Functional Polymer Blends and Nanocomposites

Related Titles

Biopolymers Based Advanced Materials
ISBN: 978-0-6482205-4-1 (e-book)
ISBN: 978-0-6482205-5-8 (hardcover)

Polymers in Oil and Gas Industry
ISBN: 978-0-6482205-0-3 (e-book)
ISBN: 978-0-6482205-1-0 (softcover)

Functional Nanomaterials and Nanotechnologies: Applications for Energy & Environment
ISBN: 978-0-6482205-2-7 (e-book)
ISBN: 978-0-6482205-3-4 (softcover)

Functional Polymer Blends and Nanocomposites

Dr. Vikas Mittal
Editor and Lead Author

CWP
Central West Publishing

This edition has been published by Central West Publishing, Australia

For more information about the books published by Central West Publishing, please visit https://centralwestpublishing.com

Disclaimer
Every effort has been made by the publisher, editor and authors while preparing this book, however, no warranties are made regarding the accuracy and completeness of the content. The publisher, editor and authors disclaim without any limitation all warranties as well as any implied warranties about sales, along with fitness of the content for a particular purpose. Citation of any website and other information sources does not mean any endorsement from the publisher and authors. For ascertaining the suitability of the contents contained herein for a particular lab or commercial use, consultation with the subject expert is needed. In addition, while using the information and methods contained herein, the practitioners and researchers need to be mindful for their own safety, along with the safety of others, including the professional parties and premises for whom they have professional responsibility. To the fullest extent of law, the publisher, editor and authors are not liable in all circumstances (special, incidental, and consequential) for any injury and/or damage to persons and property, along with any potential loss of profit and other commercial damages due to the use of any methods, products, guidelines, procedures contained in the material herein.

NATIONAL LIBRARY OF AUSTRALIA

A catalogue record for this book is available from the National Library of Australia

ISBN (print): 978-0-6482205-7-2
ISBN (e-book): 978-0-6482205-6-5

Contents

Preface

Polymer blends and nanocomposites represent functional materials generated for combining useful properties of their components. Effective mixing of the components in these materials usually results in synergistic enhancements in mechanical, thermal, flow, barrier and flammability performances. This tuning and control on the material properties leads to the applications of these functional blends and composites in many applications. A great degree of advancements in these materials have been achieved in the recent years on all aspects related to material combinations, synthesis routes, organic-inorganic phase compatibilization methodologies, etc. This text aims to present a number of such blend and nanocomposite systems. For this reason, a majority of the chapters presented in this text focus on experimental details as well as accurate material characterizations to present deep insights into controlling structure-property correlations. In addition, overviews on many commercially important nanocomposite systems have also been included.

In Chapter 1, the effect of particle compatibilization of polyethylene and polystyrene blends on the composite properties was elucidated by using a variety of 0-D and 2-D nanomaterials, viz. silica, silicate and thermally reduced graphene, in the absence of any compatibilizer. In order to explore the development of polyethylene based functional materials, especially for low temperature flexibility, blends of polyethylene with a variety of copolymers have been generated and analyzed in Chapter 2. In Chapter 3, morphology, phase miscibility, interfacial interactions and reduced low temperature brittleness of polyethylene-chlorinated polyethylene blends and polyethylene-chlorinated polyethylene-graphene oxide nanocomposites have been studied. In Chapter 4, the approach of reactive compatibilization using blend nanocomposites has been further explored to achieve significant enhancement in the composite properties at a low filler content. Blend nanocomposites of polypropylene and polystyrene using graphene as filler have been prepared in Chapter 5. In Chapter 6, efforts have been made to gain insights into the various aspects related to the composites of polyethylene with a variety of fillers such as psyllium husk (0-D), sepiolite (1-D) and thermally reduced graphene (2-D) as well as their hybrids. In Chapter 7, the influence of a variety of polymeric surface modifications physically adsorbed on the surface of sepiolite filler was studied on the thermal, mechanical and rheological properties of high density polyethylene nanocomposites. Chapter 8 aims to review recent studies reporting the development of polyethylene/graphene nanocomposites with a focus on the composite properties, especially morphological, mechanical and thermal properties. Chapter 9 reviews the recent reports on the synthesis and characterization of polyimide/graphene nanocomposites, whereas the mechanical, thermal and electrical properties of PPS nanocomposites reinforced with graphite, graphene and other derivatives have been reviewed in Chapter 10.

The book would not have been successfully accomplished without the continuous support of the chapter contributors. The book is dedicated to my family for their unswerving motivation and encouragement.

Vikas MITTAL
Australia

1

Filler Distribution and Properties of Polyethylene/ Polystyrene Blend Composites with 0-D and 2-D Nanomaterials

1.1 Introduction

Blending or alloying of polymers is an increasingly common practice for the development of new compounds having improved and diverse properties, however, phases of the mostly immiscible systems need to be efficiently mixed [1-4]. Polyolefins, especially polyethylene (PE), have outstanding properties including low cost, recyclability, good processability, non-toxicity and biocompatibility, but suffer from poor barrier properties, shrinkage and relatively low stiffness [5-10]. To overcome these limitations, blends of polyolefins with stiff polymers like polystyrene (PS) represent effective alternatives, however, their phase domains due to unfavorable enthalpy of mixing need efficient mixing [11,12]. Significant efforts have been devoted in order to modify the macrophase distribution of polymers by lowering interfacial tension and improving adhesion using compatibilizers like block copolymers [12-17]. The drawbacks associated with the usage of polymeric compatibilizers are their low molecular weight, hindrance to polymer crystallization and weakening of mechanical properties of the blends [18]. Some studies have also dealt with compatibilization of polyolefin and PS blends by using reactive blending [11,19], however, the processing requires efficient control on the reaction as no detrimental side reactions can be tolerated.

Polymers reinforced with nano-fillers, such as layered silicates, carbon nanotubes (CNTs) and graphene nano-sheets have attracted much research attention due to significant enhancements in the polymer performance [20-31]. Among these fillers, layered silicates have

Vikas Mittal[a,], Ali U. Chaudhry[a,**], Fakhruddin Patwary[a], Nadejda Matsko[b]*
[a]The Petroleum Institute (part of Khailfa University of Science and Technology), Abu Dhabi, UAE; [b]Graz Centre for Electron Microscopy, Graz, Austria
**Current address: Bletchington, Wellington County, Australia; **Current address: Texas A&M University, Qatar*

also been used to improve the blend miscibility by enhancing weak interfacial adhesion between the phases in PE/PS [32], polyamide/polypropylene (PP) [33], elastomers/PP [34], PP/ethylene propylene rubber [35] blends, among others [36]. Along with the morphological variations leading to effective phase distribution, layered silicate particles also exhibited reinforcing effect on the blend properties. Also, graphene sheets are one-atom-thick two-dimensional layers of sp^2-bonded carbon and exhibit a variety of remarkable properties such as electrical and thermal conductivity, gas impermeability and mechanical properties, etc. [37,38]. Recently, Cao *et al.* used the amphiphilic nature of graphene oxide (GO) and reported its compatibilizing behavior towards immiscible polymer blend of polyamide/polyphenylene oxide. The authors observed remarkable reduction in the size of microphase and enhanced mechanical properties. The behavior was due to coupling effect of GO owing to the presence of carboxylic acid at the edges as well as phenol, hydroxyl and epoxide groups at basal planes [39]. In addition, there is a need to further evaluate the role of particles like graphene on the compatibilization and properties of immiscible blends, in comparison with other 0-D and 2-D particles like silica and alumino-silicates.

Various mechanisms have been described in the literature for morphologically stabilized blends of immiscible polymers using particles with different shapes, sizes and surface characteristics. For instance, one of the mechanisms is the accumulation of filler at the interface of dispersed phases which suppresses the coalescence and phase separation [40]. It was reported that the particles at the interface act as a bridge which was observed by gel behavior at low frequency during rheological measurements. The gel behavior was the consequence of the formation of particle network between the viscous fluids having different phases [41]. Thus, it is important to tune the localization of the filler in the polymer phases. Pawar *et al.* [42] also reported co-continuous morphology of poly(3-hydroxybutyrate-co-3-hydroxyvalerate) (PHBV) and poly(butylene adipate-co-terephthalate) (PBAT) in the blend nanocomposites, though the filler (graphene and vermiculite platelets) was localized in the tougher PBAT phase, as shown in Figure 1.1 [42]. Reduction of microphase size of the polymer phases on particle incorporation is another common mechanism of stabilization [39]. Tanaka *et al.* reported for the stabilization behavior of binary immiscible polymer mixtures containing monodisperse glass spheres that coarsening of phases was affected by the mobile glass particles due to the dynamic interactions between

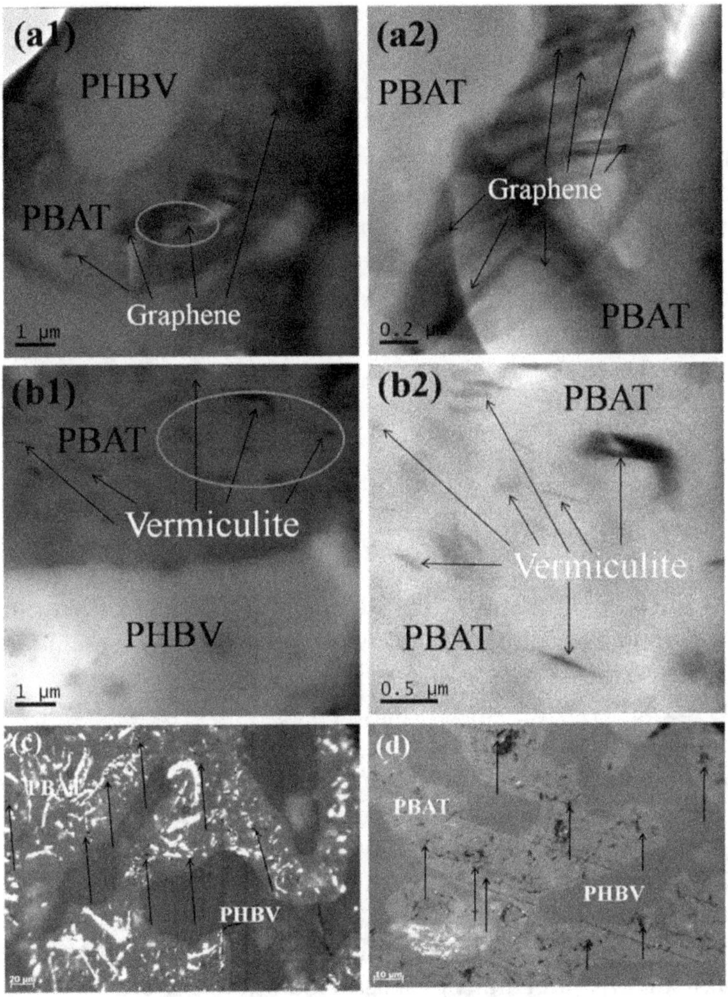

Figure 1.1 TEM micrographs of PHBV/PBAT composites (a1 and a2) with 5 wt% graphene and (b1 and b2) with 5 wt% vermiculite. Optical images of PHBV/PBAT composites (c) with 5 wt% graphene and (d) with 5 wt% vermiculite. Reproduced from Reference 42 with permission from Springer.

phase separation and wetting [43]. Similar results were also confirmed by using simulation of the motion of spherical particles in binary systems [44]. Thermodynamic parameter such as interaction parameter has also been reported to play a significant role in lowering the total Gibbs free energy (ΔG_{mix}) of blends containing the inorganic solid filler phase depending on the interaction and preferential adsorption of the components [45,46]. Further, surface features such as attached functional groups with polymer or filler have also been reported to help in compatibilization and reducing surface tension between the immiscible phases [47].

In this study, the effect of particle compatibilization of PE and PS blends on the composite properties was elucidated by using a variety of 0-D and 2-D nanomaterials, viz. silica, silicate and thermally reduced graphene, in the absence of any compatibilizer. The effect of the nano-filler morphology on its distribution and subsequent composite properties was quantified as a function of blend composition.

1.2 Experimental

1.2.1 Materials

High density polyethylene BB2581 was received from Abu Dhabi Polymers Company Limited (Borouge), UAE. It was in the form of white pellets with specific gravity 0.958. Polystyrene (PS) with average molecular weight (M_w) of 400,000 was procured from Sigma. The polymers were used as obtained. Synthetic silicon dioxide powder (ZEOFREE® 5161 S) and synthetic alumino-silicate (ZEOLEX® 23) were supplied by J. M. Huber Private Limited, India. Thermally reduced graphene was produced by thermal exfoliation of precursor graphite oxide using modified Hummer's method, as reported earlier [31,48].

1.2.2 Generation of Blends and Blend Composites

Melt mixing using mini-twin conical screw extruder (MiniLab HAAKE Rheomex CTW5, Germany) was carried out to generate PS/PE blends and nanocomposites. The polymers and fillers were dried at 60° C overnight under vacuum so as to remove any surface adsorbed moisture which can lead to unwanted degradation during the melt compounding process. Blends of three compositions were generated: 25 wt% PE + 75 wt% PS, 50 wt% PE + 50 wt% PS and 75 wt% PE + 25 wt% PS, which were correspondingly designated as $PE_{25}PS_{75}$, $PE_{50}PS_{50}$ and $PE_{75}PS_{25}$, respectively. The filler content was fixed to 7 wt% in the composites. Blends and blend composites were prepared by pre-mixing of the polymers with fillers and melt mixing these together at a temperature of 205 °C and rotational speed of 80 rpm for 7 min. To study the effect of processing conditions on the filler distribution and resulting properties, the composites were also prepared by (i) melting PE and PS, followed by filler addition, (ii) melting PE, followed by addition of graphene and PS, (iii) melting PS, followed by addition of graphene and PE, (iv) melt mixing the components together using a rotational speed of 120 rpm or using a mixing time of 15 min. Tetrahedron MTP-10 hot press was used to compression mold the blend and composite sheets. A temperature of 205 °C and a pressure of 800 bar were used (holding pressure 290 bar). The samples were cooled to 50 °C at a rate of 5 °C/min. Sheets of 12*12 cm with a thickness of 1.5 mm were produced and disc, dumbbell and bar-shaped test specimens for rheological and mechanical characterization were stamped out from these compression molded sheets.

1.2.3 Materials Characterization

Differential scanning calorimetric (DSC) measurements on the polymer blends and nanocomposites were performed using a Netzsch DSC under nitrogen atmosphere. The scans were obtained from 30-250-30 °C using heating and cooling rates of 5 °C/min. Thermal properties of the samples were analyzed using Netzsch thermogravimetric analyzer (TGA) using nitrogen as carrier gas. The scans were obtained from 50 to 700 °C at a heating rate of 10 °C/min.

Universal testing machine Testometric M-350 10CT was used to analyze the mechanical performance of the samples. The dumbbell-shaped samples of 53 mm length, 4 mm width and 2 mm thickness (gauge length was 26 mm) conforming to standard ASTM-638-V were used. A loading rate of 1 mm/min was used and the tests were carried out at room temperature. Win Test Analysis software was used for the calculation of tensile modulus and other properties. An average of five values was calculated.

DMA (dynamic mechanical analyzer) RSA3 of TA Instruments was used to study the change in modulus with temperature (25-120 °C). Dynamic temperature ramp test was used at a frequency of 5 rad/s on bar-shaped samples, using a ramp rate of 3 °C/min and 1% strain for all samples. AR 2000 rheometer from TA Instruments was used to characterize rheological properties such as storage modulus (G′), loss modulus (G″), viscosity (η′) and elasticity (η″), along with complex modulus (G*) and complex viscosity (η*). Disc shaped samples of 25 mm diameter and 2 mm thickness were characterized at 205 °C using a gap opening of 1 mm. The samples were subjected to a strain sweep at 205 °C and angular frequency of 1 rad/s. Subsequently, the frequency sweep analysis (dynamic testing) of all blends and blend nanocomposite samples was carried out at 1% strain from ω = 0.01 to 100 rad/s.

For the microscopy analysis, the samples were mounted in special holders that at the same time fit in the microtome and were also suitable for the examination of the block face by atomic force microscopy (AFM). Ultrathin sections (30-90 nm) as well as block faces of the composite samples were obtained using a Leica Ultracut E microtome (Leica, Austria) equipped with a diamond knife (Diatome, Switzerland) at -120 °C. The so-obtained sections for transmission electron microscopy (TEM) analysis were collected on 400 mesh formvar electron microscopy grids (coated with holey carbon film), and were examined in a Philips CM 20 (Philips/FEI, Germany) electron microscope at 200 kV at room temperature without additional staining. TEM image processing was performed using DigitalMicrograph software (Gatan, USA). All energy-dispersive X-ray spectroscopy (EDXS) experiments on blends and composites were performed in STEM mode with a probe current of 20 nA and a beam diameter of 100 nm. EDXS spectra were collected using a HPGe detector (solid angle 0.13 sr). For the calculation of elemental distribution images, jump ratio method was used and EDXS spectra have been visualized using DigitalMicrograph software (Gatan, USA). The block faces of specimens after cryo-ultramicrotomy were investigated at ambient conditions using Dimension 3100 AFM/SPM (Veeco, USA) atomic force microscope. AFM images were collected in tapping mode using silicon nitride cantilevers with natural frequencies in the 300 kHz range (force constant = 20 N m^{-1}, tip radius = 10 nm (NT-MDT, Russia)). AFM image processing was performed using Nanoscope v720 software (Veeco, USA). The block faces of specimens after cryo-ultramicrotomy were also investigated at ambient conditions using ZEISS Axioplan light microscope (Zeiss, Germany) equipped with Zeiss Axio Cam ICc 1 CCD camera in reflected polarized light.

1.3 Results and Discussion

In the current study, different fillers with 0-D and 2-D nanostructure were added to PE/PS blends and their effect on the morphology and physical properties of the composites was studied as a function of blend composition. Initially, all the blend composites were generated using one-step melt mixing process where all the component were added to the compounder together. In the later part of the study, variations in the mixing protocols were made to study the corresponding effect on the filler distribution and composite properties. The effect of mixing time and rotational speed in the compounder was also analyzed. However, the compounding temperature was not exceeded beyond 205 °C so as to avoid any extensive thermal degradation of the polymers. The filler extent was also kept constant at 7 wt% so as to relate the variation in the performance of the composites to the type of the filler and blend composition. Also, the filler fraction of 7 wt% was chosen, as this was reported to an optimum content in an earlier study [49].

Figure 1.2 demonstrates the TGA thermograms of polymer blends and blend composites with silica, silicate and graphene, as a function of blend composition. As observed in Figure 1.2a, the blends exhibited two step degradation behavior, where the low temperature step

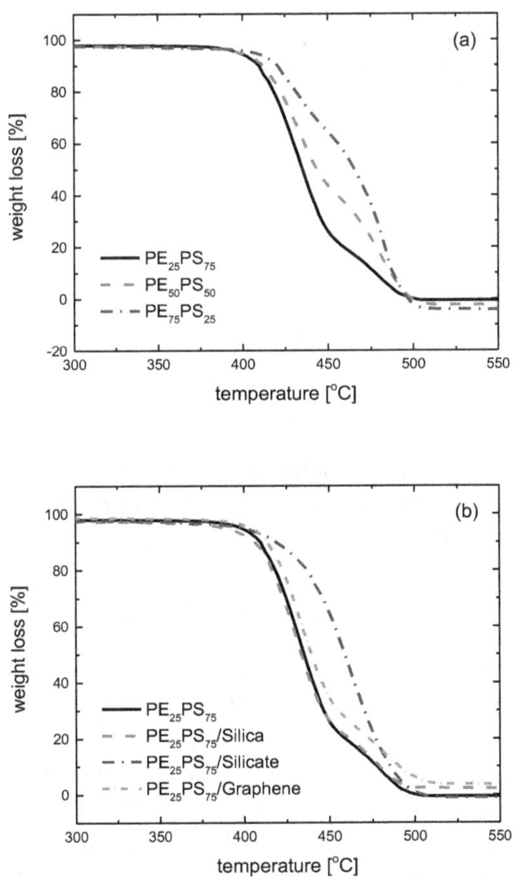

Figure 1.2 (continued on next page)

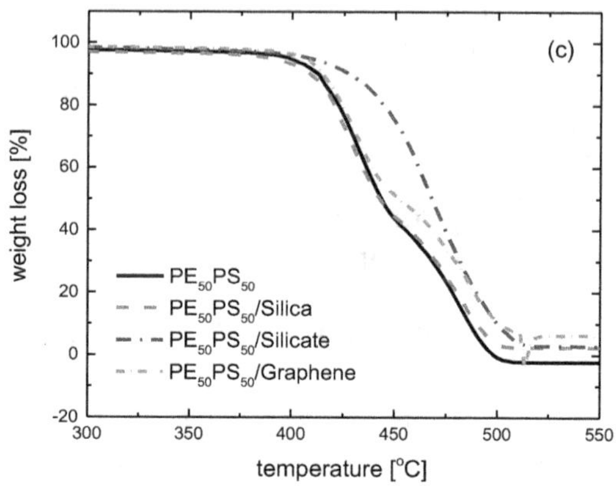

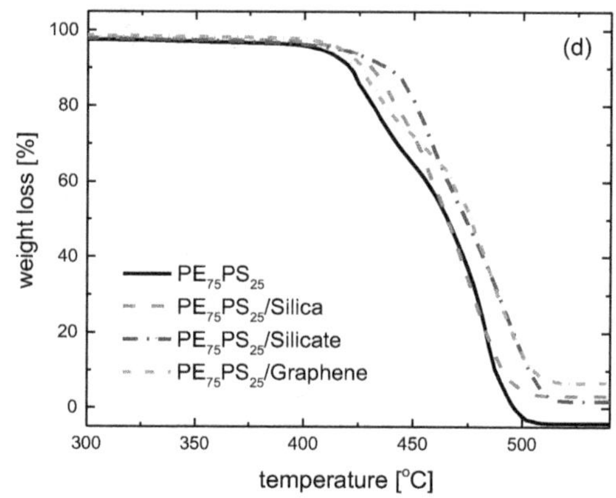

Figure 1.2 TGA thermograms of polymer blends (a) and silica, silicate and graphene composites with polymer blends with compositions of (b) $PE_{25}PS_{75}$, (c) $PE_{50}PS_{50}$ and (d) $PE_{75}PS_{25}$.

step corresponded to PS phase, whereas the high temperature degradation step was attributed to PE. Presence of the individual degradation patterns of PE and PS also indicated the immiscibility of the phases. On increasing the amount of PE in the blend, both onset and peak degradation temperature were observed to shift to higher values, thus, indicating the enhancement of thermal stability on the addition of PE. The thermal stability of the polymer was observed to largely improve in the composites with silicate and graphene fillers. The silicate filler provided maximum enhancement in the onset and peak degradation temperatures at all blend compositions. Addition of silica did not impact the polymer degradation temperatures. In addition, the blend composites with silica and graphene also exhibited two step degradation behavior as the pure blends owing to different PE and PS phases. In contrast, the degradation patterns of composites with silicate filler did not exhibit two step degradation pattern, and rather a smooth single step degradation behavior was observed. Probably, silicate provided effective phase compatibilization between PS and PE, thus, leading to a reduction in their domain sizes in the composites.

Figure 1.3 and Table 1.1 also present the calorimetric properties of the blends and nanocomposites. Second heating curves, demonstrated in Figure 1.3, were used for the analysis of peak melting temperature as well as melt enthalpy. In these blends, no melt transition

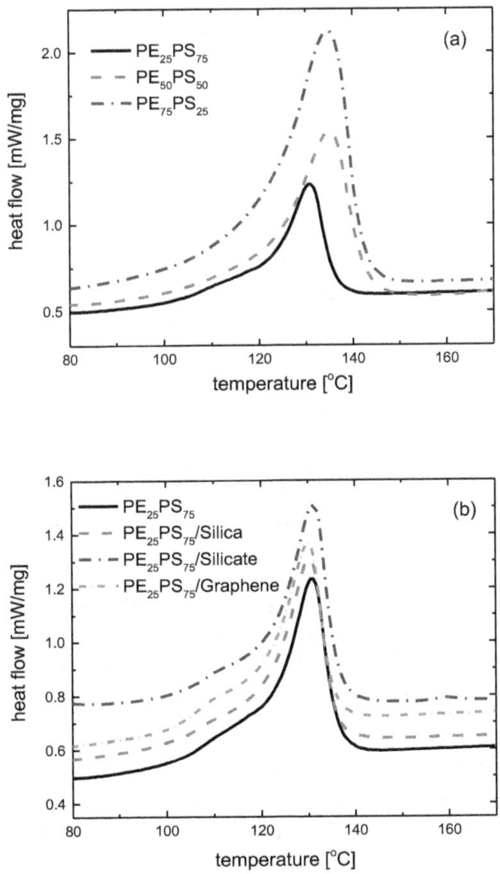

Figure 1.3 (continued on next page)

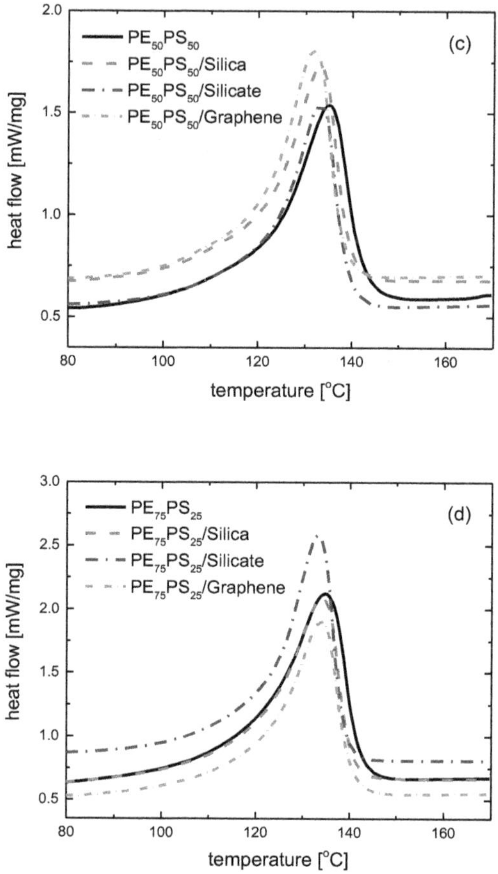

Figure 1.3 DSC melting thermograms of polymer blends (a) and silica, silicate and graphene composites with polymer blends with compositions of (b) $PE_{25}PS_{75}$, (c) $PE_{50}PS_{50}$ and (d) $PE_{75}PS_{25}$.

was observed for PS due its amorphous nature. On the other hand, the peak melting temperature of the PE was observed to decrease from 134.6 °C for 75% PE content to 130.6 °C when the PE extent was 25%, indicating possible hindrance to the PE crystallization by the majority PS phase. The decrease in the melt enthalpy (normalized to PE content) also indicated reduced PE crystallinity with decreasing PE content in the blends. For instance, melt enthalpy for the blend with 75% PE was 156 J/g, which was reduced to 132 J/g in the 25% PE blend. Comparing with the value of 293 J/g for the theoretical melt enthalpy of 100% crystalline PE sample[50], degree of crystallinity of

Table 1.1 Calorimetric properties of the PE/PS blends and composites

	Peak melting temperature T_m, °C	Melt enthalpy ΔH_m, J/g	Peak crystallization temperature T_c, °C	Crystallization enthalpy ΔH_c, J/g
$PE_{25}PS_{75}$	130.9	132	115.7	128
$PE_{50}PS_{50}$	133.1	134	115.4	136
$PE_{75}PS_{25}$	134.6	156	115.7	151
$PE_{25}PS_{75}$/Silica	130.2	163	116.5	168
$PE_{25}PS_{75}$/Silicate	131.3	163	115.4	168
$PE_{25}PS_{75}$/Graphene	130.9	159	115.2	151
$PE_{50}PS_{50}$/Silica	133.1	140	115.2	146
$PE_{50}PS_{50}$/Silicate	133.2	137	115.5	142
$PE_{50}PS_{50}$/Graphene	133.1	144	116.0	148
$PE_{75}PS_{25}$/Silica	134.1	148	116.5	145
$PE_{75}PS_{25}$/Silicate	133.2	141	115.7	145
$PE_{75}PS_{25}$/Graphene	134.0	124	115.2	123

53% and 45% respectively was observed. Similar trend was also exhibited by the crystallization enthalpy of the PE phase in the blends (Table 1.1). The melt enthalpy was significantly affected by the filler addition, which was further a function of blend composition. In the blend with 25% PE, addition of filler resulted in significant enhancement of melt enthalpy, though peak melting temperature was not affected. The peak crystallization temperature was also observed to be similar to the blend indicating the absence of any nucleation effect by the filler. In addition, all fillers exhibited similar enhancement in the melt enthalpy of PE. The blend with 50% PE also exhibited increment in the melt enthalpy with the addition of filler, though the extent of enhancement was reduced in comparison with the composite of the blend with 25% PE content. The blend with 75% PE exhibited a reduction in the melt enthalpy on incorporation of fillers, with the maximum reduction observed in the case of graphene. Thus, depending on the blend compositions, the fillers had a variable effect on the PE crystallinity, though there was no indication of the nucleation effect by the fillers. The observed effect would have resulted due to the specific filler dispersion in the blend composites generated as a function of blend composition. As the PE crystallinity is significantly affected on addition of the fillers, it is also envisaged that the filler particles may have been predominantly localized in the PE phase. The blend and composites with 75% PS also exhibited an endothermic signal near 110 °C (Figure 1.3), probably indicating the glass transition temperature of polystyrene. The signal was not observed in the other materials owing to constraining of the PS phase inside PE, thus, hindering the movement of PS chains.

Figure 1.4 shows the TEM micrographs of the $PE_{50}PS_{50}$ composites with silica, silicate and graphene fillers. In the images, the darker phase represented PE, whereas lighter phase contained PS. The filler was observed to be specifically localized in the PE phase, with PS phase completely devoid of it. Similar morphology was observed for all filler types and supported the findings from calorimetric analysis where the observed effect on PE crystallinity was suspected to occur due to specific localization of filler in PE. In certain instances, the filler particles were also present at the interface of the PE and PS phases. As PE chains are not expected to have any specific interaction with the fillers, thus, the observed composite morphology would not result due to thermodynamic mixing of the phases. In contrast, the kinetic factors seem to have resulted in the observed morphology, due to the processing method used to generate the nanocomposites. As both PS and PE were added together in the compounder along with the filler, thus, the early melting of PE phase would have resulted in its mixing with the filler before the PS melt could be formed. Thus, the PS melt would have phase mixed with PE-filler mixture, rather than PE and filler individually. Thus, in the scenario of formation of the melt from both PE and PS phases before the addition of filler, the distribution of filler in both phases can be expected. Also, it should be noted that no significant interface stabilization occurred in these composites due to the filler largely embedded in the PE phase. However, the domains of PE and PS were reduced in the composites as compared to pure blends, which indicated that the filler particles still exhibited stabilization effect, even though most of the filler was present in the PE phase. Also, the phases did not coalesce with time, further confirming the phase stability in the composites. Similar effect was also observed earlier by Pawar *et al.* [42], where specific localization of filler in PBAT phase in the PHBV/PBAT blend composite led to the synergistic enhancement of the composite properties. In fact, the observed morphology of filler localization in the flexible phase (PE) is also beneficial from the perspective of physical properties of the composites, as distribution of the filler in the stiffer phase (PS) would have resulted in even further decrease in the toughness of the composites. In addition, though TGA analysis had signaled the possibility of better phase

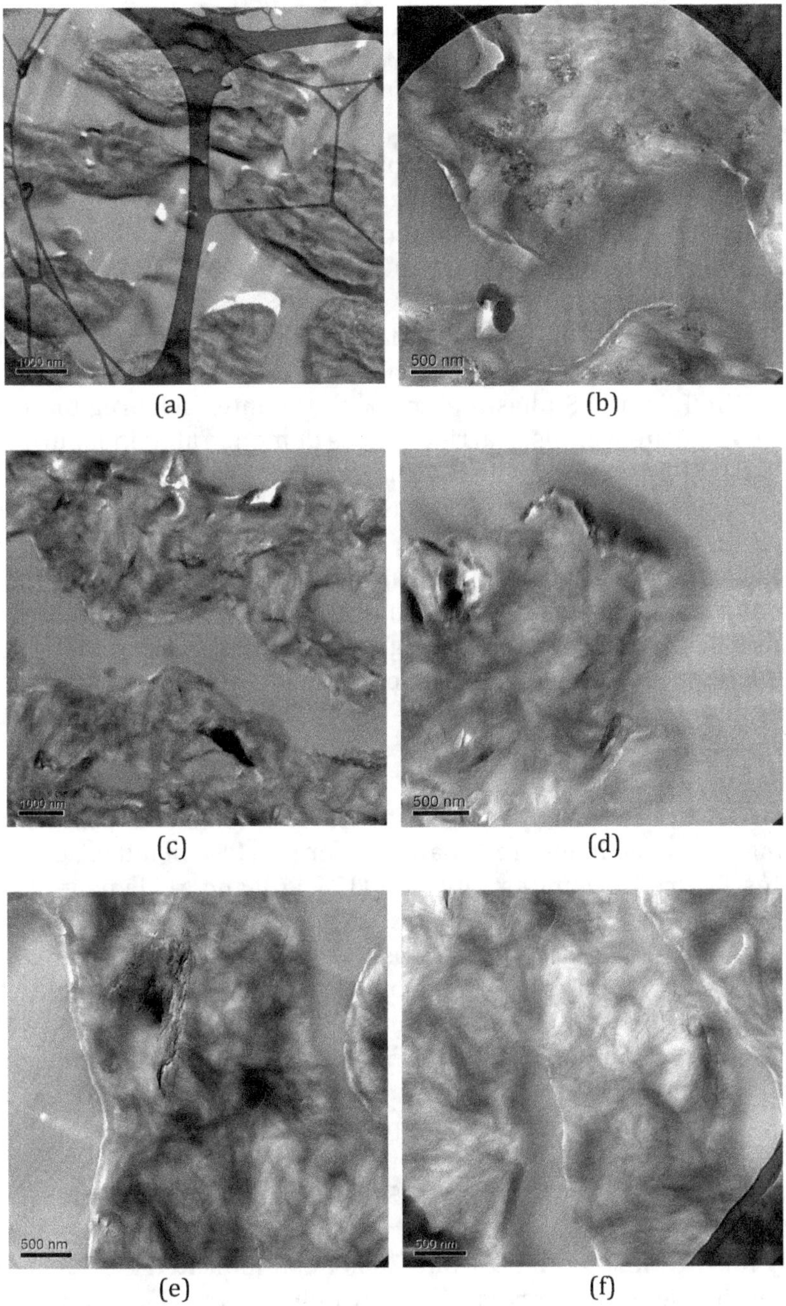

Figure 1.4 TEM micrographs of $PE_{50}PS_{50}$ composites at two magnifications depicting the phase distribution; (a),(b) using silica as filler, (c),(d) using silicate as filler and (e),(f) using graphene as filler.

mixing in the case of silicate composites due to the presence of a single degradation step, however, morphological analysis did not exhibit any difference in the morphology of the composites with different fillers.

Figure 1.5 depicts further morphological analysis of the $PE_{50}PS_{50}$ composites with silica and graphene. Figures 1.5a and b depict the TEM and PCR (plasmon to carbon ratio) images of the $PE_{50}PS_{50}$ composite with silica. As observed in Figure 1.4, the TEM image confirmed the smaller domain sizes of PS and PE phases. In addition, the distribution of silica solely in PE phase was also confirmed in the corresponding PCR image (seen as white spherical domains). EDX analysis of the PE and PS phases along with the interfacial region in the silica composite was also carried out, as demonstrated in Figure 1.5c and d. The PE phase exhibited a sharp signal for Si, thus, confirming the presence of silica in the PE phase. On the other hand, no Si signal was observed in the PS phase. The interfacial region also exhibited a weak signal of Si, thus, also indicating the presence of a fraction of filler at the interface between PE and PS. Figure 1.5e and f also present the similar TEM and PCR images of the $PE_{50}PS_{50}$ composite with graphene. The PCR image clearly confirmed the distribution of graphene platelets inside the PE phase with a fraction also stabilizing the interface between PE and PS phases.

Morphological analysis of the composites generated from the blends with other compositions was also carried out to confirm the findings reported in Figure 1.4 and 1.5 for $PE_{50}PS_{50}$ blend based composites. Figure 1.6 demonstrates the TEM, PCR and AFM images of the $PE_{75}PS_{25}$ blend based composites with silica and graphene. The $PE_{50}PS_{50}$ blend had a co-continuous morphology, whereas the blend $PE_{75}PS_{25}$ exhibited dispersed morphology of the PS phase inside PE. Similar filler distribution was observed in the composites. The filler particles were localized in the PE phase, with a fraction of filler present at the interface, thus, confirming no change in the nature of component dispersion with blend composition.

Table 1.2 demonstrates the tensile properties of the blends and nanocomposites. The blend with 75% PE exhibited the tensile modulus of 1284 MPa, which was enhanced to 1435 and 1593 MPa respectively for the blend with 50% and 25% PE. It indicated that the modulus of the blend was correspondingly enhanced as the extent of the stiffer phase was increased. The addition of filler in the blends increased the tensile modulus significantly. Comparing among all the fillers, graphene resulted in the largest extent of enhancement in the

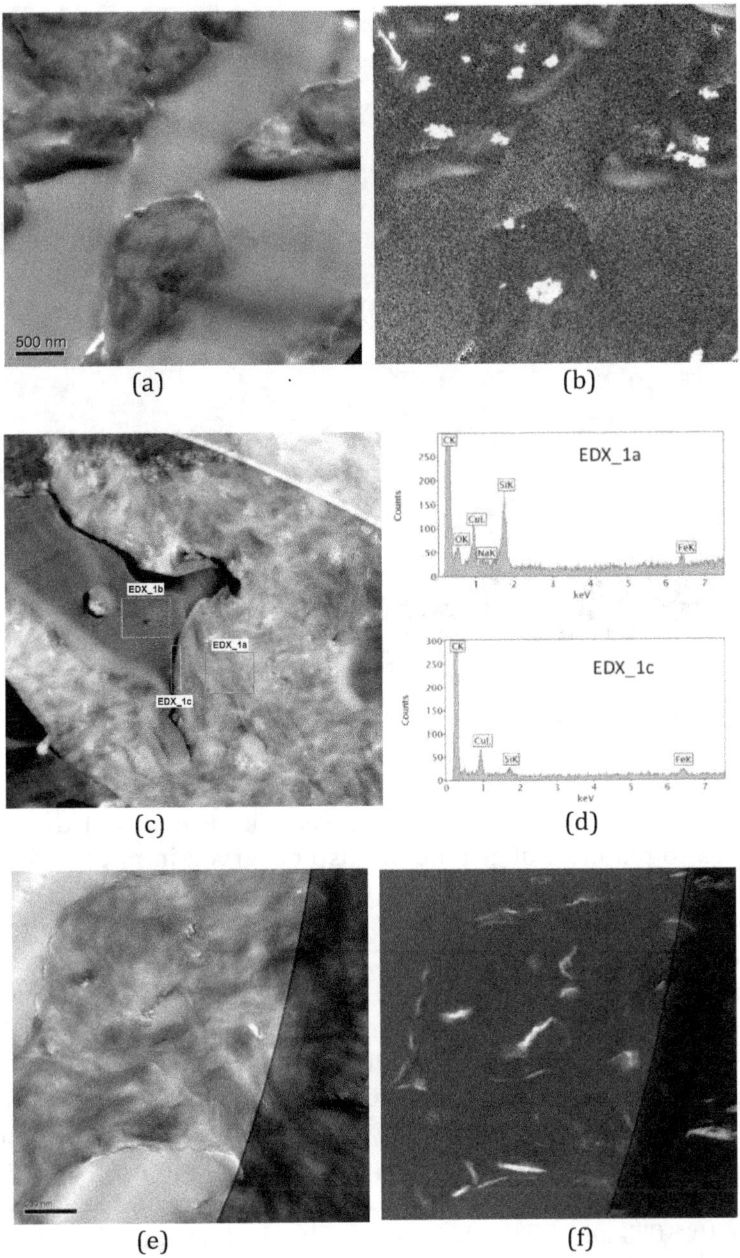

Figure 1.5 (a) TEM and (b) corresponding PCR image; (c) TEM and (d) corresponding EDX analysis for determining the distribution of silica particles in the composite for $PE_{50}PS_{50}$/Silica composite; (e) TEM and (f) corresponding PCR image for $PE_{50}PS_{50}$/Graphene composite.

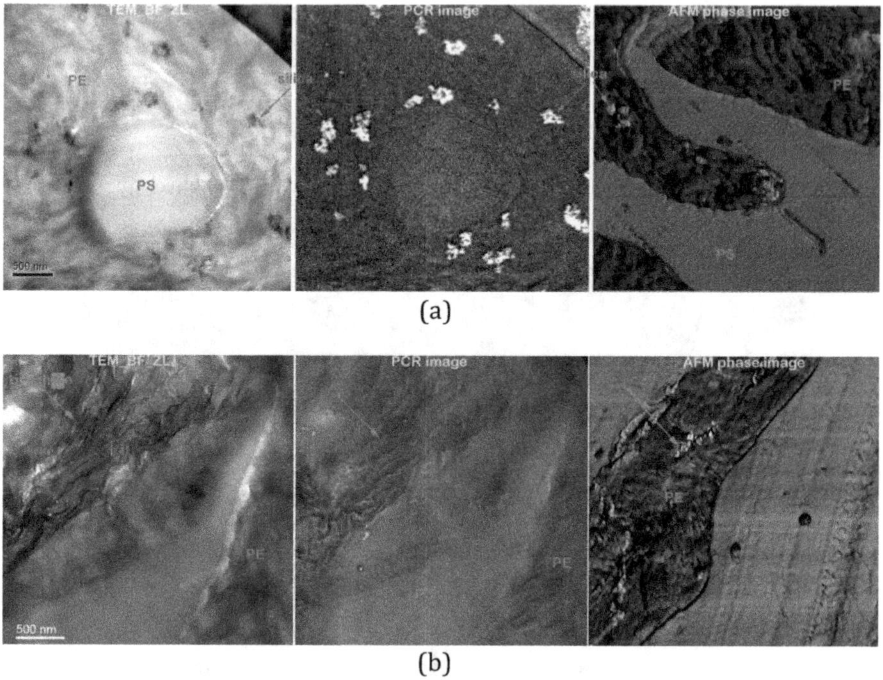

Figure 1.6 Morphological analysis of (a) $PE_{75}PS_{25}$/Silica and (b) $PE_{75}PS_{25}$/Graphene composites.

modulus, followed by silicate and silica. It indicated that the larger aspect ratio platelets of graphene (also observed in microscopy images) contributed to effective stress transfer from the polymer chains. Overall, an increase of 17% in the modulus was observed for the blend composites with 25% PE, whereas the increment was 29% and 36% for the blend composites with 50% and 75% PE content. It has also to be noted that the maximum enhancement in the modulus of the blend composite with 75% PE occurred even after reduction of crystallinity in the composites on addition of filler. This indicated that the filler may have higher extent of dispersion in this case due to larger fraction of available PE phase. Similarly, the peak stress and stress at break was also observed to enhance for the blend composites with 50% and 75% PE content. The peak strain of the composites was also largely retained when the PE content in the blend was high. It also indicated that the specific localization of the filler in the PE phase did not largely impact the flexibility. Thus, the blend composite with 75% PE exhibited more optimal enhancement in the mechanical properties, followed by the blend composites with 50% PE.

Table 1.2 Mechanical properties of the PE/PS blends and composites

	Tensile modulus[a], MPa	Peak stress[b], MPa	Peak strain[c], %	Stress at break[d], MPa
$PE_{25}PS_{75}$	1593	26	1.5	26
$PE_{50}PS_{50}$	1435	17	1.6	15
$PE_{75}PS_{25}$	1284	15	2.5	12
$PE_{25}PS_{75}$/Silica	1719	21	1.4	20
$PE_{25}PS_{75}$/Silicate	1760	21	1.1	20
$PE_{25}PS_{75}$/Graphene	1864	25	1.4	24
$PE_{50}PS_{50}$/Silica	1561	20	1.2	19
$PE_{50}PS_{50}$/Silicate	1682	20	1.1	18
$PE_{50}PS_{50}$/Graphene	1853	19	1.2	18
$PE_{75}PS_{25}$/Silica	1272	13	2.5	13
$PE_{75}PS_{25}$/Silicate	1423	14	2.3	14
$PE_{75}PS_{25}$/Graphene	1749	17	2.0	17

[a]Relative probable error 3%; [b]relative probable error 3%; [c]relative probable error 5%; [d]relative probable error 10%.

Figure 1.7 shows the dynamic mechanical performance of the blends and nanocomposites as a function of temperature (room temperature till 120 °C). Similar to the tensile modulus, the storage modulus at room temperature was observed to be higher for the blends with higher extent of PS. On increasing the temperature, the storage modulus decreased gradually, though the trend between the blends was retained. The blends exhibited a sharp decrease in the modulus around 100 °C, which is the glass transition temperature of PS. The reduction was most prominent in the $PE_{25}PS_{75}$ blend owing to the large fraction of PS in the blend. Thus, beyond this temperature, both polymer phases were in rubbery state, thereby, resulting in lower storage modulus. Due to this, no difference in the performance of different blends was observed beyond this temperature. The storage modulus was observed to increase in the blend composites with 50% and 75% PE content, with maximum extent of increment achieved on the incorporation of graphene, similar to the findings from tensile analysis. However, the blend composite with 25% PE did not exhibit any increment in the modulus on filler addition, in fact, the modulus was observed to decrease to some extent. Tensile analysis also exhibited least extent of modulus enhancement for this case. As the filler was contained in the 25% PE fraction in this case, probably some filler aggregation may have taken place. As the blends, all the composites exhibited sharp decline in the modulus around 100 °C, though this transition was observed to shift to higher temperatures in the case of graphene containing composites. This transition, owing to the glass transition temperature of PS, is also depicted in Figure 1.7e. The transition in the blend with 75% PS was sharp and well defined. In the graphene based composite generated with this blend, the transition was observed to shift to a higher temperature, as observed earlier. Thus, it confirmed that the addition of graphene enhanced the thermal resistance of the polymer due to effective heat transfer and conductivity of graphene platelets. The transition was observed to become broad as the amount of PS was decreased in the blends. Similar to the case with 75% PS, the blend composites of graphene with 50% PS and 25% PS also exhibited the shift of transition temperature

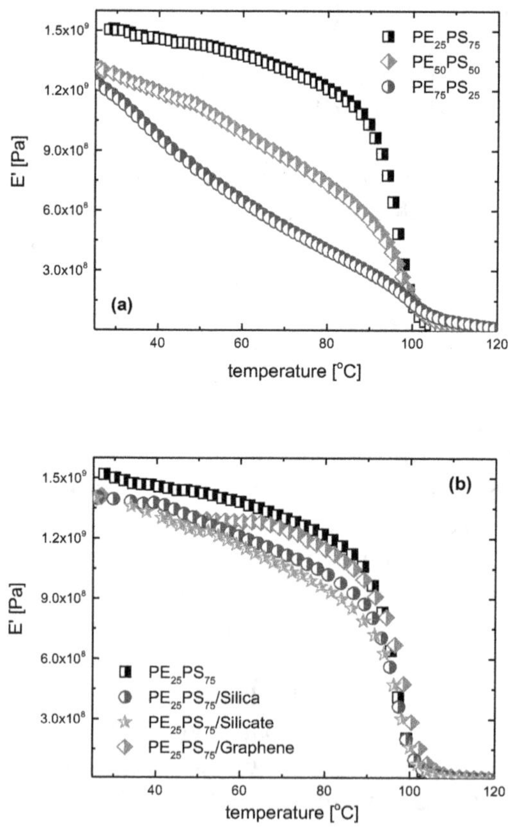

Figure 1.7 (continued on next page)

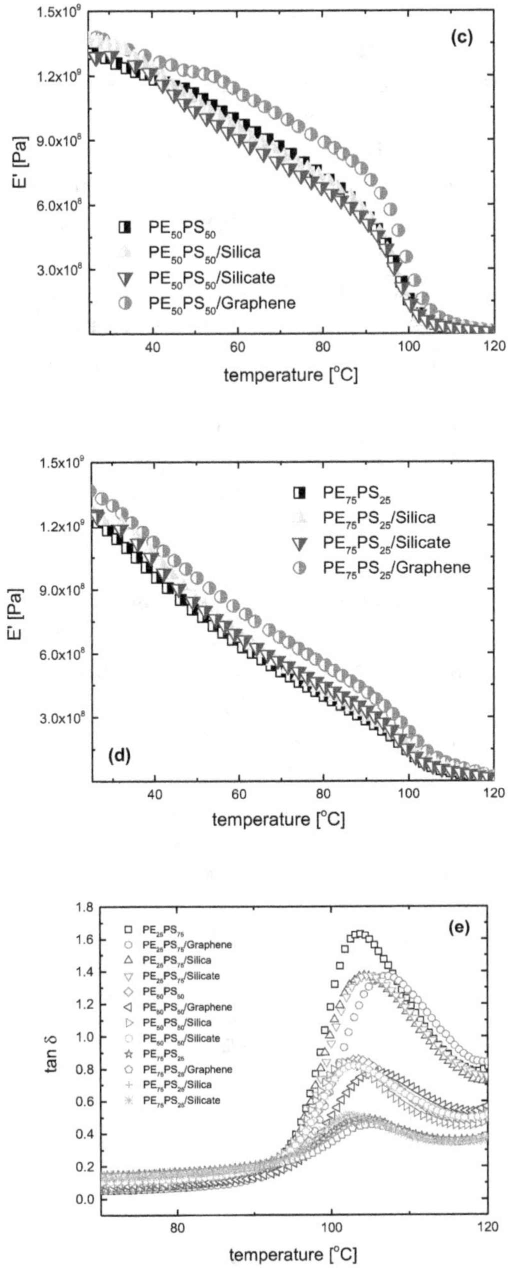

Figure 1.7 E' vs T curves of polymer blends (a) and silica, silicate and graphene composites with polymer blends with compositions of (b) $PE_{25}PS_{75}$, (c) $PE_{50}PS_{50}$ and (d) $PE_{75}PS_{25}$; (d) tan delta vs. T curves.

of PS to a higher temperature, and the extent of enhancement was also a function of PS content in the composites. Therefore, both mechanical and thermal properties were significantly affected by large aspect ratio graphene platelets, as compared to silicate and silica fillers. Even though the silicate filler had also large aspect ratio platelets, however, reinforcement effect was not as high as graphene platelets. Higher polarity of silicate platelets may have led to lower degree of their dispersion in the nanocomposites as compared to graphene platelets.

Figure 1.8 also demonstrates the rheological performance of the blends and nanocomposites as a function of angular frequency at a temperature of 205 °C. All the samples exhibited gradual decline in the extent of modulus enhancement as a function of angular frequency due to shear thinning effect. At the melt shearing conditions, the blend with 75% PE exhibited largest complex modulus, probably due to the more viscous nature of the PE fraction at these conditions. The blend with 50% PE also exhibited slightly higher complex modulus due to this reason. Even at higher angular frequency, the differences between the samples were maintained indicating composition dependent behavior of the blends. In the composites, the addition of filler resulted in enhancement of the modulus, with graphene exhibiting highest extent of increment, similar to earlier findings. The complex viscosity of the composites also increased with the addition of filler, as shown in Figure 1.9. Similar to the modulus, addition of graphene resulted in maximum extent of viscosity increase. However, the composites were still processable using the same conditions as pure blends.

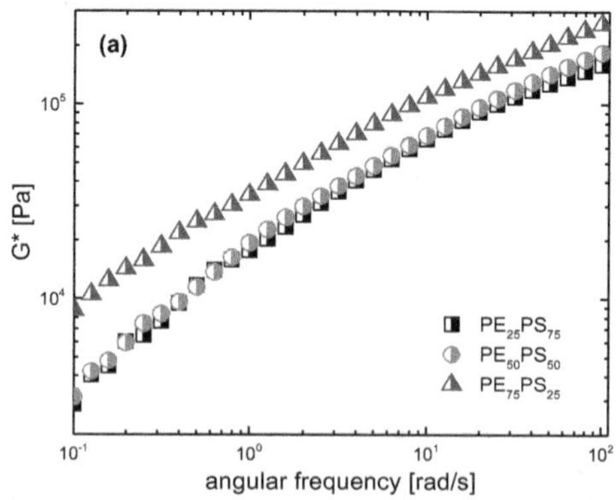

Figure 1.8 (continued on next page)

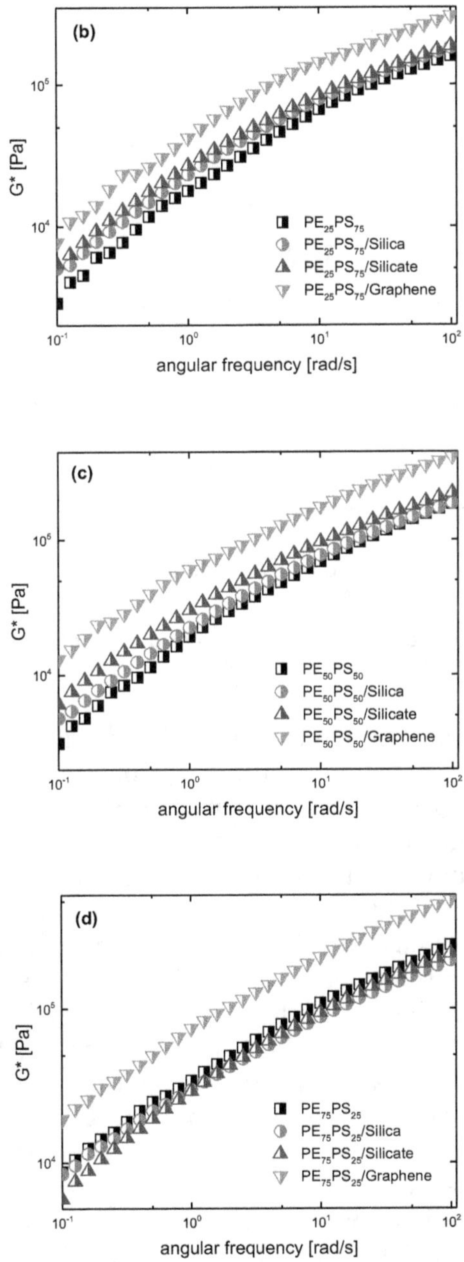

Figure 1.8 Complex modulus of polymer blends (a) and silica, silicate and graphene composites with polymer blends with compositions of (b) $PE_{25}PS_{75}$, (c) $PE_{50}PS_{50}$ and (d) $PE_{75}PS_{25}$.

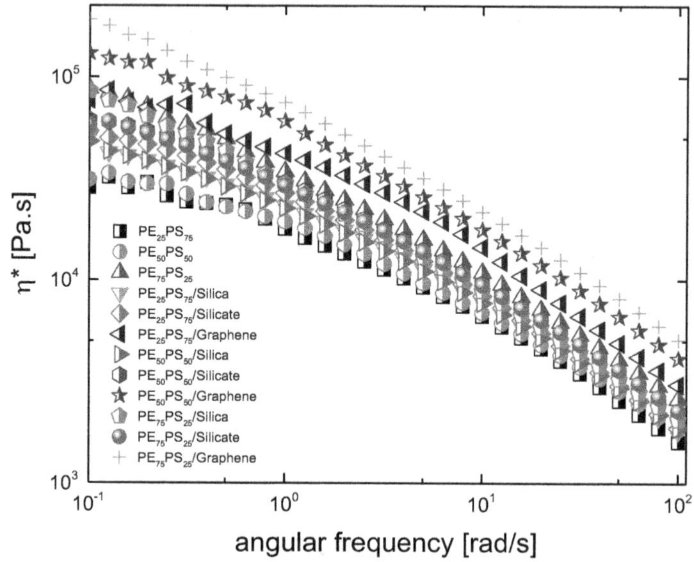

Figure 1.9 Complex viscosity of polymer blends and composites as a function of angular frequency.

The miscibility of the blends was further studied using Cole-Cole viscosity plot which develops relationships between real (η') and imaginary (η'') parts of complex viscosity [51-53]. A smooth, semi-circular shape of the graph would suggest miscible systems with homogenous phase. The deviation from this behavior indicates phase segregation due to immiscibility of the components. Though these analyses have been traditionally developed for polymer blends, however, some insights can also be gained about the phase miscibility in blend composites. Figure 1.10 shows the Cole-Cole plots for the materials. In the case of blend with 25% PE content, the viscosity plot did not exhibit semi-circular shape. In comparison, the plots in the composites exhibited nearly semi-circular shape indicating good degree of miscibility, though mixing at molecular level was absent. Thus, it indicated that the presence of filler would have resulted in phase stabilization and reduction of phase domains, as suggested earlier. In the case of blend composites with 50% and 75% PE, the silicate based composites exhibited specifically semi-circular shape of the viscosity plot. It also corroborated with the earlier findings from the TGA analysis, where a single degradation pattern was observed in the thermo-

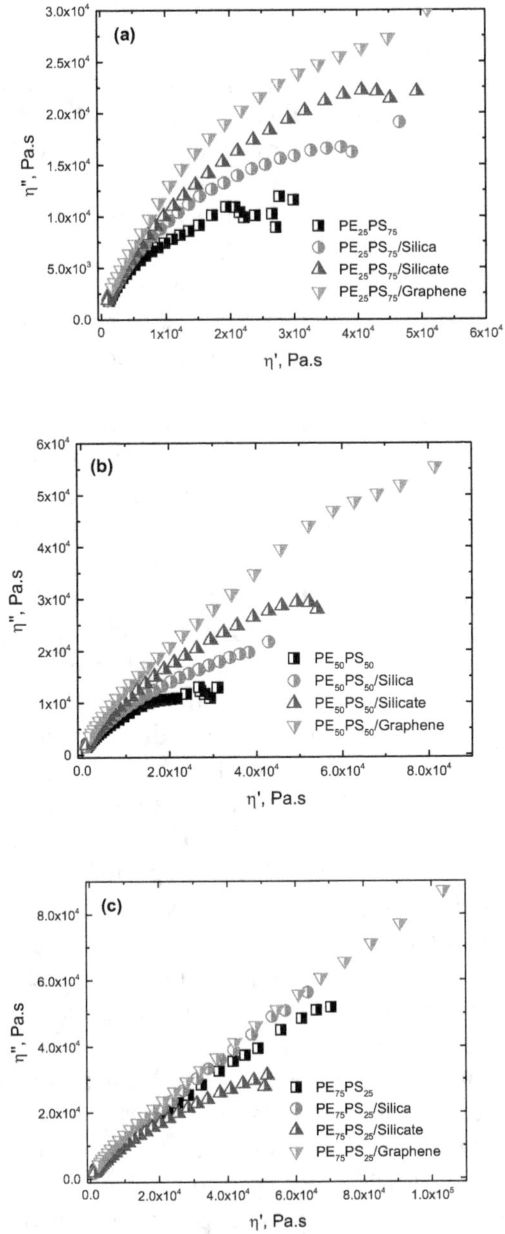

Figure 1.10 Analysis of phase miscibility thorough Cole-Cole plots (viscosity plots between η' and η'') plots for the silica, silicate and graphene hybrids with polymer blends with compositions of (a) $PE_{25}PS_{75}$, (b) $PE_{50}PS_{50}$ and (c) $PE_{75}PS_{25}$.

grams, thereby, confirming enhancement in the phase intermixing and reduction of domain sizes.

To gain further insights into the development of morphology of the composites during compounding, different processing protocols were also employed. As observed earlier, the fillers were observed to localize in the polymer phase forming the melt earlier (i.e. PE). Thus, to confirm this further, the composites were also prepared by (i) melting PE and PS, followed by filler addition, (ii) melting PE, followed by addition of graphene and PS, (iii) melting PS, followed by addition of graphene and PE, (iv) melt mixing the components together using a rotational speed of 120 rpm (7 min mixing time) or using a mixing time of 15 min (rotational speed of 80 rpm). For instance, Figure 1.11 shows the light microscopy micrographs of the $PE_{50}PS_{50}$/Graphene composite when (a) PE and PS were melted first followed by graphene addition and (b) PS was melted first followed by addition of graphene and PE. The bright phase in the micrographs represents PS, whereas the darker phase is PE. In the first case, when the polymer phases were already in the melt form before the filler was added, the filler particles did not have any specific localization, as these were observed to be uniformly distributed (observed as black domains) in both PE and PS phases in the micrograph. On the other hand, in the second case, where the filler was added to the PS melt before PE was added to the compounder, the filler thus had opportunity to mix specifically with the PS phase. This resulted in predominant localization of the filler in PS. As suggested earlier, this confirmed that the observed morphology in the composites evolved due to the kinetic mixing protocol employed to generate the nanocomposites. In other words, the filler was observed to localize in the polymer which formed the melt first.

In addition to the morphology, the effect of the different processing protocols was also studied on the composite properties. Table 1.3 demonstrates the effect of different processing conditions on the mechanical properties of $PE_{50}PS_{50}$/Graphene composite. For comparison, bottom row in the table also shows the mechanical properties of the composite generated using the earlier processing method of one step mixing at 205 °C at 80 rpm for 7 min.

As evident from the table, the composite generated with earlier processing method exhibited the best mechanical performance. Mixing all the components together at 120 rpm as well as for 15 min probably led to some degree of thermal degradation of the polymer, thus, resulting in reduced modulus. On the other hand, localizing filler

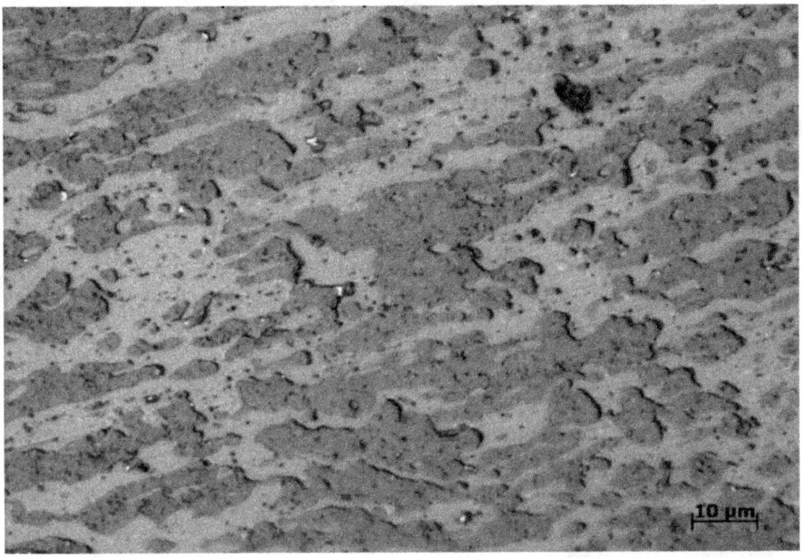

(a)

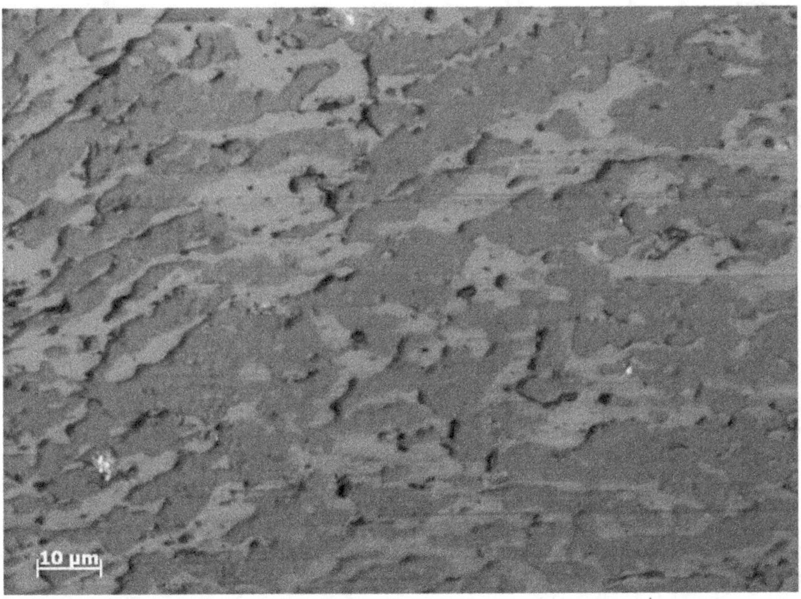

(b)

Figure 1.11 Light microscopy images of $PE_{50}PS_{50}$/Graphene composites when (a) PE and PS were melted followed by graphene addition and (b) PS was melted followed by addition of graphene and PE.

in PS was also not optimal. Even the trial where PE was melted and added with graphene before PS did not lead to similar mechanical performance as the composite where all the components are added together, which will also result in similar morphology. Longer duration of compounding in the former case would have caused some polymer degradation. Localizing filler in both PS and PS phases led to probably some aggregation or phase separation as the mechanical properties were observed to deteriorate significantly. Thus, the findings from the mechanical performance indicated that the earlier

Table 1.3 Effect of processing conditions on the mechanical performance of $PE_{50}PS_{50}$/Graphene composite

Composite	Tensile modulus[a], MPa	Peak stress[b], MPa	Peak strain[c], %	Stress at break[d], MPa
melt PE+PS followed by graphene addition and mixing for 7 min at 205 °C and 80 rpm	1034	3	0.4	3
melt PE, add graphene, followed by PS addition and mixing for 7 min at 205 °C and 80 rpm	1778	17	1.3	16
melt PS, add graphene, followed by PE addition and mixing for 7 min at 205 °C and 80 rpm	1654	13	1.1	13
add PE, PS and graphene together followed by mixing for 7 min at 205 °C and 120 rpm	1642	11	1.0	11
add PE, PS and graphene together followed by mixing for 15 min at 205 °C and 80 rpm	1746	18	1.3	18
add PE, PS and graphene together followed by mixing for 7 min at 205 °C and 80 rpm	1853	19	1.2	18

[a]Relative probable error 3%; [b]relative probable error 3%; [c]relative probable error 5%; [d]relative probable error 10%.

processing method resulted in optimum morphology with significant extent of property enhancement. In addition, the method was also efficient due to single step addition of the components, thus, reducing the mixing time as well as exposure of the polymer to high temperature for longer durations.

Table 1.4 Effect of processing conditions on the calorimetric performance of $PE_{50}PS_{50}$/Graphene composite

	Peak melting temperature T_m, °C	Melt enthalpy ΔH_m, J/g	Peak crystallization temperature T_c, °C	Crystallization enthalpy ΔH_c, J/g
melt PE+PS followed by graphene addition and mixing for 7 min at 205 °C and 80 rpm	130.2	129	111.9	112
melt PE, add graphene, followed by PS addition and mixing for 7 min at 205 °C and 80 rpm	131.9	138	112.1	133
melt PS, add graphene, followed by PE addition and mixing for 7 min at 205 °C and 80 rpm	133.7	140	111.8	138
add PE, PS and graphene together followed by mixing for 7 min at 205 °C and 120 rpm	131.5	140	111.1	133
add PE, PS and graphene together followed by mixing for 15 min at 205 °C and 80 rpm	132.2	146	114.4	138
add PE, PS and graphene together followed by mixing for 7 min at 205 °C and 80 rpm	133.1	144	116.0	148

Table 1.4 also shows the effect of processing conditions on the calorimetric performance of the $PE_{50}PS_{50}$/Graphene composite. The bottom row depicts the properties of the composites generated using earlier processing method. Similar to the mechanical performance, the composite generated using earlier processing protocol also exhibited optimal calorimetric response with high melt enthalpy and peak melting point. The peak crystallization point of the composite was also the highest among the composites, thus, indicating earlier initiation of crystallization, which might be because of the localization of filler in the semi-crystalline PE phase. The thermal degradation behavior of the composites was also similar, with no significant difference in the onset and peak degradation temperatures of the composites generated using different processing methods.

Figure 1.12 also demonstrates the evolution of complex modulus of the $PE_{50}PS_{50}$/Graphene composite as a function of different processing conditions. Similar to the mechanical and calorimetric properties, the complex modulus of the earlier developed composite was either same or higher than the other composites. Thus, these findings confirmed that the earlier followed processing methodology of one step addition of the polymers and filler phase was efficient due to faster compounding (lower energy consumption) and lower risk of polymer degradation, which thus led to the development of stable morphology with optimal composite properties.

Overall, it was observed that the morphology of the composites was significantly affected by processing method employed to generate these composites. The methodology involving single step mixing of the components was suitable to generate optimal morphology of filler stabilization of the phases and thus corresponding extent of property enhancement. In addition, the longer mixing time as well as higher rotational speed also deteriorated the composite performance.

1.4 Conclusions

Blend composites have evolved as materials of significant interest due to the potential of synergistic combination of component properties. The blends are generally created to achieve toughness, whereas the filler phase imparts stiffness to the system, thereby leading to simultaneous enhancement of these properties, which otherwise is challenging to achieve in ether blends or composites alone. These combinations of blends and fillers also lead to occasional cost saving

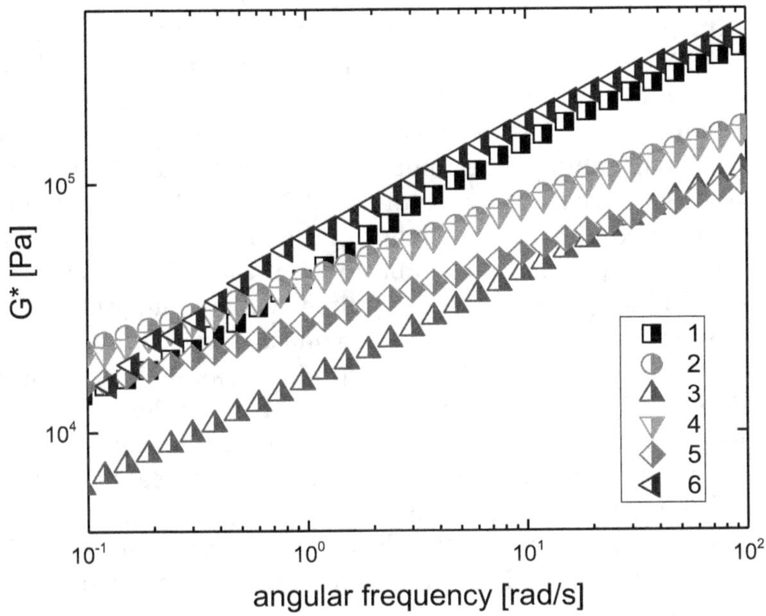

Figure 1.12 Complex modulus of the $PE_{50}PS_{50}$/Graphene composites as a function of different processing methodologies.
1: melt PE+PS followed by graphene addition and mixing for 7 min at 205 °C and 80 rpm
2: melt PE, add graphene, followed by PS addition and mixing for 7 min at 205 °C and 80 rpm
3: melt PS, add graphene, followed by PE addition and mixing for 7 min at 205 °C and 80 rpm
4: add PE, PS and graphene together followed by mixing for 7 min at 205 °C and 120 rpm
5: add PE, PS and graphene together followed by mixing for 15 min at 205 °C and 80 rpm
6: add PE, PS and graphene together followed by mixing for 7 min at 205 °C and 80 rpm

by reducing the overall cost of the composites, without sacrificing the properties. In the current study, a functional PE/PS blend system has been successfully used to provide optimal enhancement in the properties when incorporated with different fillers. One step melt mixing process could be used for generating the composites, which resulted in the kinetic embedding of the filler particles in PE phase due to its earlier melting during compounding. A fraction of the filler was also observed at the interface between PE and PS phases. In addition, the filler particles were also observed to cause reduction in the phase domains, thus, enhancing the phase intermixing between immiscible PE and PS. Different fillers like silica, silicate and graphene exhibited similar morphology.

The blends with higher fraction of PE were observed to have higher thermal stability. The composites also exhibited enhancement in the thermal stability on the addition of filler, with silicate filler imparting the maximum enhancement in the onset and peak degradation temperatures. The blends and most of the composites (except silicate containing composites) exhibited two step degradation pattern in the TGA thermograms owing to phase immiscibility. The calorimetric analysis also revealed reduction in peak melting point and crystallinity of PE on enhancing the amount of PS in the blends. In the composites, especially generated with blends containing 25 and 50% PE, the melt enthalpy of PE was observed to increase, though there was no difference in the peak meeting and crystallization temperatures, as compared to the blends. The blend and composites with 75% PS content also exhibited the glass transition of PS at slightly higher temperature of 110 °C.

The tensile modulus in the blend composites increased with filler with maximum extent of increase provided by graphene platelets as well as in the blends containing 75% PE. High aspect ratio platelets of graphene thus were beneficial in generating good degree of stress transfer. Though the silicate also contained large aspect ratio platelets, however, its high polarity may have resulted in lower degree of filler dispersion in the polymer matrix. The storage modulus of the composites also increased with the addition of filler, especially in the composite generated using the blends with 50% and 75% PE. The transition temperature of PS was also observed to shift to higher temperatures with filler as well as PS content. The complex modulus also increased with filler addition to the composites as a function of angular frequency, with maximum enhancement exhibited by graphene. Variation in the processing methodologies led to significant changes in the composite morphology. In the case of pre-melting of PS phase, the filler was also observed to localize in PS, thus, confirming the kinetic mixing of filler in the compounder. The morphology generated earlier by one step mixing of the components at optimal processing parameters (80 rpm rotational speed and 7 min mixing time) resulted in the highest extent of enhancement in the composite performance, as modulus, melt enthalpy as well as shear viscosity were observed to be maximum in this case. The generated composites, thus, open the opportunities for the development of phase stabilized polymer blend composites using a variety of fillers.

References

1. Fekete, E., Foldes, E., and Pukanszky, B. (2005) Effect of molecular interactions on the miscibility and structure of polymer blends. *European Polymer Journal*, **41**, 727-736.
2. Lipatov, Y. S. (2002) Polymer blends and interpenetrating polymer networks at the interface with solids. *Progress in Polymer Science*, **27**, 1721-1801.
3. Litmanovich, A. D., Plate, N. A., and Kudryavtsev, Y. V. (2002) Reactions in polymer blends: interchain effects and theoretical problems. *Progress in Polymer Science*, **27**, 915-970.
4. Sundararaj, U., and Macosko, C. W. (1995) Drop breakup and coalescence in polymer blends: The effects of concentration and compatibilization. *Macromolecules*, **28**, 2647-2657.
5. Katz, H. S., and Milewski, J. V. (1987) *Handbook of Fillers for Plastics*, Van Nostrand, USA.
6. Rothon, R. (1995) *Particulate-Filled Polymer Composites*, Longman, UK.
7. Brydson, J. A. (1975) *Plastic Materials*, Newnes-Butterworths, UK.
8. Karger-Kocsis, J. (1995) *Polypropylene Structure, Blends and Composites*, Chapman & Hall, UK.
9. Elias, H.-G. (2003) *Anwendungen von Polymeren*, Wiley-VCH, Germany.
10. Robertson, G. L. (1993) *Food Packaging - Principles and Practice*, volume 6 of Packaging and Converting Technology, Marcel Dekker, Inc, USA.
11. Diaz, M. F., Barbosa, S. E., and Capiati, N. J. (2007) Reactive compatibilization of PE/PS blends. Effect of copolymer chain length on interfacial adhesion and mechanical behavior. *Polymer*, **48**, 1058-1065.
12. Xu, S.-A., and Cham, C.-M. (1998) Polystyrene/high density polyethylene blends compatibilized by a tri-block copolymer I. Properties and morphology. *Polymer Journal*, **30**, 552-558.
13. Ismail, H. H., and Nasir, M. (2002) The effect of various compatibilizers on mechanical properties of polystyrene/polypropylene blend. *Polymer Testing*, **21**(2), 163-170.
14. D'orazio, L., Guarino, R., Mancarella, C., Martuscelli, E., and Cecchin, G. (1997) Effects of the addition of a graft copolymer of propylene with styrene. *Journal of Applied Polymer Science*, **65**(8), 1539-1553.
15. Aziz, A. A., Akil, H. M., Jamaludin, S. M. S., and Ramli, N. A. M. (2011)

The effect of multiple compatibilizers on the impact properties of polypropylene/polystyrene (PP/PS) blend. *Polymer-Plastics Technology and Engineering*, **50**(8), 768-775.

16. Kallel, T., Massardier-Nageotte, V., Jaziri, M., Gerard, J.-F., and Elleuch, B. (2003) Compatibilization of PE/PS and PE/PP blends. I. Effect of processing conditions and formulation. *Journal of Applied Polymer Science*, **90**(9), 2475-2484.
17. Galloway, J. A., Jeon, H. K., Bell, J. R., and Macosko, C. W. (2005) Block copolymer compatibilization of cocontinuous polymer blends. *Polymer*, **46**, 183-191.
18. Osman, M. A., Mittal, V., and Suter, U. W. (2007) Poly(propylene)-layered silicate nanocomposites: Gas permeation properties and clay exfoliation. *Macromolecular Chemistry and Physics*, **208**, 68-75.
19. van Ballegooie, P., and Rudin, A. (1988) Reactive extrusion of polystyrene/polyethylene blends. *Polymer Engineering & Science*, **28**(21), 1434-1442.
20. Lan, T., Kaviratna, P. D., and Pinnavaia, T. J. (1994) On the nature of polyimide-clay hybrid composites. *Chemistry of Materials*, **6**, 573-575.
21. Chin, I.-J., Thurn-Albrecht, T., Kim, H.-C., Russell, T. P., and Wang, J. (2001) On exfoliation of montmorillonite in epoxy. *Polymer*, **42**, 5947-5952.
22. Lim, S. K., Kim, J. W., Chin, I.-J., Kwon, Y. K., and Choi, H. J. (2002) Preparation and interaction characteristics of organically modified montmorillonite nanocomposite with miscible polymer blend of poly(Ethylene Oxide) and poly(Methyl Methacrylate). *Chemistry of Materials*, **14**, 1989-1994.
23. Wang, Z., and Pinnavaia, T. J. (1998) Nanolayer reinforcement of elastomeric polyurethane. *Chemistry of Materials*, **10**, 3769-3771.
24. Yano, K., Usuki, A., and Okada, A. (1997) Synthesis and properties of polyimide-clay hybrid films. *Journal of Polymer Science, Part A: Polymer Chemistry*, **35**, 2289-2294.
25. Shi, H., Lan, T., and Pinnavaia, T. J. (1996) Interfacial effects on the reinforcement properties of polymer-organoclay nanocomposites. *Chemistry of Materials*, **8**, 1584-1587.
26. Kuilla, T., Bhadra, S., Yao, D., Kim, N. H., Bose, S., and Lee, J. H. (2010) Recent advances in graphene based polymer composites. *Progress in Polymer Science*, **35**, 1350-1375.
27. Cai, D., and Song, M. (2010) Recent advance in functionalized graphene/polymer nanocomposites. *Journal of Materials Chemistry*, **20**, 7906-7915.
28. Steurer, P., Wissert, R., Thomann, R., and Muelhaupt, R. (2009) Functionalized graphenes and thermoplastic nanocomposites based upon expanded graphite oxide. *Macromolecular Rapid Communications*, **30**, 316-327.

29. Nguyen, D. A., Lee, Y. R., Raghu, A. V., Jeong, H. M., Shin, C. M., and Kim, B. K. (2009) Morphological and physical properties of a thermoplastic polyurethane reinforced with functionalized graphene sheet. *Polymer International*, **58**, 412-417.
30. Fang, M., Wang, K., Lu, H., Yang, Y., and Nutt, S. (2009) Covalent polymer functionalization of graphene nanosheets and mechanical properties of composites. *Journal of Materials Chemistry*, **19**, 7098-7105.
31. Chaudhry, A. U., and Mittal, V. (2013) High-density polyethylene nanocomposites using masterbatches of chlorinated polyethylene/graphene oxide. *Polymer Engineering and Science*, **53**, 78-88.
32. Sinha Ray, S., Pouliot, S., Bousmina, M., and Utracki, L. A. (2004) Role of organically modified layered silicate as an active interfacial modifier in immiscible polystyrene/polypropylene blends. *Polymer*, **45**(25), 8403-8413.
33. Wahit, M. U., Hassan, A., Rahmat, A. R., Lim, J. W., and Ishak, Z. A. M. (2006) Effect of organoclay and ethylene-octene copolymer inclusion on the morphology and mechanical properties of polyamide/polypropylene blends. *Journal of Reinforced Plastics and Composites*, **25**(9), 933-955.
34. Austin, J. R., and Kontopoulou, M. (2006) Effect of organoclay content on the rheology, morphology, and physical properties of polyolefin elastomers and their blends with polypropylene. *Polymer Engineering and Science*, **46**(11), 1491-1501.
35. Yang, H., Zhang, X., Qu, C., Li, B., Zhang, L., Zhang, Q., and Fu, Q. (2007) Largely improved toughness of PP/EPDM blends by adding nano-SiO_2 particles. *Polymer*, **48**(3), 860-869.
36. Si, M., Araki, T., Ade, H., Kilcoyne, A. L. D., Fisher, R., Sokolov, J. C., and Rafailovich, M. H. (2006) Compatibilizing bulk polymer blends by using organoclays. *Macromolecules*, **39**(14), 4793-4801.
37. Kim, J., Cote, L. J., Kim, F., Yuan, W., Shull, K. R., and Huang, J. (2010) Graphene oxide sheets at interfaces. *Journal of the American Chemical Society*, **132**(23), 8180-8186.
38. Kim, H., Abdala, A. A., and Macosko, C. W. (2010) Graphene/polymer nanocomposites. *Macromolecules*, **43**(16), 6515-6530.
39. Cao, Y., Zhang, J., Feng, J., and Wu, P. (2011) Compatibilization of immiscible polymer blends using graphene oxide sheets. *ACS Nano*, **5**(7), 5920-5927.
40. Elias, L., Fenouillot, F., Majeste, J. C., Alcouffe, P., and Cassagnau, P. (2008) Immiscible polymer blends stabilized with nano-silica particles: Rheology and effective interfacial tension. *Polymer*, **49**(20), 4378-4385.
41. Thareja, P., and Velankar, S. (2007) Particle-induced bridging in immiscible polymer blends. *Rheologica Acta*, **46**(3), 405-412.
42. Pawar, S. P., Misra, A., Bose, S., Chatterjee, K., and Mittal, V. (2015)

Enzymatically degradable and flexible bio-nanocomposites derived from PHBV and PBAT blend: Assessing thermal, morphological, mechanical, and biodegradation properties. *Colloid and Polymer Science*, **293**, 2921-2930.

43. Tanaka, H., Lovinger, A. J., and Davis, D. D. (1994) Pattern evolution caused by dynamic coupling between wetting and phase separation in binary liquid mixture containing glass particles. *Physical Review Letters*, **72**(16), 2581-2584.
44. Ginzburg, V. V., Qiu, F., Paniconi, M., Peng, G., Jasnow, D., and Balazs, A. C. (1999) Simulation of hard particles in a phase-separating binary mixture. *Physical Review Letters*, **82**(20), 4026-4029.
45. Nesterov, A. E. (1999) Compatibilizing effect of a filler in binary polymer mixtures. *Polymer*, **40**(5), 1347-1349.
46. Nesterov, A. E. (2001) Effect of an interface with solid on the component distribution in separated phases of binary polymer mixtures. *European Polymer Journal*, **37**(2), 281-285.
47. Hong, J. S., Namkung, H., Ahn, K. H., Lee, S. J., and Kim, C. (2006) The role of organically modified layered silicate in the breakup and coalescence of droplets in PBT/PE blends. *Polymer*, **47**(11), 3967-3975.
48. McAllister, M. J., Li, J. L., Adamson, D. H., Schniepp, H. C., Abdala, A. A., Liu, J., Herrera-Alonso, M., Milius, D. L., Car, R., Prud'homme, R. K., and Aksay, I. A. (2007) Single sheet functionalized graphene by oxidation and thermal expansion of graphite. *Chemistry of Materials*, **19**, 4396.
49. Mittal, V., and Chaudhry, A. U. (2015) Polymer-graphene nanocomposites: Effect of polymer matrix and filler amount on properties. *Macromolecular Materials and Engineering*, **300**(5), 510-521.
50. Joshi, M., Butola, B. S., Simon, G., and Kukaleva, N. (2006) Rheological and viscoelastic behavior of HDPE/octamethyl-POSS nanocomposites. *Macromolecules*, **39**, 1839-1849.
51. Cho, K., Lee, B. H., Hwang, K. M., Lee, H., and Choe, S. (1998) Rheological and mechanical properties in polyethylene blends. *Polymer Engineering and Science*, **38**, 1969-1975.
52. Kim, H. K., Rana, D., Kwag, H., and Choe, S. (2001) Melt rheology of ethylene 1-octene copolymer blends synthesized by Ziegler-Natta and metallocene catalysts. *Korea Polymer Journal*, **8**, 34-43.
53. Kwag, H., Rana, D., Cho, K., Rhee, J., Woo, T., Lee, B. H., and Choe, S. (2000) Binary blends of metallocene polyethylene with conventional polyolefins: Rheological and morphological properties. *Polymer Engineering and Science*, **40**, 1672-1681.

2

Polyethylene-Copolymer Blends: Morphology and Properties

2.1 Introduction

Polymer blends are generated by the mixing of two or more polymers together in order to combine the characteristics of individual polymers. The compatibilization of the phases in the blends is also achieved by a number of physical or chemical means. Polymer blends are widely used materials in the modern polymer industry owing to their wider range of properties such as performance combinations, aid in processing, low capital costs as compared with development of new monomers and polymers of required properties, etc. As a result, polymer blends find application in numerous fields such as adhesion, colloidal stability and design of composite materials [1]. The dispersed phase in the many blend systems can also acquire a specific morphology which is beneficial for applications such as barrier performance achieved by the dispersed phase acquiring the platy morphology. Both solution as well as melt mixing techniques to generate the polymer blends have found their application, though melt mixing is more environmentally friendly method due to the absence of any additional solvent. On the other hand, the melt mixing method requires the use of high temperature and shear for the generation of polymer blends, which may degrade the heat sensitive polymers, thus, requiring optimization of the mixing conditions so as to achieve good mixing, however, also avoiding extensive thermal degradation. The studies on the polymer blends have significantly increased in recent years leading to the generation of functional blend systems with superior properties [2-10]. Both miscible and immiscible blends have been studied along with new pathways to achieve compatibilization

Vikas Mittal[a,], M. Suria[a,**], Nadejda Matsko[b]*
[a]The Petroleum Institute (part of Khalifa University of Science and Technology), Abu Dhabi, UAE; [b]Graz Centre of Electron Microscopy, Graz, Austria
**Current address: Bletchington, Wellington County, Australia; **Current address: University of Waterloo, Canada*

of the phases in the immiscible blends.

Polyolefins represent the largest group of industrial thermoplastics in terms of production and have superior rank among commodity plastics owing to their use in a large number of applications. The number of individual polyolefin members is limited and the number of applications based on such materials is constantly increasing. Thus, various polyolefin-based blends and composites are required to meet the increasing demand of materials with required properties and processing suitable for specific applications [11-13]. For instance, Wang *et al.* [7] reported polyolefin based blends with styrene polymers and engineering plastics and developed effective strategy to compatibilize the blends by using a polyolefin based multiphase compatibilizer. Chlorinated polyethylene (CPE) blends with different polymers have also been commonly reported for many purposes such as improvements in the toughening properties of host polymers [14], as a compatibilization agent through enhancement of interfacial interactions between two polymers [15,16], improved processing properties [17], ameliorated adhesion properties [18], enhanced ignition resistance properties [19], and modified gas transport properties [20]. In a recent study, Mittal and Matsko [21] studied the morphology of polyethylene-CPE blends for the purpose of generating materials for reduced low temperature brittleness. CPE materials of two different chlorination levels (25% and 35%) were employed at 5 wt% content. Figure 2.1 shows the AFM height images of the PE and PE-CPE block faces prepared by microtoming at -140 °C. Surface topography analysis (surface height or roughness variation) at 3 different places on the sample surface was also performed. It is noteworthy that all parameters related to hardness of conventional PE, such as the elastic constants, tensile strength, and crushing strength increased with decreasing temperature. Also, the elastic region of the stress-strain curve was extended to higher stress values with decreasing temperature. These results were consistent with the experience that cryo-sectioning of polycrystalline PE becomes more difficult at lower temperature. Incorporation of the amorphous CPE (CPE35) to PE, however, resulted in completely different behavior. Higher height variation in its micrograph indicated higher degree of compression which represents compressibility and flexibility of polymer materials during sectioning. The extent of plasticization was also higher in the case of amorphous CPE35 polymer where the height variation was significantly large as compared with PE blend with CPE25.

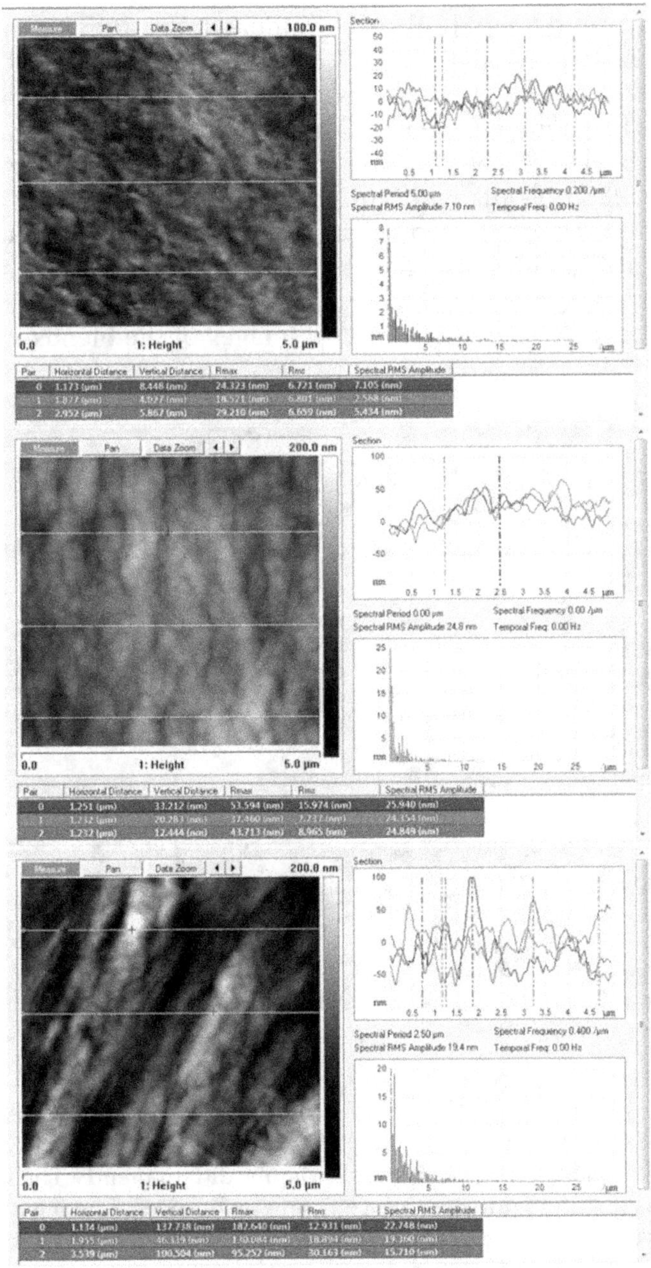

Figure 2.1 AFM height images and corresponding height corrugation analysis for PE (top image), PE+5%CPE25 (middle row) and PE+5%CPE35 (bottom row) when sectioned at -140 °C. Reproduced from Reference 21 with permission from Springer.

Chaudhry and Mittal [22] also expanded their study on the blends of high density polyethylene (HDPE) with CPE in the weight fraction range from 1% to 30%. The rheological analysis of the data using Cole-Cole, Han-Chuang and van Gurp plots confirmed the miscibility of CPE25 blends (except for 30% CPE25 blend at lower frequency) whereas CPE35 blends with 10-30% CPE content were immiscible. Highest increase in the rheological properties (complex moduli) was observed at 2% CPE content. The optical microscopy also confirmed the better mixing of CPE25 polymers in HDPE than CPE35. For instance, Figure 2.2 shows the optical images of the blends with both

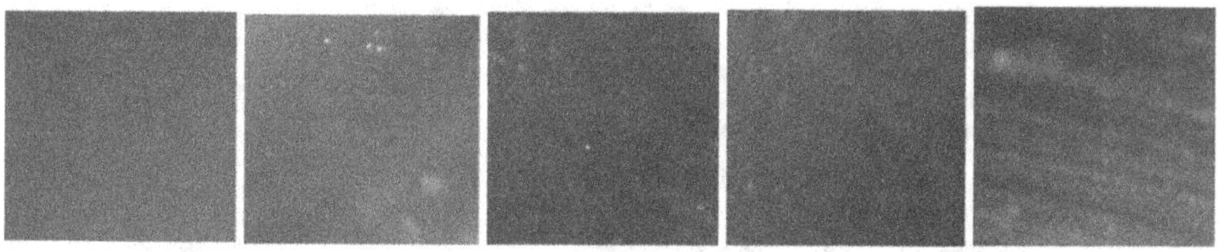

Left to right: (a)-(e)

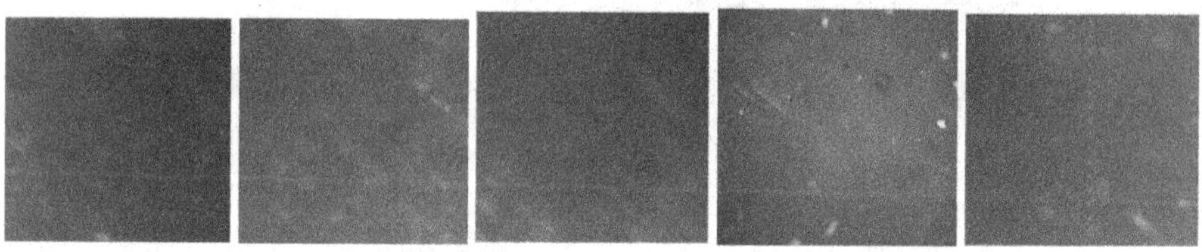

Left to right: (f)-(j)

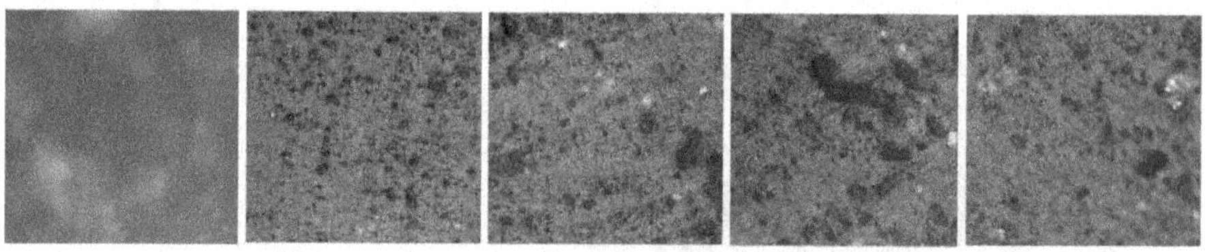

Left to right: (k)-(o)

Figure 2.2 Optical micrographs of (a) pure HDPE, (b) 1% CPE25, (c) 2% CPE25, (d) 5% CPE25, (e) 10% CPE25, (f) 15% CPE25, (g) 20% CPE25, (h) 30% CPE25, (i) 1% CPE35, (j) 2% CPE35, (k) 5% CPE35, (l) 10% CPE35, (m) 15% CPE35, (n) 20% CPE35, and (o) 30% CPE35 blends. The width of the images reads 500 μm. Reproduced from Reference 22 with permission from Wiley.

CPE25 and CPE35 as a function of blend composition. The CPE25 blends with 1%, 2%, and 5% CPE concentration (Figure 2.2b-d) were similar to pure HDPE (Figure 2.2a), thus, indicating molecular mixing of the two phases. The blends with 10%, 15%, 20%, and 30% CPE25

(Figure 2.2e-h, respectively) had small whitish regions indicating CPE25 phase, but these were observed to be dispersed homogenously. In the case of CPE35 blends, the lower concentrations of CPE did lead to good mixing (Figure 2.2i-k for 1%, 2%, and 5%, respectively). However, beyond 5%, the immiscibility of the two phases was extensive and large domains of CPE35 (as black areas in the bright field images of Figure 2.2l-o for 10%, 15%, 20%, and 30% CPE35 content) were clearly visible. The mechanical properties of the CPE25 blends were also superior than the corresponding CPE35 blends especially at higher CPE concentration where effects of immiscibility as well as matrix plasticization played a role.

In order to further explore the development of polyethylene based functional materials, especially for low temperature flexibility, blends of polyethylene with a variety of copolymers have been generated in the current study. The copolymers employed for this purpose were polyethylene-*g*-maleic anhydride copolymer (PE-MA), ethylene methacrylic acid copolymer with zinc ion (EMAZ) and ethylene vinyl acetate-*g*-maleic anhydride (EVA-MA). The amount of the copolymers was also varied in polyethylene in order to study the effect on miscibility and resulting blend properties. In addition, the effect of copolymer molecular weight was also analyzed in the blends by employing PE-MA copolymers of two different molecular weights. Melt mixing process was employed for the generation of blends due to its industrial acceptability. The compounding parameters were also optimized to ensure effective mixing without thermal degradation of the polymers.

2.2 Experimental

2.2.1 Materials

HDPE (BB2581) of specific gravity 0.958 was supplied by Abu Dhabi Polymers Company Limited (Borouge), UAE. Various copolymers of different specifications and functionalities were procured from suppliers, as mentioned in Table 2.1. The polymer materials were used as obtained.

2.2.2 Generation of Blends

Polymer blends were prepared by melt mixing of the copolymers with HDPE using mini twin conical screw extruder (MiniLab HAAKE

Table 2.1 Specifications of HDPE and copolymers

Polymer/ Compatibilizer	Supplier and trade name	MFI, g/10 min (2.16 kg, 190 °C)	Density, g/cm³	Peak melting temperature, °C
PE	Borouge BB2581	0.35	0.958	147
EVA-MA	Dupont Fusabond C190	16	0.950	71
EMAZ	Dupont Surlyn 9020	1	0.96	85
PE-MA1	Aldrich PE-*g*-MA	500 cP viscosity (140 °C)	--	--
PE-MA2	Aldrich PE-*g*-MA	1700-4500 cP viscosity (140 °C)	--	--

Rheomex CTW5, Germany). A mixing temperature of 185 °C for 7 min at 80 rpm with batch size of 5 g was used. The screw length and screw diameter were 109.5 mm and 5/14 mm conical, respectively. Blends of copolymers in weight percentages 5%, 10%, and 15%, were generated. The disc and dumbbell-shaped test samples were prepared by mini injection molding machine (HAAKE MiniJet, Germany) at a processing temperature of 185 °C. The injection pressure was 700 bar for 6 s, whereas holding pressure was 400 bar for 3 s. The temperature of the mold was kept at 55 °C.

2.2.3 Characterization Techniques

Thermal properties of the blends were recorded using Netzsch thermogravimetric analyzer (TGA). Nitrogen was used as a carrier gas and the scans were obtained from 50 to 700 °C at a heating rate of 10 °C/min. Calorimetric properties of blends were recorded on a Netzsch DSC under nitrogen atmosphere. The scans were obtained from 30-180-30 °C using heating and cooling rates of 10 °C/min and 5 °C/min, respectively. The heat enthalpies, useful to estimate the extent of crystallinity, were recorded in a narrow error range (±0.1%), which were also confirmed by repeated runs.

AR 2000 rheometer from TA Instruments was used to characterize the rheological properties of the blends such as storage modulus (G'), loss modulus (G''), and complex viscosity (η^*). Disc shaped samples of 25 mm diameter and 2 mm thickness were measured at 185 °C using a gap opening of 1.2 mm. Strain sweep scans were recorded at ω = 1 rad/s from 0.1 to 100% strain and the samples were observed to be shear stable up to 10% strain. For the comparison, frequency sweep scans (dynamic testing) of all the samples were recorded at 2% strain from ω = 0.1 to 100 rad/s.

The mechanical testing of the blends was performed on universal testing machine (Testometric, UK). The dumbbell-shaped samples with 53 mm length, 4 mm width, and 2 mm thickness were used. A loading rate of 5 mm/min was employed and the tests were carried out at room temperature. WinTest Analysis software was used for the calculation of tensile modulus and yield stress properties of the blends. An average of five values was calculated.

For the microscopy analysis, the samples were mounted in special holders that at the same time fit in the microtome and were also suitable for the examination of the block face by atomic force microscopy (AFM). The block faces of specimens after cryo-ultramicrotomy were investigated at ambient conditions using Dimension 3100 AFM/SPM (Veeco, USA) atomic force microscope. AFM images were collected in tapping mode using silicon nitride cantilevers with natural frequencies in the 300 kHz range (force constant = 20 $N.m^{-1}$, tip radius = 10 nm (NT-MDT, Russia)). AFM image processing was performed using Nanoscope v720 software (Veeco, USA). The block faces of specimens after cryo-ultramicrotomy were also investigated at ambient conditions using ZEISS Axioplan light microscope (Zeiss, Germany) equipped with Zeiss Axio Cam ICc 1 CCD camera in reflected polarized light.

Wide-angle X-ray diffraction (WAXRD) analysis of the blends was performed using a Panalytical powder diffractometer (X'Pert PRO) using Cu Kα radiation (λ = 1.5406 Å) in reflection mode. A zero-background holder was used to minimize the noise. The samples were step-scanned from 2Θ = 10-70° at room temperature using a step size of 2Θ = 0.02° and a step time of 10 s. Time studies were also carried out by measuring the same samples after 90 days so as to ascertain any change in the crystal structure.

2.3 Results and Discussion

The choice of the copolymers in the study was made on the basis of functional groups. The nonpolar component of the compatibilizers was based on polyethylene in order to attain compatibility with the matrix polymer. The miscibility of these copolymers with PE can also subsequently lead to the compatibilization of PE with reinforcing fillers through the polar functional groups. Many studies have reported the blending of these copolymers with PE, however, no systematic study was performed where the effect of these copolymers was studied on the morphology as well as properties as a function of their molecular architecture and content in the blend. For instance, the effect of maleic anhydride on the PE microstructure and properties was studied through EVA-MA as well as PE-MA copolymers. In addition, the ionomer EMAZ was used to study the effect of ionic species on the polymer matrix. The melt flow indices of the compatibilizers varied between 1 and 20, thus, also indicating a wide variation in their molecular weight characteristics. As mentioned earlier, PE-MA of two different molecular weight characteristics were employed to study the molecular weight effect on the blend morphology and properties. Comparison with the MFI of PE of 0.35 also revealed that the copolymers had much lower molecular weight than PE. In addition, the copolymers were semi-crystalline in nature with significantly lower crystallinity and peak melting points than pure PE.

Figure 2.3 demonstrates the TGA thermograms of PE blends with PE-MA1, PE-MA2, EMAZ and EVA-MA as a function of blend composition. Owing to their low molecular weight, no enhancement in the thermal stability of PE was expected on the addition of copolymers, however, it was necessary to ensure that the thermal stability of the PE is not significantly reduced on the addition of these copolymers, which would subsequently affect the high temperature applications. At 5% and 10% content of PE-MA1 in the blend, slight reduction in the onset and peak degradation temperature of PE was observed. Further enhancing the content of PE-MA1 in the blend led to significant reduction in the onset of degradation temperature of PE, as observed in Figure 2.3a. However, all the blends were still stable till 350 °C. In addition, the degradation pattern of the polymer was not impacted on the addition of PE-MA1 copolymer as confirmed by the single degradation step for PE and blends. Addition of higher molecular weight PE-MA2 to PE exhibited better thermal performance as compared to PE-MA1. No change in the onset and peak degradation temperature of PE was observed even after the incorporation of 15% of PE-MA2, thus, confirming superior thermal performance of PE/PE-MA2 blends. Similar to PE-MA1 blends, the PE-MA2 blends also exhibited single degradation pattern. In the case of EMAZ based blends, the thermal degradation of PE was retained till 10% EMAZ content in the blend. However, a reduction in the onset and peak degradation temperature was observed on further enhancing the EMAZ content to 15%. Similar to PE-MA1, it indicated that the optimal amount of copolymer for addition in PE was 10%, beyond which the thermal performance reduction was significant. Similarly, addition of 5% EVA-MA copolymer to PE also exhibited no effect on the thermal degradation of PE. However, addition of 10% and 15% copolymer to PE correspondingly reduced the onset and peak degradation temperature of PE. Thus, it can be concluded that even though all the blends were still stable up to 350 °C, 10% amount in PE was optimal as it largely retained the thermal performance.

Table 2.2 and Figures 2.4-2.5 represent the calorimetric behavior of PE blends with PE-MA1, PE-MA2, EMAZ and EVA-MA. In the blends

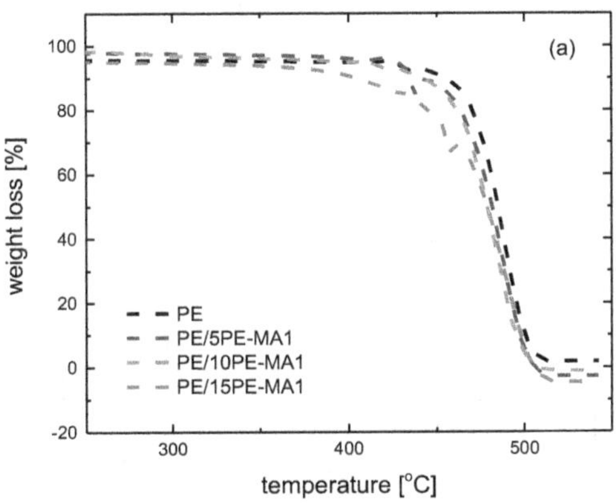

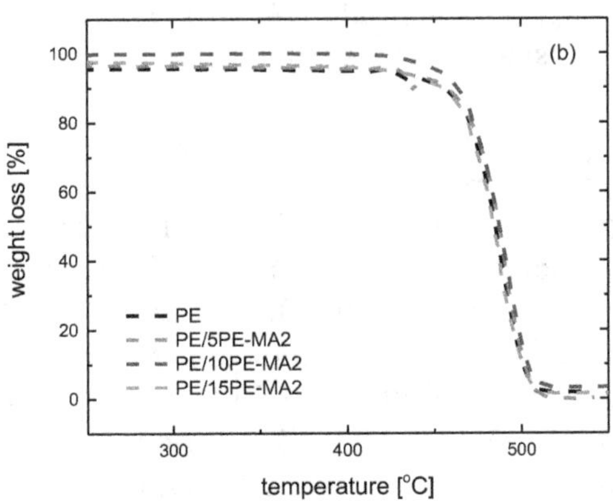

Figure 2.3 (continued on next page)

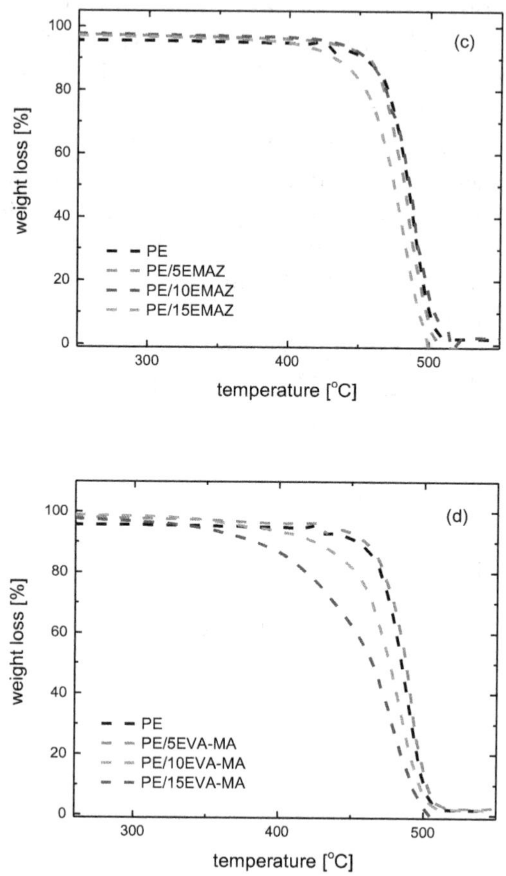

Figure 2.3 TGA thermograms of (a) PE-MA1, (b) PE-MA2, (c) EMAZ and (d) EVA-MA blends (increasing copolymer content when moving downwards).

with EMAZ and EVA-MA, a slight hump at the corresponding copolymer melting temperatures of 85 °C and 71 °C was also observed in the melt thermograms of the blends, as seen in Figure 2.4c and d. The magnitude of the copolymer peak was also observed to increase with increasing copolymer amount in the blend, thus, indicating phase immiscibility on enhancing the copolymer content. In an earlier study [23], the melt transition attributed to PE was reported to exhibit dual peaks with PE-MA, EVA-MA and EMAZ copolymers at 5 wt% content, in the presence of filler. In the current study, this behavior was not observed, though melting transitions corresponding to EMAZ and EVA-MA were observed in the thermograms. PE-MA based blends were observed to have a single melt transition peak, thus, indicating better phase miscibility. It has also to be noted that the observed changes in the current study would have resulted due to the use of different processing conditions in two studies. Though better phase intermixing was achieved in the current study with PE-MA copolymers, however, it can be expected that the copolymers were not molecularly or thermodynamically mixed with PE owing to the absence of any interacting force. The blends were observed to retain their microstructure as no change in the calorimetric behavior was observed

Table 2.2 Calorimetric properties of PE blends

Material	Peak melting temperature, °C	Melt enthalpy, J/g	Peak crystallization temperature, °C
PE	138.4	154	117.0
PE/5PE-MA1	137.8	148	116.1
PE/10PE-MA1	136.6	136	115.7
PE/15PE-MA1	135.0	130	115.4
PE/5PE-MA2	136.7	148	116.6
PE/10PE-MA2	136.1	139	116.2
PE/15PE-MA2	135.1	120	116.2
PE/5EMAZ	138.2	132	115.8
PE/10EMAZ	137.2	120	115.2
PE/15EMAZ	136.6	116	115.1
PE/5EVA-MA	138.3	129	115.5
PE/10EVA-MA	136.9	121	115.1
PE/15EVA-MA	135.1	114	114.4

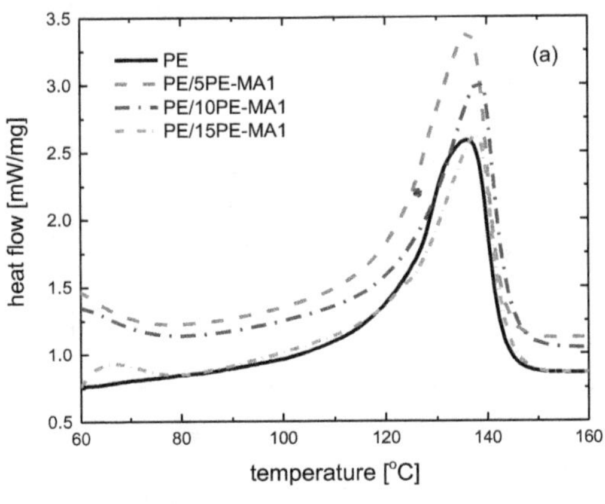

Figure 2.4 (continued on next page)

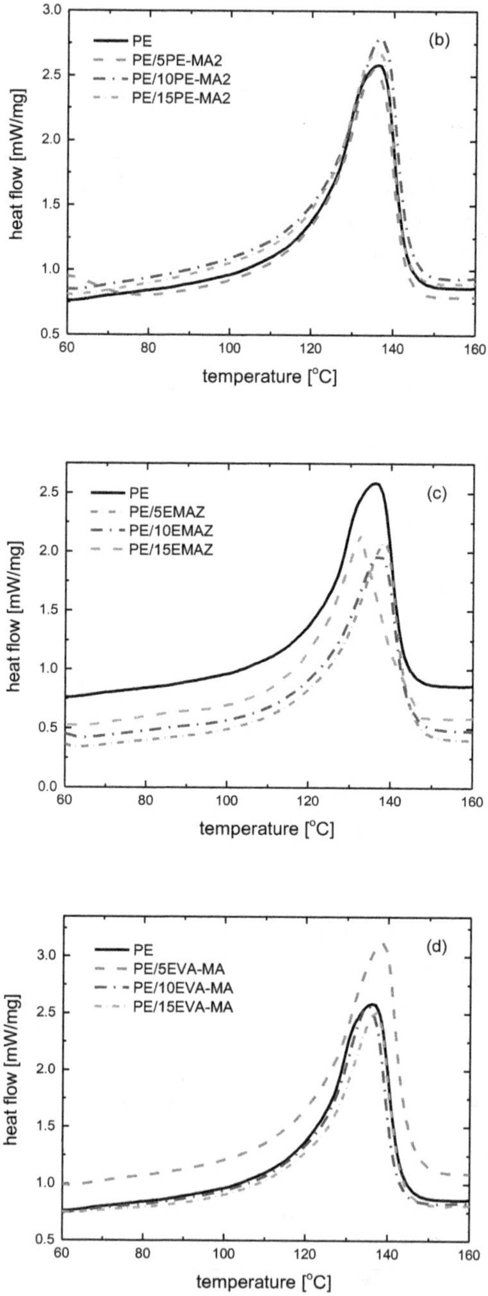

Figure 2.4 Melting thermograms of (a) PE/PE-MA1, (b) PE/PE-MA2, (c) PE/EMAZ and (d) PE/EVA-MA blends.

even after few processing cycles.

All copolymers led to a slight reduction in the peak melting temperature of PE. In addition, the extent of reduction was a function of the copolymer content in the blend. Similar to the peak melting point, the melt enthalpy of PE was also observed to decrease with the addition of copolymers, thus, indicating the hindrance to polymer crystallization in the presence of copolymer chains. Maximum reduction in the melt enthalpy was observed in the case of EVA-MA and EMAZ copolymers. For instance, the blends with 15% EMAZ and EVA-MA exhibited the melt enthalpy values of 116 J/g and 114 J/g, in comparison with 154 J/g for pure PE. On the other hand, PE-MA1 and PE-MA2 based blends with 15% copolymer content exhibited the melt enthalpy values of 130 J/g and 120 J/g, respectively. As both EMAZ and

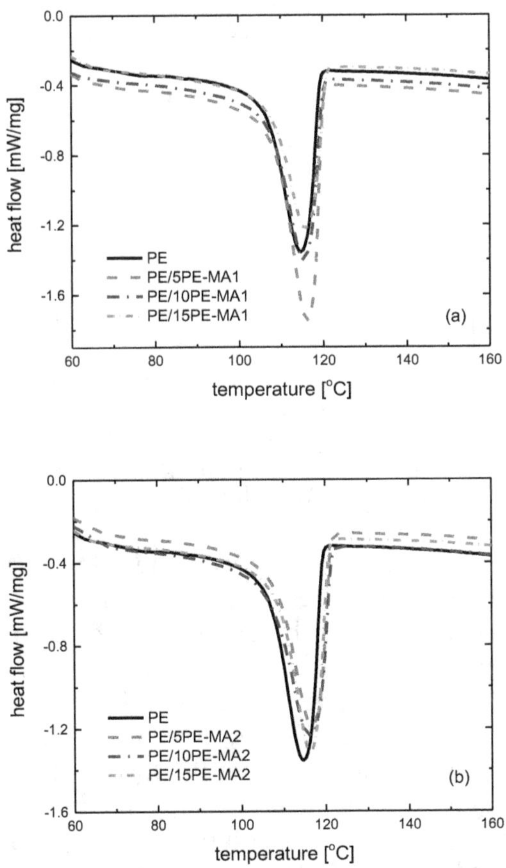

Figure 2.5 (continued on next page)

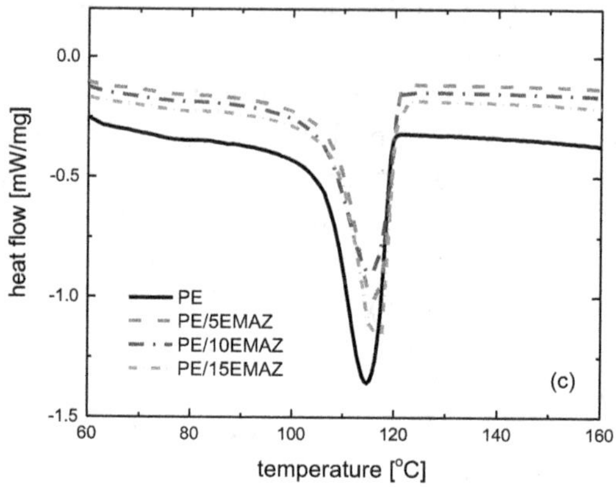

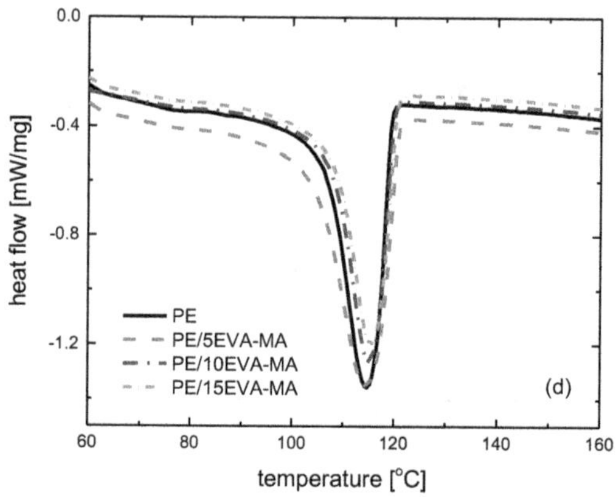

Figure 2.5 Crystallization thermograms of (a) PE/PE-MA1, (b) PE/PE-MA2, (c) PE/EMAZ and (d) PE/EVA-MA blends.

EVA-MA copolymers were also observed to have lower degree of miscibility with PE, it would have resulted in higher extent of hindrance to polymer crystallization, thereby decreasing the degree of crystallinity of PE. The peak crystallization temperature of PE though decreased slightly from 117 °C to 114-116 °C in the blends, but the amount of copolymer did not affect it, thus, indicating no nucleation effect. The DSC studies carried out after 90 days of blend preparation also revealed no change in the melt enthalpy and peak melting/crystallization temperatures. As both polymers in the blend are above their respective glass transition temperatures (T_g) at room temperature, thus, no contraction/expansion of the polymer pairs would be expected. In the case of polymer blends with one component above and other below T_g at room temperature, the contraction/expansion phenomenon has been reported to cause significant changes in the microstructure and crystallinity of the phases in the blends, also called physical ageing [24].

The microstructure of the blends was also studied by X-ray diffraction, as shown in Figure 2.6. The incorporation of copolymers did not affect the major diffraction peaks attributed to PE crystals, as also observed earlier from DSC analysis. However, small changes in the 17-22° 2Θ angle range were observed in the case of EMAZ and EVA-MA containing blends, which may also confirm the findings from DSC about more significant changes in the crystallinity and morphology due to EVA-MA and EMAZ as compared to both PE-MA1 and PE-MA2

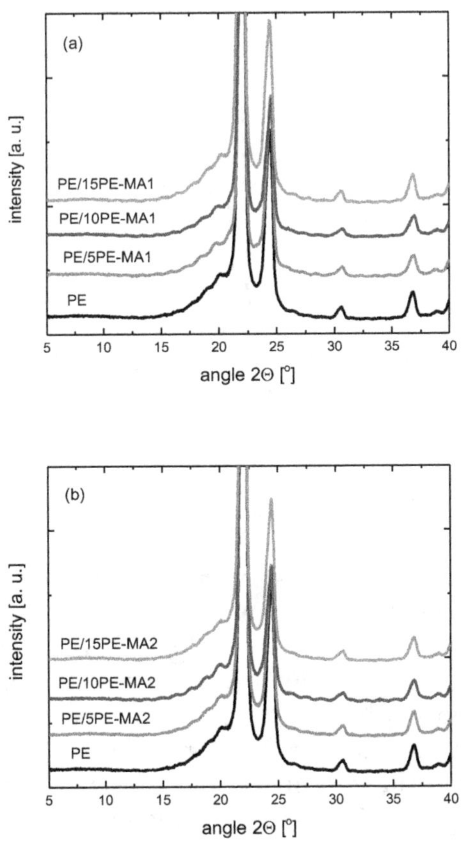

Figure 2.6 (continued on next page)

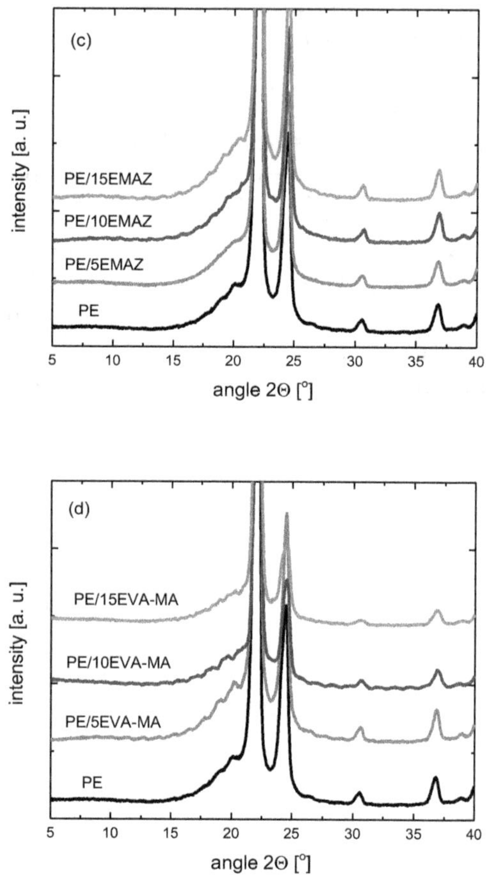

Figure 2.6 X-ray diffractograms of (a) PE/PE-MA1, (b) PE/PE-MA2, (c) PE/EMAZ and (d) PE/EVA-MA blends.

copolymers, however, these observations were required to be further confirmed through microscopy analysis.

Further insights into the morphology of the blends were also gained through microscopy. Figure 2.7 shows the light microscopy images of the blends with 15 wt% copolymer content at two different magnifications. At the microscale, the morphology of the blends was observed to be uniform with no obvious phase distribution of the PE and copolymer phases. Similar morphology was also observed for other blend compositions. Further morphological features about the blends were observed in the atomic force microscopy analysis of the blends, as demonstrated in Figures 2.8 to 2.11. Specifically, Figure 2.8

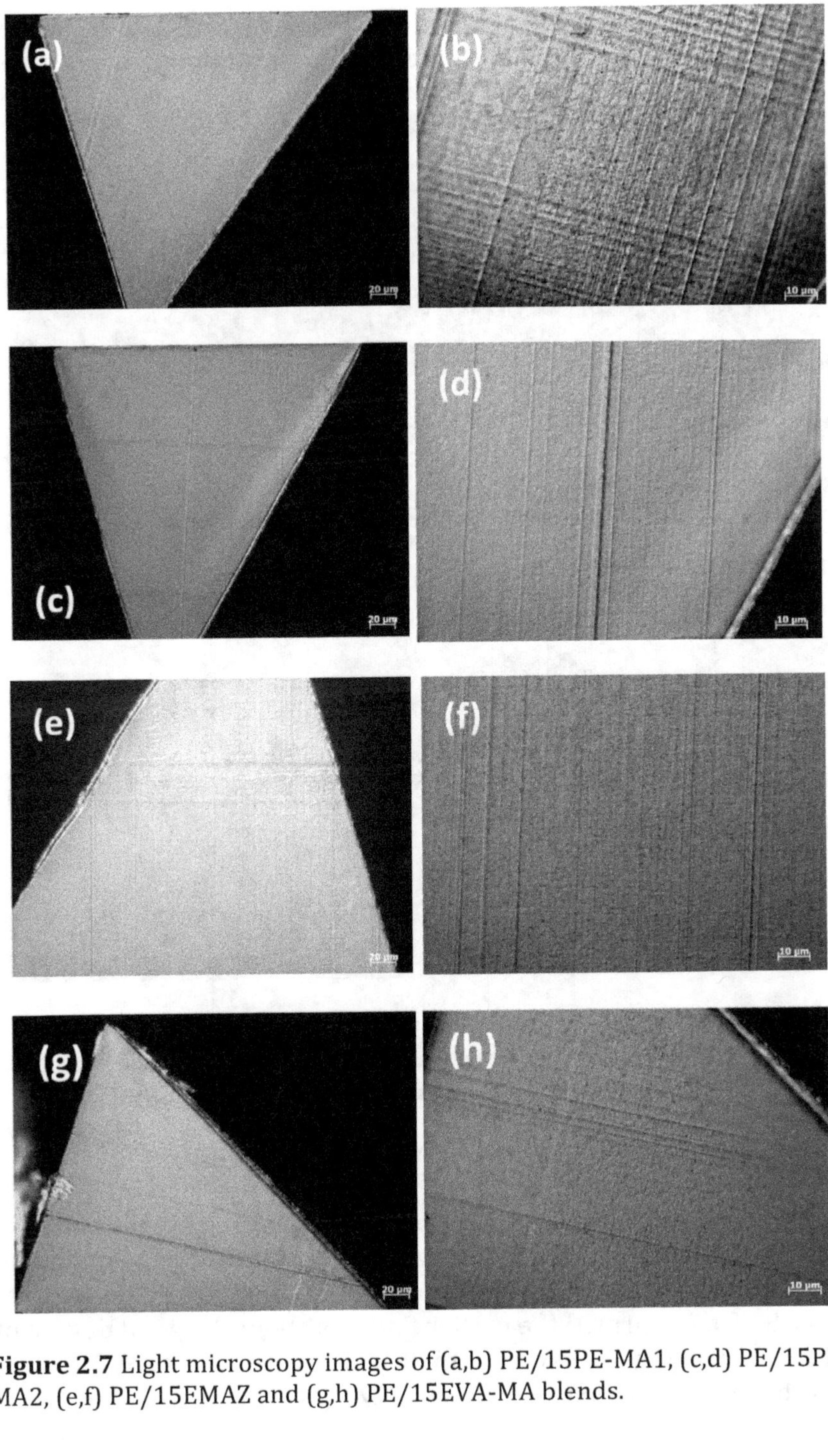

Figure 2.7 Light microscopy images of (a,b) PE/15PE-MA1, (c,d) PE/15PE-MA2, (e,f) PE/15EMAZ and (g,h) PE/15EVA-MA blends.

represents the height (Figure 2.8a) and phase (Figure 2.8b-d) images of the PE/15PE-MA1 blend at different magnifications. As observed

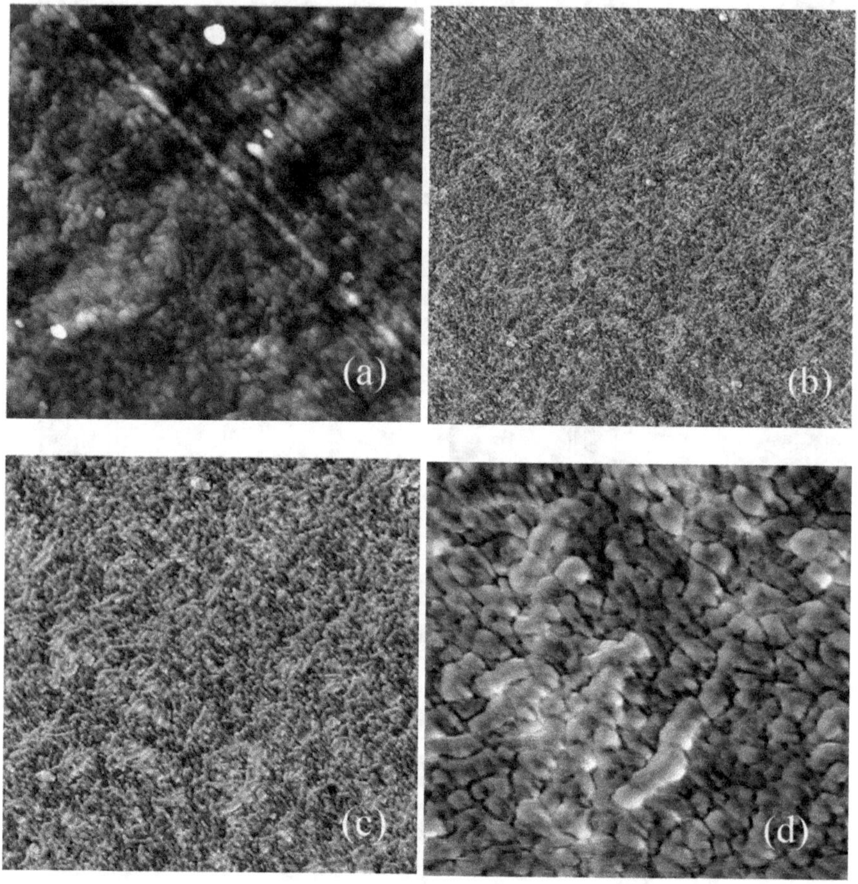

Figure 2.8 AFM images of the PE/15PE-MA1 blend: (a) height image (the width of the image is 10 μm); (b-d) phase images with image width as 20, 10 and 3 μm respectively. Fine dispersion of mixture of hard (bright phase representing crystalline region) and soft (dark phase representing amorphous region) regions is clearly visible.

in the phase images, the PE-MA1 blend exhibited uniform morphology with fine mixture of hard (bright phase representing crystalline region) and soft (dark phase representing amorphous region) regions. It also confirmed the earlier findings about the good degree of phase intermixing of PE-MA1 with PE. Similarly, Figures 2.9-2.11 present the height and phase images of PE/15PE-MA2, PE/15EMAZ and

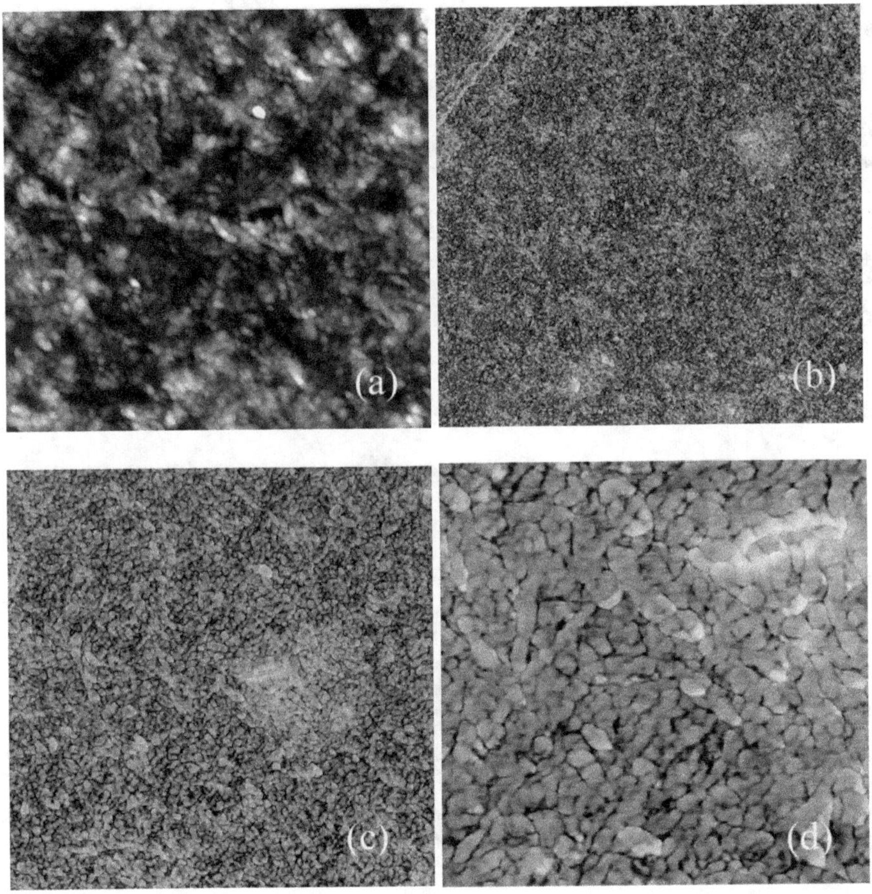

Figure 2.9 AFM images of the PE/15PE-MA2 blend: (a) height image (the width of the image is 10 μm); (b-d) phase images with image width as 20, 10 and 3 μm respectively. Fine dispersion of mixture of hard (bright phase representing crystalline region) and soft (dark phase representing amorphous region) regions is clearly visible.

and PE/15EVA-MA blends. The PE-MA2 also exhibited similar uniform phase morphology, thus, confirming that both PE-MA1 and PE-MA2 resulted in homogenous samples, even at 15 wt% copolymer content. As observed earlier in DSC and X-ray diffraction, the phase morphology of PE was observed to be significantly affected in the EMAZ and EVA-MA based blends. EVA-MA was observed to exhibit maximum distortion of the phase morphology, which also corroborated with the maximum decrease in melt enthalpy for this material. The hard and soft phases in the EMAZ and EVA-MA based blends did

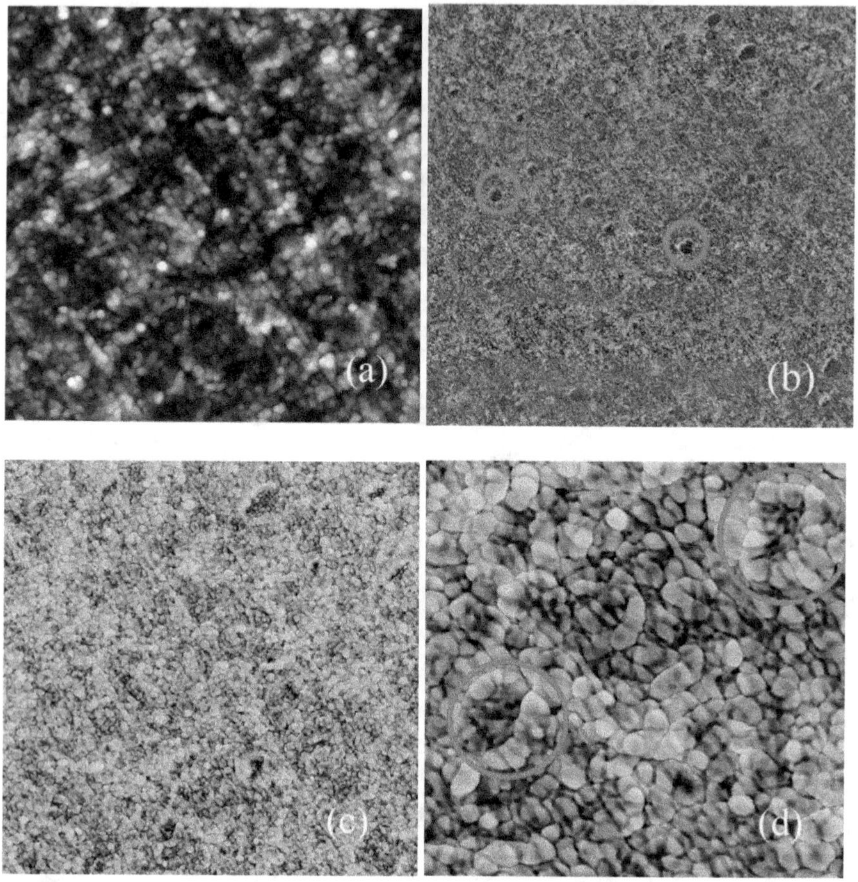

Figure 2.10 AFM images of the PE/15EMAZ blend: (a) height image (the width of the image is 10 μm); (b-d) phase images with image width as 20, 10 and 3 μm respectively. The fine dispersion of mixture of hard (bright phase representing crystalline region) and soft (dark phase representing amorphous region) regions was disturbed representing immiscibility. In addition, the presence of "eye-like" structures with soft central part and dense borders was observed.

not show compatibility with each other. In addition, the presence of "eye-like" structures with soft central part and dense borders were observed due to the immiscible copolymer phase present as large dispersed domains and surrounded by semi-crystalline PE chains.

Network structure of the polymer blends was evaluated with shear rheology and the storage and loss moduli of the samples as a function of angular frequency are presented in Figures 2.12 and 2.13.

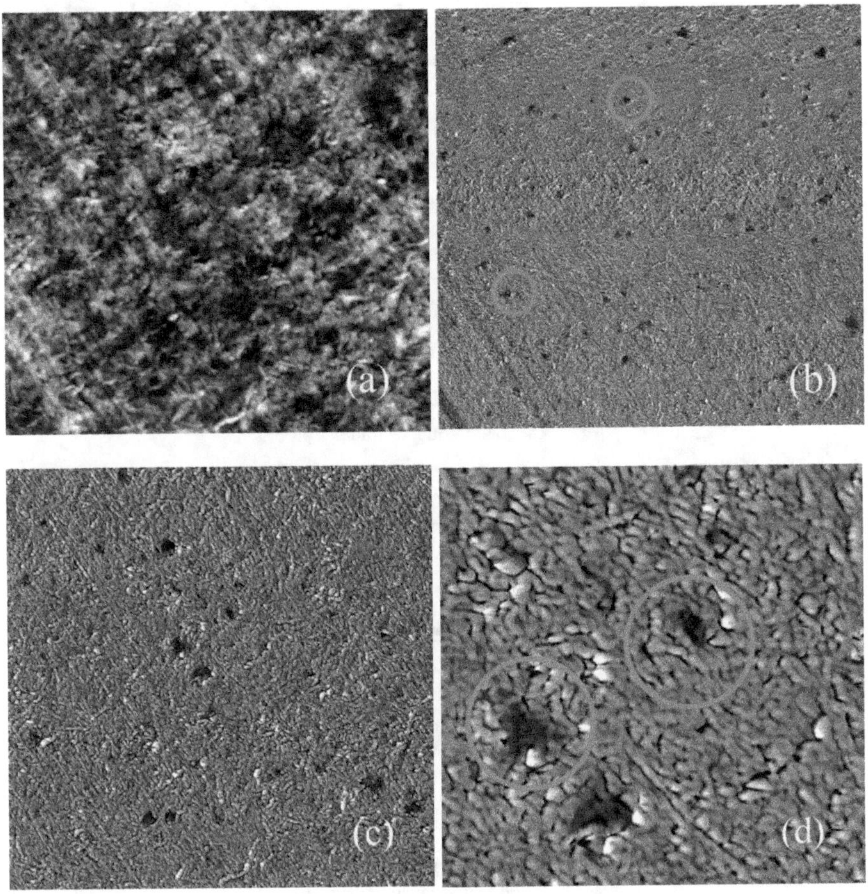

Figure 2.11 AFM images of the PE/15EVA-MA blend: (a) height image (the width of the image is 10 μm); (b-d) phase images with image width as 20, 10 and 3 μm respectively. The fine dispersion of mixture of hard (bright phase representing crystalline region) and soft (dark phase representing amorphous region) regions was disturbed representing immiscibility. In addition, the presence of "eye-like" structures with soft central part and dense borders was observed.

Strain sweep was conducted and samples were found to be safe up to 10% strain. Frequency sweep of the samples was performed with controlled shear strain of 2% using frequency range of 0.1-100 rad/s. All the samples exhibited gradual decline in the extent of modulus enhancement as a function of angular frequency due to the shear thinning effect. In the case of PE-MA1 based blends, the storage modulus of the blend with 5% copolymer content exhibited higher value than

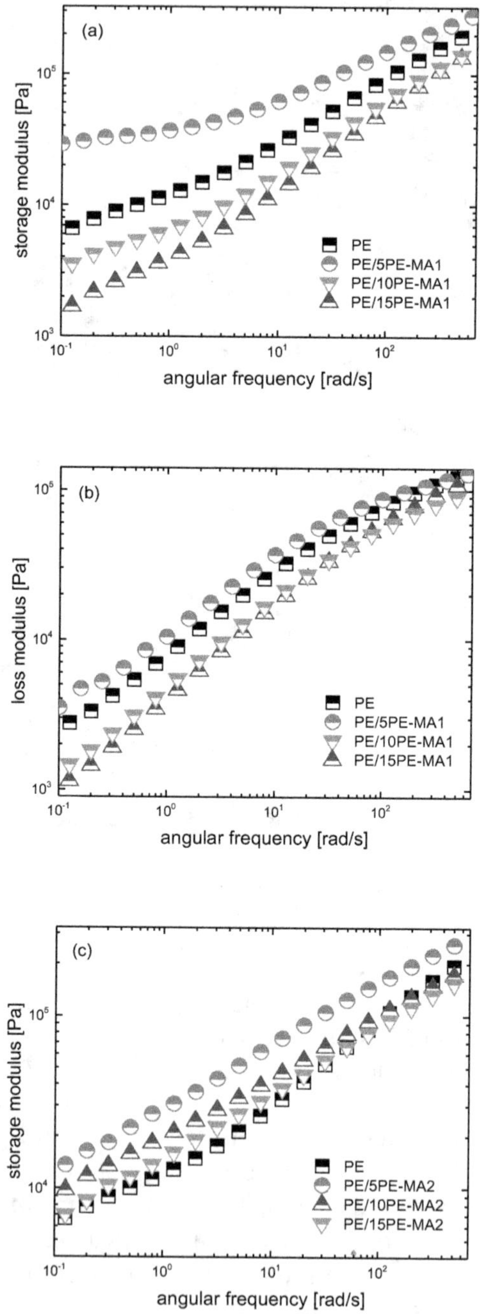

Figure 2.12 (continued on next page)

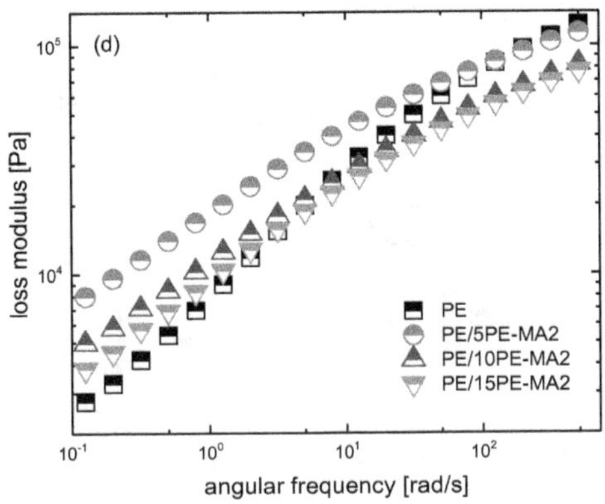

Figure 2.12 (a) Storage (G') and (b) loss (G") modulus of the PE/15PE-MA1 blend; and (c) storage (G') and (d) loss (G") modulus of the PE/15PE-MA2 blend, as a function of angular frequency as well as copolymer concentration in the blends.

the pure polymer over the complete angular frequency range, though the difference was observed to decrease at higher angular frequency. It indicated that the good degree of intermixing of the phases may have enhanced the shear behavior of the polymer. On further increasing the PE-MA1 content, the modulus was observed to decrease due to matrix plasticization in the presence of copolymer. Even though the blends with 10% and 15% PE-MA1 had lower storage modulus than PE, however, the difference in the magnitude was not large. The loss modulus of the PE-MA1 based blends exhibited the similar trend as storage modulus with the modulus curves almost converging into

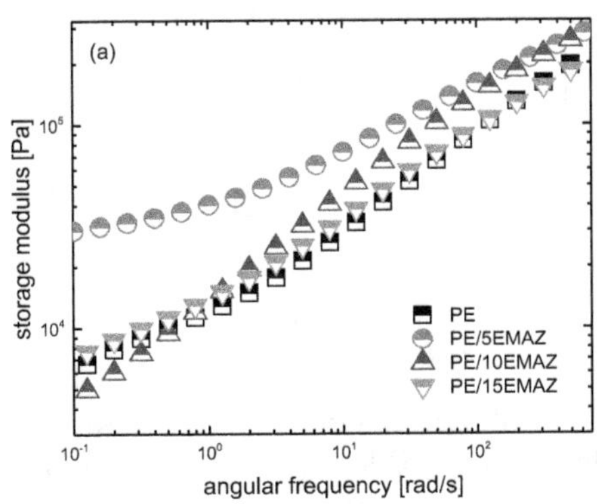

Figure 2.13 (continued on next page)

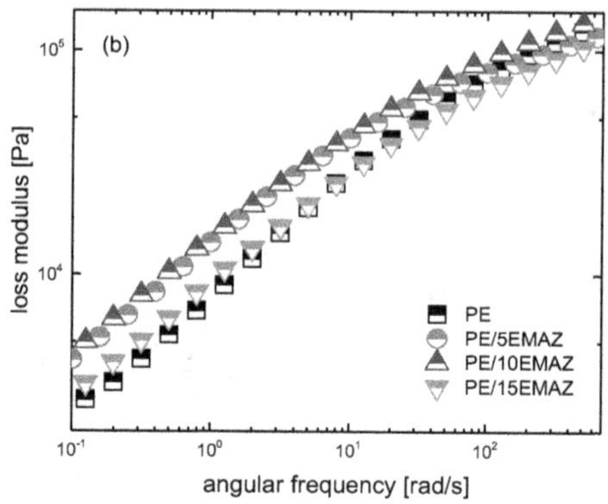

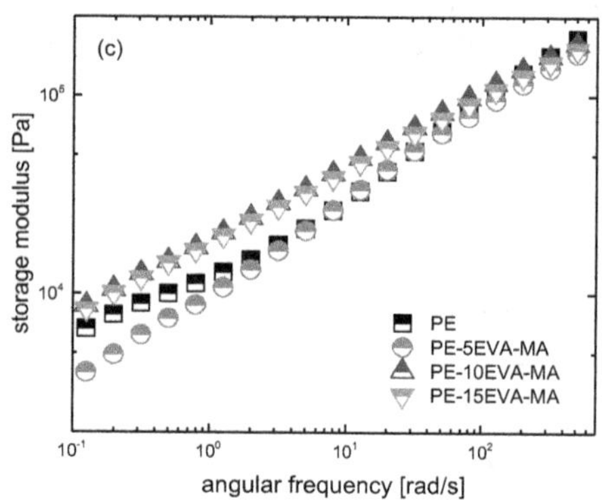

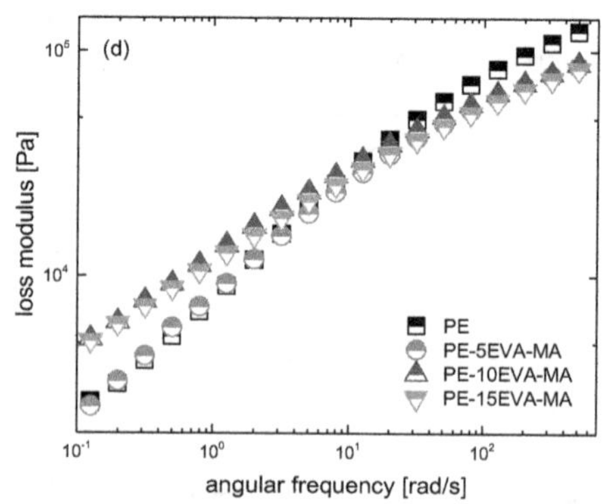

Figure 2.13 (a) Storage (G') and (b) loss (G'') modulus of the PE/15EMAZ blend; and (c) storage (G') and (d) loss (G'') modulus of the PE/15EVA-MA blend, as a function of angular frequency as well as copolymer concentration in the blends.

a single curve at higher angular frequency. In addition, the loss modulus of the blends was lower than the storage modulus, especially for 5% copolymer blend, thus, indicating that the elastic behavior was dominant in the blends at the measurement conditions. In the case of PE-MA2, the modulus was also observed to improve with the addition of the copolymer, though the extent of increase was not significant. Similar to the PE-MA1 case, the blend with 5% PE-MA2 had the highest modulus. The modulus of the blends with 10% and 15% copolymer was also higher than pure PE in this case, especially at lower angular frequency. Similar trends were also observed for the EMAZ and EVA-MA based blends, where the storage modulus of the blends was slightly improved as compared to the pure polymer. Overall, the PE-MA1 blends were observed to exhibit the highest degree of matrix plasticization due to the lower molecular weight of PE-MA1. Also, there was no significant difference in the bulk rheological performance of the blends due to the differences in the phase morphology at the nano-scale, probably due to the measurements in the melt state.

Figure 2.14 also demonstrates the complex viscosity of the blends. Contrary to the modulus, the viscosity of the blends decreased on increasing angular frequency. Due to the increase in frequency, polymer structure shows temporary network of entanglements which leads to more flexibility (lower viscosity). In many blends, this allowed more deformation energy to be stored resulting in elastic dominance which leads to increased modulus. Thus, it simultaneously reduced the contribution of lost deformation energy and, hence, viscous behavior. Similar to storage modulus, some increments were

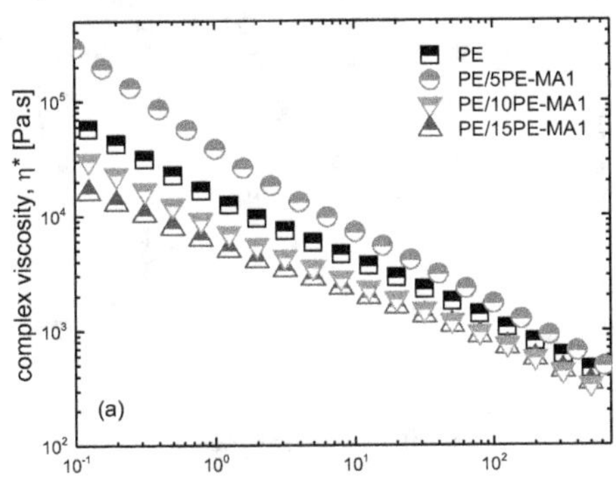

Figure 2.14 (continued on next page)

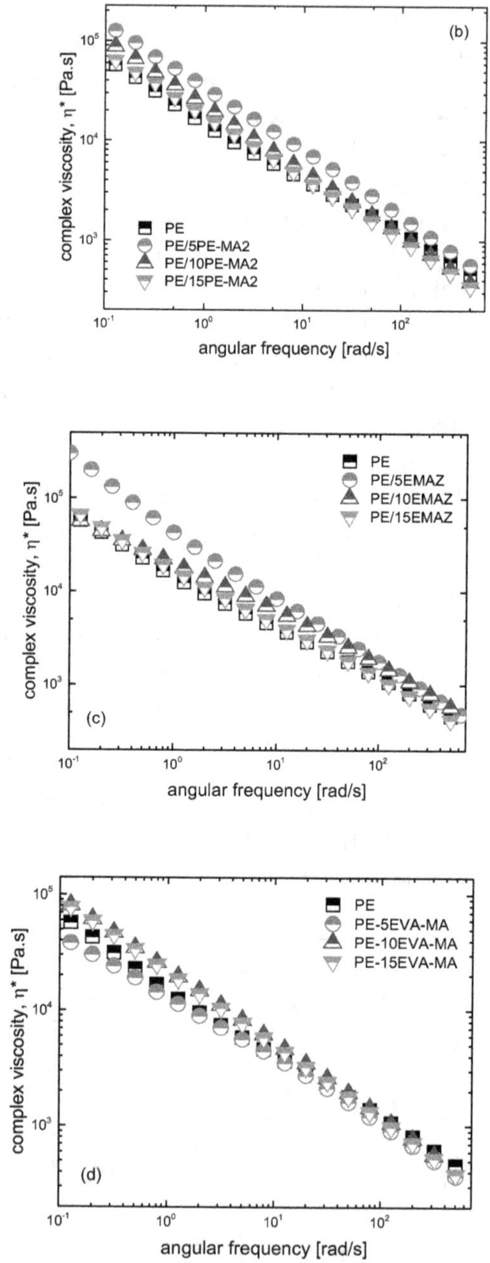

Figure 2.14 Complex viscosity of (a) PE/PE-MA1, (b) PE/PE-MA2, (c) PE/EMAZ, and (d) PE/EVA-MA blends, as a function of angular frequency as well as copolymer concentration in the blends.

observed in the viscosity of the blends with the addition of copolymers. Even though the viscosity of the blends increased in many instances as compared to the pure polymer, however, the extent of increase was not significant and blends could still be processed using the same conditions as pure PE.

The rheological behavior of the blends was also analyzed using the criteria reported for compatibility of polymer blends by Han and Chuang [25,26] as shown in Figure 2.15. As G' is plotted vs. G", such analysis generates composition independent correlation in the case of compatible blends (indicated by the merging of the curves into a common curve), whereas the correlation is composition dependent for incompatible blends. For the PE-MA1 blends, the curves almost converged into single curve at higher modulus (or higher frequency)

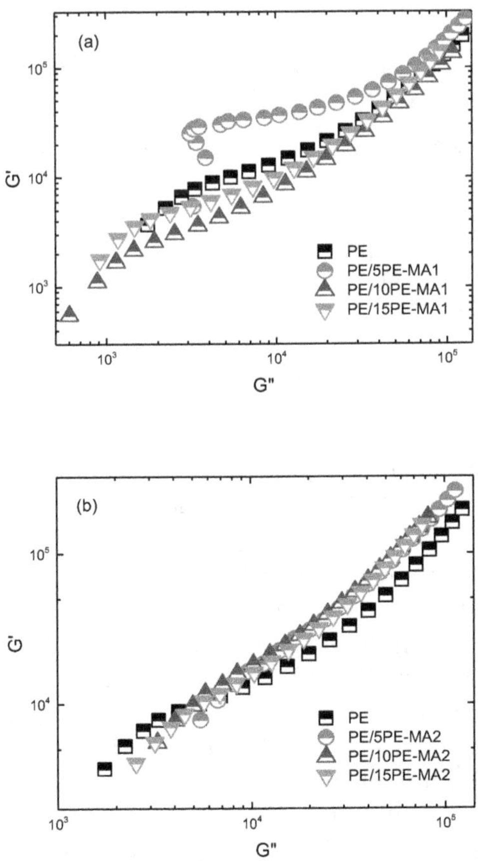

Figure 2.15 (continued on next page)

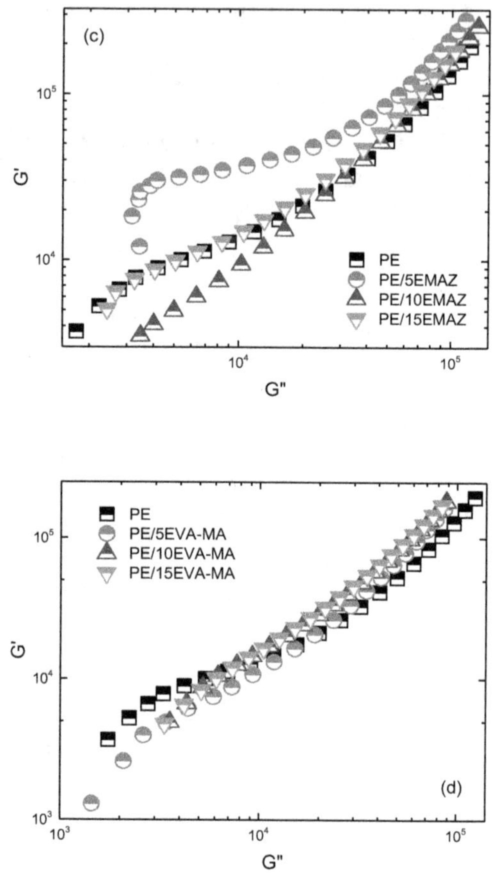

Figure 2.15 Han-Chuang (G' vs. G") plots of (a) PE/PE-MA1, (b) PE/PE-MA2, (c) PE/EMAZ, and (d) PE/EVA-MA blends.

values, however, deviations were observed at lower modulus values. Thus, it also indicated that the phases were not molecularly miscible, even though good degree of intermixing was observed for these blends earlier. Similarly, PE-MA2 based blends also exhibited somewhat concentration independent behavior due to good phase intermixing, as observed earlier. The blends with EMAZ and EVA-MA copolymers exhibited higher extent of concertation dependent behavior than PE-MA based blends at both low and high modulus values due to larger degree of phase immiscibility, which also corroborated with the earlier observations from calorimetric and morphological analysis. Similar insights about the phase miscibility were also gained through Cole-Cole viscosity plots as well as van Gurp plots (by plotting η' vs. η" as well as complex modulus, G* with phase angle δ).

Figure 2.16 demonstrates the tensile modulus of the blends as a function of copolymer content in the blends. The modulus of pure polymer was measured to be 1172 MPa. In all the blends, the modulus was observed to decrease as a function of copolymer content owing to reduced crystallinity and matrix plasticization. Largest decrease in the modulus was exhibited by the EVA-MA blends. For instance, the modulus of the blend with 15% EVA-MA was measured as 846 MPa, in comparison with 915 MPa, 936 MPa and 915 MPa for the blends with 15% content of EMAZ, PE-MA2 and PE-MA1. It indicated that the observed immiscible phase morphology as well as largest extent of crystallinity reduction in the PE/EVA-MA blends resulted in corresponding reduction in tensile modulus. Though a lower extent of phase immiscibility was also observed for EMAZ containing blends, however, these blends exhibited almost similar performance as the PE-MA based blends. Overall, though the modulus of PE expectedly decreased with the incorporation of a variety of copolymers, however, the blends still exhibited appreciable modulus. Figure 2.17 also demonstrates the effect of copolymer content on the yield and break stress as well as corresponding yield and break strain. Similar to

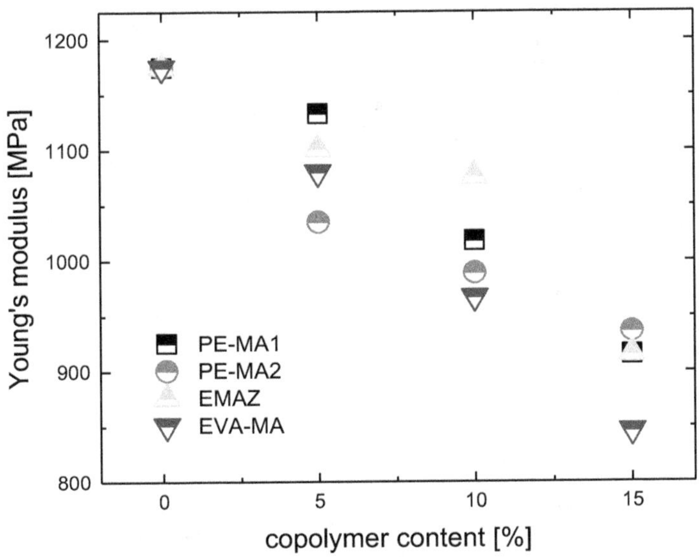

Figure 2.16 Tensile modulus of the blends as a function of copolymer content. The modulus of the blends varied in the range of ±2%.

tensile modulus, the yield stress of the blends decreased with increasing the copolymer content. Among the different copolymers, EVA-MA based blends exhibited largest extent of reduction in the yield stress, thus, re-confirming the effect of reduced crystallinity, plasticization as well as phase immiscibility. The blend with 15% EVA-MA content exhibited a reduction of >25% in the yield stress. Due to enhanced flexibility in the blends, the yield strain was also correspondingly increased with copolymer content. From a value of 6% for pure PE, the yield strain in the blends with 15% copolymer content was observed in the range of 11-12%, thus, exhibiting extensive plasticization of the matrix. Similarly, the stress at break and strain at break in the blends were also affected by the copolymer content, though a large probable error existed in the values of these quantities. As mentioned earlier,

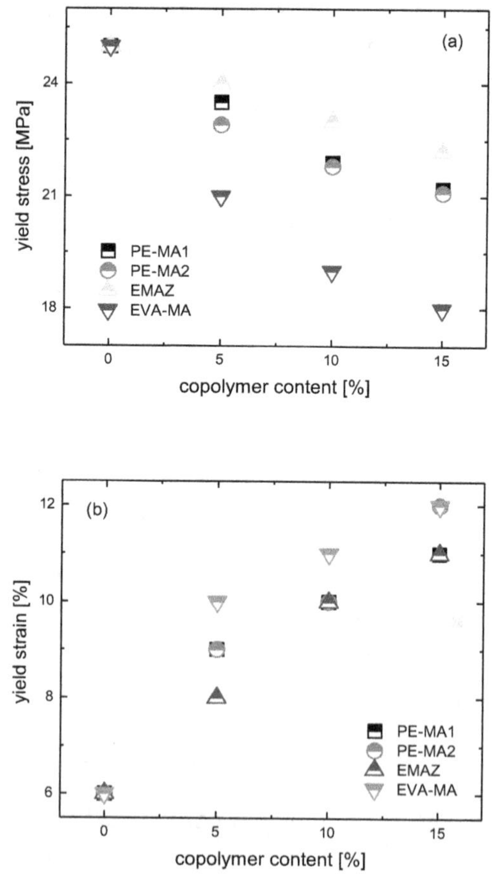

Figure 2.17 (continued on next page)

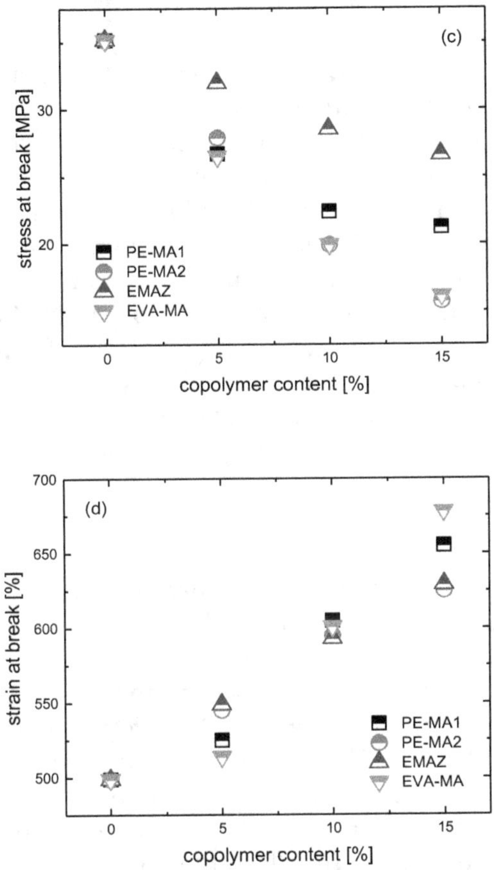

Figure 2.17 (a) Yield stress, (b) yield strain (c) stress at break and (d) strain at break of the blends as a function of copolymer content. The yield stress and yield strain varied in the range of ±2%, whereas the stress at break and strain at break had the relative probable error of ±5%.

the tensile properties of the blends in an earlier study with CPE25 and CPE35 were also significantly affected by the type as well as amount of the CPE used [22]. For instance, the tensile modulus as well as the peak strength of the CPE25 blends had lower decrease in magnitude as compared with CPE35 blends. The decrease was observed only after 5% CPE content and the difference between the two systems enhanced on enhancing the CPE content. These findings were also similar to the tensile properties observed in the current study. Thus, the factors affecting the mechanical performance of the blends are the nature of copolymer, matrix plasticization, effect on crystallinity as well as extent of compatibility of the copolymer with the matrix polymer.

The generated blends provide systematic insights into the effect of a variety of copolymers on the thermal, mechanical, rheological and morphological properties of PE. In addition, the effect could also be quantified as a function of copolymer content. These findings enable the user to choose the appropriate blend system and composition to target the required set of properties. In addition, the insights into the phase miscibility of the copolymers in PE would also benefit the design of advanced composites where the compatibilization of the filler phase with the polymer is necessary. Though these findings are applicable to PE based systems, however, blends and composites based on other polyolefins can also benefit from these, owing to their similar nature.

2.4 Conclusions

In the current study, development of polyethylene based functional blends with a variety of copolymers has been demonstrated. The blends of polyethylene-*g*-maleic anhydride copolymer (PE-MA), ethylene methacrylic acid copolymer with zinc ion (EMAZ) and ethylene vinyl acetate-*g*-maleic anhydride (EVA-MA) copolymer in the weight fraction of 5%, 10% and 15% were generated. The thermal degradation of the PE was affected by the addition of the copolymers (except PE-MA2), as the onset and peak degradation temperatures were reduced, though the blends were still stable up to 350 °C. The incorporation of copolymers also reduced the peak melting point of PE in the blends, which was also a function of copolymer content. The melt enthalpy of PE was also observed to decrease owing to the hindrance to PE chains, with largest reduction in enthalpy exhibited by EVA-MA. The EVA-MA and EMAZ based blends exhibited phase immiscibility due to the presence of separate melt transition of these copolymers in the DSC thermograms of the blend. The X-ray diffraction analysis also indicated some changes in the diffraction peaks in the EVA-MA and EMAZ copolymer based blends, though major diffraction peaks of PE were not affected in intensity as well as peak position. The morphological analysis of the blends confirmed the earlier finding of phase immiscibility in EVA-MA based blends. The EMAZ based blends also exhibited some phase inhomogeneity, whereas the PE-MA based blends had uniform morphology. The rheological performance of the blends exhibited generally higher modulus in the 5% copolymer blends and no obvious effect of phase morphology was observed on the properties in melt. The tensile modulus of the blends was a function of reduced crystallinity, matrix plasticization as well as phase morphology. As a result, the EVA-MA based blends exhibited lower tensile values than other copolymers. Overall, even though the mechanical properties of the blends deteriorated after the inclusion of copolymers, these were still appreciable. The generated blends lead to valuable insights about the properties and morphology of PE based blends with different copolymers, which are subsequently useful for the design of advanced composites where compatibilization with the filler is additionally required.

References

1. Lipatov, Y. S. (2002) Polymer blends and interpenetrating polymer networks at the interface with solids. *Progress in Polymer Science*, **27**, 1721-1801.
2. Fekete, E., Foldes, E., and Pukanszky, B. (2005) Effect of molecular interactions on the miscibility and structure of polymer blends. *European Polymer Journal*, **41**, 727-736.
3. Litmanovich, A. D., Plate, N. A., and Kudryavtsev, Y. V. (2002) Reactions in polymer blends: Interchain effects and theoretical problems. *Progress in Polymer Science*, **27**, 915-970.
4. Sundararaj, U., and Macosko, C. W. (1995) Drop breakup and coalescence in polymer blends: The effects of concentration and compatibilization. *Macromolecules*, **28**, 2647-2657.
5. Macosko, C. W., Guegan, P., Khandpur, A. K., Nakayama, A., Marechal, P., and Inoue, T. (1996) Compatibilizers for melt blending: Premade block copolymers, *Macromolecules*, **29**, 5590-5598.
6. Rameshwaram, J. K., Yang, Y.-S., and Jeon, H. S. (2005) Structure-property relationships of nanocomposite-like polymer blends with ultrahigh viscosity ratios. *Polymer*, **46**, 5569-5579.
7. Wang, D., Li, Y., Xie, X.-M., and Guo, B.-H. (2011) Compatibilization and morphology development of immiscible ternary polymer blends. *Polymer*, **52**, 191-200.
8. Higgins, J. S., Tambasco, M., and Lipson, J. E. G. (2005) Polymer blends; stretching what we can learn through the combination of experiment and theory. *Progress in Polymer Science*, **30**, 832-843.
9. Menyhard, A., Varga, J., Liber, A., and Belina, G. (2005) Polymer blends based on the β-modification of polypropylene. *European Polymer Journal*, **41**, 669-677.

10. Gopakumar, T. G., Ponrathnam, S., Lele, A., Rajan, C. R., and Fradet, A. (1999) In situ compatibilisation of poly(phenylene sulphide)/wholly aromatic thermotropic liquid crystalline polymer blends by reactive extrusion: Morphology, thermal and mechanical properties. *Polymer*, **40**, 357-364.
11. Jancar, J. (1999) *Mineral Fillers in Thermoplastics I: Raw Materials and Processing*, Springer-Verlag, Germany.
12. Mittal, V. (2012) *Functional Polymer Blends: Synthesis, Properties and Performance*, CRC Press, USA.
13. Hippi, U. (2008) *Polyolefin Composites*, John Wiley & Sons, USA.
14. Zhou, L., Wang, X., Lin, Y., Yang, J., and Wu, Q. (2003) Comparison of the toughening mechanisms of poly(vinyl chloride)/chlorinated polyethylene and poly(vinyl chloride)/acrylonitrile-butadiene-styrene copolymer blends. *Journal of Applied Polymer Science*, **90**, 916-924.
15. Eastwood, E. A., and Dadmun, M. D. (2002) Compatibilization of poly(vinyl chloride) and polyolefin elastomer blends with multiblock/blocky chlorinated polyethylenes. *Polymer*, **43**, 6707-6717.
16. He, P., Huang, H., Xiao, W., Huang, S., and Cheng, S. (1997) A study on PVC/LLDPE blends with solid-state-chlorinated polyethylene as compatibilizer. *Journal of Applied Polymer Science*, **64**, 2535-2541.
17. Stoeva, S. (2006) Study of poly(vinyl chloride) blends with solid-state chlorinated polyethylene. *Journal of Applied Polymer Science*, **101**, 2602-2613.
18. Waddington, S., and Briggs, D. (1991) Adhesion Mechanisms between polymer coatings and polypropylene studied by XPS and SIMS. *Polymer Communications*, **32**, 506-508.
19. Ogoe, S. A., Liou, D. W., Ainsworth, O. C., Bethea, J. R., Flores, D. R., Ellebracht, S. R., Muskopf, J. W., Werling, C. L., Wilson, E. J., Cocup, J. W., and Allison, J. L. (1996) Blends of Polycarbonate and Chlorinated Polyethylene, US patent 5,525,651.
20. Barrie, J. A., and Webb, W. D. (1989) Gas transport in miscible blends of ethylene-vinyl acetate copolymer and chlorinated polyethylene. *Polymer*, **30**, 327-332.
21. Mittal, V., and Matsko, N. (2013) PE-CPE blends and their graphene oxide nanocomposites with reduced low temperature brittleness. *Colloid and Polymer Science*, **291**(8), 1949-1961.
22. Chaudhry, A. U., and Mittal, V. (2014) Blends of high-density polyethylene with chlorinated polyethylene: Morphology, thermal, rheological, and mechanical properties. *Polymer Engineering and Science*, **54**(1), 85-95.
23. Mittal, V., and Chaudhry, A. U. (2015) Effect of amphiphilic compatibilizers on the filler dispersion and properties of polyethylene-thermally reduced graphene nanocomposites. *Journal of Applied Polymer Science*, **132**(35), doi:10.1002/app.42484.

24. Matsko, N. B., and Mittal, V. (2014) Thermotropic dynamic processes in multiphase polymer systems by (cryo-)AFM. *Soft Matter*, **10**, 5478-5488.
25. Chuang, H. K., and Han, C. D. (1984) Rheological behavior of polymer blends. *Journal of Applied Polymer Science*, **29**, 2205-2229.
26. Han, C. D., and Chuang, H. K. (1985) Criteria for rheological compatibility of polymer blends. *Journal of Applied Polymer Science*, **30**(11), 4431-4454.

3

PE-CPE Blends and PE-CPE-Graphene Oxide Nanocomposites: Morphology, Phase Miscibility, Interfacial Interactions and Reduced Low Temperature Brittleness

3.1 Introduction

Processes like polymer blending or incorporation of specific fillers to the polymer matrices are commonly performed to acquire new polymeric materials with altered properties [1,2]. Polymer blends have advantages such as property profile combinations, aid in processing, low capital costs as compared to development of new monomers and polymers of required properties. On the other hand, composites fulfill requirements for different applications where cost and weight reduction, dimensional stability, opacity, heat stability, and processability are required. The properties of the resulting materials depend on a number of factors like interaction between the phases, dispersion of one phase into the other, compatibilization, processing conditions, composition, etc.

Polyolefins, for example polyethylene (PE), have wide range of properties including low cost, ease of recycling, good processability, non-toxicity, bio-compatibility, and good chemical resistance. The number of individual polyolefin members is limited and the number of applications based on such materials is constantly increasing. Various polyolefin based blends and composites are thus required to meet the increasing demand of materials with required properties and processing suitable for specific applications [1,3].

Chlorinated polyethylene (CPE) of different chlorination levels is generally blended with polymers for different purposes such as toughening of matrix polymers [4], enhancing the interfacial interactions between two polymers [5,6], improving processing [7] and

Vikas Mittal[a,], Gisha E. Luckachan[a], and Nadejda Matsko[b]*
[a]The Petroleum Institute (part of Khalifa University of Science and Technology), Abu Dhabi, UAE; [b]Graz Centre of Electron Microscopy, Graz, Austria
**Current address: Bletchington, Wellington County, Australia*

adhesion properties [8], enhancing ignition resistance [9] and modifying gas transport properties [10]. In reported studies on PE+CPE systems, CPE with 36% chlorination content was blended with high density polyethylene and low density polyethylene [11] and the mechanical properties were observed to depend on morphological changes occurring at higher concentration of CPE36 in the blends [11,12]. Blends of CPE have also been reported with other polymer systems, however, studies on the miscibility, structural properties and performance of commercially important blends of different types of chlorinated polyethylene with polyolefins like polyethylene are still missing.

Graphene has been a filler of choice in studies on polymer nanocomposites due to its excellent electrical and mechanical properties, which are significantly better than other filler materials [13-18]. Graphene (and graphene oxide (GO)) can be produced from graphite by different methods such as thermal expansion of chemically intercalated graphite, micromechanical exfoliation of graphite, chemical vapor decomposition and chemical reduction method [15]. The compatibility of the graphene derivatives with the polymer matrices is of utmost importance to achieve optimum enhancement in the polymer properties. Graphene has very low surface energy owing to the presence of less number of functional groups (such as carboxyl, epoxide, and hydroxyl, etc.) on the surface which leads to lower compatibility with polar polymer matrices [13]. On the other hand, dispersion of polar graphene oxide in non-polar polymers like polyethylene also suffers due to absence of positive interactions between them. Addition of amphiphilic compatibilizer having polar and non-polar groups is used to compatibilize filler and host polymer, thus, resulting in improved filler dispersion. For example, maleic anhydride was reported to compatibilize nanocomposites of low density polyethylene and exfoliated graphite nanoplatelets [19]. Kim *et al.* [20] also reported the importance of functionalized polyethylene as well as solution mixing method for exfoliation of graphene in the polyethylene matrix. CPE has also been reported to be efficient compatibilizer in composites. For example, CPE was used to comaptibilize poly(vinyl chloride) and wood flour composites (chlorine content ranged from 25% to 42%), and improvements in mechanical and rheological properties were reported [21]. In a recent study by the authors of this study, CPE polymers of 25% and 35% chlorination levels were also used to comaptibilize PE-graphene oxide system and rheological and mechanical properties of the nanocomposites were observed to depend on the degree of chlorination as well as CPE content [22].

Due to the immense potential of graphene in enhancing the material characteristics, further analysis of its nanocomposites with commercially important polymers is required in terms of characterization of miscibility of the phases and changes in the crystalline structure of the matrix polymer. In addition, it is also needed to study physical and chemical interactions between the compatibilizer and filler in comparison to interactions between polymer and compatibilizer phases as well as polymer and filler phases. To accomplish this, two types of chlorinated polyethylene polymers (with 25% and 35% chlorination content, i.e. from the thermoplastic and elastomer range respectively [23]) were used to generate blends with PE as well as PE+CPE nanocomposites with graphene oxide. A recent study on these nanocomposites reported their thermal, mechanical and rheological properties [22], however, more fundamental aspects of morphology as well as interfacial interactions have been studied in the current work in order to optimize the filler exfoliation to achieve optimum composite design. To generate masterbatches of CPE compatibilizer and graphene oxide, solution blending technique was used. The melt mixing method was used to obtain PE+CPE blends as well as composites of masterbatches with PE. The morphology of the blends and composites was specifically characterized using microscopy (electron and atomic force microscopy), shear rheology, thermogravimetric analysis (TGA) coupled with mass spectrometer, as well as Raman and infra-red (IR) spectroscopy.

3.2 Experimental

3.2.1 Materials

High density polyethylene BB2581 was kindly supplied by Abu Dhabi Polymers Company Limited (Borouge), UAE. Chlorinated polyethylene grades Weipren® 6025 (25% chlorine content, named as CPE25) and CPE 135A (35% chorine content, named as CPE35) were obtained from Lianda Corporation, USA and Weifang Xuran Chemicals, China respectively. The CPE25 was in the form of white granules (specific gravity 1.1-1.3), whereas the appearance of CPE35 was white powder (specific gravity 1.1-1.16). The melt flow index (190 °C, 2.16 kg) for CPE25 and CPE35 was measured to be

1.8 and 1.9 g/10 min respectively, whereas the pure polymer had a melt flow index of 0.35 g/10 min (190 °C, 2.16 kg). The heat of fusion of CPE25 and CPE35 polymers was 45 and 2 J/g respectively. The polymer materials were used as obtained. Graphene oxide was synthesized by thermal exfoliation of precursor graphite oxide using modified Hummer's method as reported earlier [22,24,25]. The density of the obtained graphene oxide was measured by tapped density tester to be 0.0161 g/ml.

3.2.2 PE+CPE Blends

Melt mixing was used to generate polymer blends of CPE25 and CPE35 with PE using mini twin conical screw extruder (MiniLab HAAKE Rheomex CTW5, Germany). A mixing temperature of 190 °C for three minutes at 60 rpm with batch size of 5 g was used. The screw length and screw diameter were 109.5 mm and 5/14 mm conical respectively. The percentage of the CPE phase was fixed at 5 wt%. Disc shaped test samples were prepared by mini injection molding machine (HAAKE MiniJet, Germany) at a processing temperature of 190 °C. The injection pressure was 700 bar for 6 s whereas holding pressure was 400 bar for 3 s. The temperature of the mold was kept at 50 °C.

3.2.3 Nanocomposite Generation

The nanocomposites were prepared by either solution mixing followed by melt mixing or direct melt mixing, as described earlier [22]. Solution mixing of CPE in graphene oxide was achieved by dissolving CPE polymer in *p*-xylene solvent (3 wt% solid content) at 100 °C under reflux and dispersing graphene oxide in *p*-xylene for 1 h at room temperature followed by mixing of the two at 100 °C. The slurry was sonicated using ultrasonic horn (70% amplitude). The mixture was cooled and kept overnight at room temperature followed by 40 °C for 24 h in order to remove any solvent residues. The so-formed CPE-graphene oxide masterbatches were melt mixed with PE at 190 °C using mini twin screw extruder using the same processing conditions used to form blends. The overall amount of graphene oxide was fixed at 0.5 wt%. Melt mixing of pure PE as well as PE-graphene oxide was also similarly performed. Disc shaped test specimens for melt rheological analysis were similarly injection molded at 190 °C, as described above.

3.2.4 Characterization of Blends and Nanocomposites

AR 2000 rheometer from TA Instruments was used to characterize the rheological properties such as storage modulus (G′), loss modulus (G″), viscosity (η′) and elasticity (η″). Disc shaped samples of 25 mm diameter and 2 mm thickness were measured at 185 °C using a gap opening of 1.2 mm. Strain sweep scans were recorded at $\omega = 1$ rad/s from 0.1 to 100% strain and the samples were observed to be shear stable up to 10% strain. For the comparison, frequency sweep scans (dynamic testing) of all the samples were recorded at 4% strain from $\omega = 0.1$ to 100 rad/s.

For the microscopy analysis, the composite samples were mounted in special holders which at the same time fit in the microtome and were suitable for the examination of the block face by atomic force microscopy. Ultrathin sections (30-90 nm) as well as block faces of the composite samples were obtained using a Leica Ultracut E microtome (Leica, Austria) equipped with a diamond knife (Diatome, Switzerland) at -120 °C. Sections for transmission electron microscopy (TEM) analysis were collected on 400 mesh formvar electron microscopy grids (coated with holey carbon film), and were examined in a Philips CM 20 (Philips/FEI, Germany) electron microscope at 200 kV at room temperature without staining. TEM image processing was performed using DigitalMicrograph software (Gatan, USA). All energy dispersive X-ray spectroscopy (EDXS) experiments were performed in STEM mode with a probe current of 20 nA and a beam diameter of 100 nm. EDXS spectra were collected using a HPGe detector (solid angle 0.13 sr). For the calculation of elemental distribution images, jump ratio method was used and EDXS spectra have been visualized using Digital Micrograph software (Gatan, USA). The block faces of specimens after cryo-ultramicrotomy were investigated at ambient conditions using Dimension 3100 AFM/SPM (Veeco, USA) atomic force microscope. AFM images were collected in tapping mode using silicon nitride cantilevers with natural frequencies in the 300 kHz range (force constant 20 N/m, tip radius 10 nm (NT-MDT, Russia)). AFM image processing was performed using Nanoscope v720 software (Veeco, USA).

IR spectra were collected on Nicolet iS10 spectrometer equipped with SmartiTR diamond ATR accessory (angle of incidence of 45°), DTGS KBr detector and KBr beam splitter. It had a diamond ATR crystal (index of refraction 2.4 at 1000 cm^{-1}) and a depth penetration of 2 μm at 1000 cm^{-1} for sample with refractive index of 1.5. Spectra were recorded by the OMNIC software in the 4000-525 cm^{-1} region with a resolution of 4 cm^{-1} from 16 scans. Raman spectra were recorded on a Jobin Yvon Horiba LabRAM spectrometer with back-scattered confocal configuration using a HeNe laser (633 nm). A long working distance objective with magnification 50x was used both to collect the scattered light and to focus the laser beam on the sample surface. CCD (charge-coupled device) detector exposure time was 10 s and an average of 2 cycles was used to increase S/N ratio. Raman shifts were calibrated using Si wafer signal at 520.7 cm^{-1}.

LINSEIS STA PT1600 simultaneous thermogravimetric analysis - differential thermal analysis (TGA-DTA) system was used to measure the weight change during heating. This system is coupled with a Phipher mass spectrometer (MS) which allowed to determine HCl (m/z=36), H_2O (m/z=18) and CO_2 (m/z=44) elimination during heating. All measurements are realized with a heating rate of 10 °C/min, under air atmosphere (flow rate 20 ml/min).

Wide angle X-ray diffraction analysis of the composites was performed on Panalytical Powder Diffractometer (X'Pert PRO) using CuKα radiation (λ = 1.5406 A°) in reflection mode. Zero-background holder was used to minimize the noise. The samples were step-scanned from 10-70° 2Θ at room temperature using a step size of 0.02° 2Θ and a step time of 10 s.

3.3 Results and Discussion

In the current study, the morphological features of PE+CPE blends and PE+GO+CPE nanocomposites, generated using chlorinated polyethylene with two different chlorination levels, were characterized. The chlorination contents were differed in order to study the impact on miscibility with polyethylene as well as dispersion of graphene oxide, thus, affecting the final morphology of these systems. For example, CPE is expected to disperse graphene oxide in PE, however, the overall miscibility in the ternary system may be affected. It was also known from earlier reported results, CPE25 was semi-crystalline in nature with a peak melting temperature of 130 °C, whereas CPE35 was amorphous as no crystalline melting peak was observed in its DSC thermogram [22]. Hence, due to different structural characteristics of the CPE polymers, different morphology was expected in blends and nanocomposites containing these polymers.

The oscillatory torque increased for all the samples as a function of angular frequency during the rheological testing due to enhanced resistance of the melt at higher rates of deformation (Figure 3.1). The build-up of torque as a function of angular frequency for PE+GO, PE+CPE25 and PE+CPE35 was higher than the PE, PE+GO+CPE25 and PE+GO+CPE35 samples. It indicated that the CPE copolymers in the blends also generated similar torque as graphene oxide platelets in the un-compatibilized composite. Thus, stiff CPE chains acted as filler particles in the matrix, however, it should also be noted that thc amount of graphene in the PE+GO composites was very low as compared to CPE amounts in PE+CPE blends. Apart from that, graphene oxide platelets were also observed to be very flexible and could be bent and folded under shear. The observed torque was much lower in PE+GO+CPE nanocomposites and the least values were obtained for PE+GO+CPE35 composite. As PE+GO as well as PE+CPE systems individually had higher torque values, the observed decrease in the composites, thus, can be traced to the interactions between CPE and graphene oxide phases. As graphene oxide

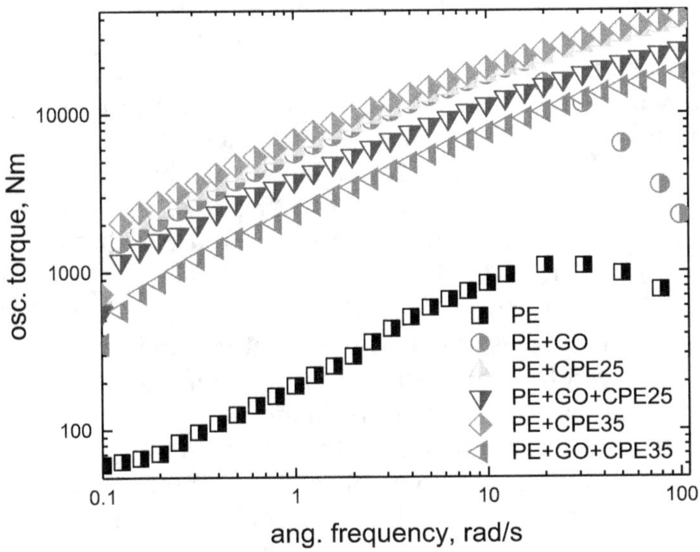

Figure 3.1 Buildup of oscillatory torque during the rheological analysis as a function of angular frequency.

has a polar surface, therefore, it is expected to interact positively with CPE chains. CPE35 due to its higher chlorination content can have even enhanced extent of these interactions. As a result, the dispersion of graphene oxide platelets under shear in the PE matrix would improve as extent of polarity of compatibilizer was increased, leading to lower torque generation due to the well dispersed graphene oxide platelets. However, such selective partitioning of CPE phase near to the graphene surface may also lead to immiscibility with the polymer matrix. PE and PE+GO were also observed to have sudden shear thinning at ~20 rad/s angular frequency. In the case of PE+GO nanocomposite, such a behavior has been reported to be due to the rupture of interface between the polymer and graphene oxide at higher frequencies [22].

Studies on the miscibility of the blends and nanocomposites were performed using Cole-Cole [26-28], van Gurp [29,30] and Han Chuang [31,32] analyses, as shown in Figure 3.2. Cole-Cole analysis develops relationships between real (η') and imaginary (η'') parts of complex viscosity. A smooth, semi-circular shape of the graph would suggest miscible phases, whereas, the deviation from this behavior indicates phase segregation due to immiscibility of the components. As observed in Figure 3.2a, the plots of PE+GO, PE+CPE25 and PE+CPE35 followed a semi-circular shape indicating the miscible phase morphology in these samples. However, the PE+GO+CPE nanocomposites showed deviation from the semi-circular path at higher viscosity (or lower frequency) values indicating presence of immiscibility in the phases. However, it is necessary to further support the findings from Cole-Cole analysis as these findings could sometimes be misleading [33]. Figure 3.2b shows the van Gurp plots representing the relationship between complex modulus (G*) and phase angle delta (δ). The findings from the Cole-Cole analysis were confirmed by van Gurp plots. In the case of PE, PE+GO, PE+CPE25 and PE+CPE35 samples, the time-temperature superposition principle was observed to hold as indicated by the merging of the curves into a common curve. On the other hand, the PE+GO+CPE samples deviated from the common curve, thus, confirming the presence of immiscibility in these samples. The nanocomposite with CPE35 compatibilizer exhibited most notable deviation indicating that the extent of immiscibility in this case was higher than the PE+GO+CPE25 composite. The curves of both PE+GO+CPE composites almost converged with each other at higher frequency (lower delta value), however, they still did not merge with the common curve. The rheological behavior of the samples was also analyzed by using the criteria reported by Han and Chuang using a plot between

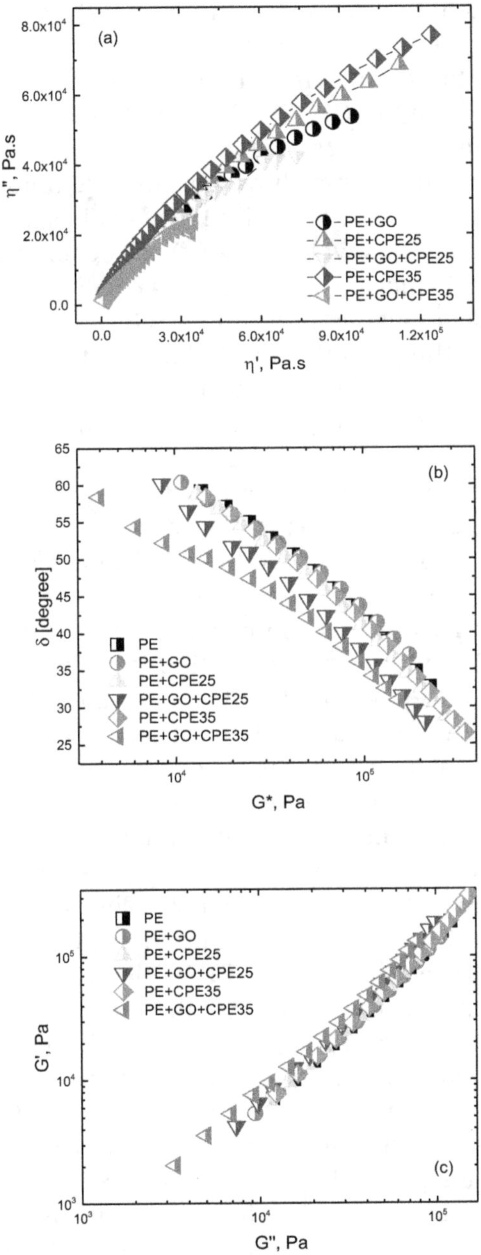

Figure 3.2 Miscibility studies on PE+CPE blends and PE+GO+CPE composites using a) Cole-Cole analysis, b) van Gurp analysis, and c) Han Chuang analysis.

G' and G", as shown in Figure 3.2c. The curves for the ternary composite systems did not merge with others indicating the presence of immiscibility in these systems. Similar to the van Gurp analysis, the nanocomposite with CPE35 compatibilizer had maximum deviation. The deviation between PE+GO+CPE25 and PE+GO+CPE35 was similarly the maximum at lower frequency, but the curves merged with each other at higher frequency, however, they retained their deviation from the other samples. The miscibility studies, thus, confirmed that the polymer blends as well as PE+GO nanocomposite were miscible with uniform phase (slope of curves ~2), however, the PE+GO+CPE nanocomposites had immiscible phase [34,35]. These also signal towards the earlier suggested notion that due to the good interaction of CPE with graphene surface, the concentration of CPE phase near to interface with graphene oxide increased leading to immiscibility with the PE matrix. CPE35 containing composite, thus, expectedly had higher extent of immiscibility as compared to the CPE25 containing composite. This also explains the different mechanical properties obtained for these composites in an earlier study [22]. CPE25 containing composite had higher modulus than PE due to less immiscibility as compared to much lower modulus in the case of CPE35 containing composite due to immiscible phases. However, it should also be noted that the CPE25 is semi-crystalline in nature, while CPE35 was amorphous. Therefore, resulting changes in the crystalline structure of the PE chains would also contribute to the changes in the mechanical response.

In the TEM micrographs of the PE polymer and PE+CPE blends in Figure 3.3, the morphology of the pure polymer was observed to change on incorporating the CPE polymer. The pure PE due to its semi-crystalline nature showed distinct morphological features, which were also present in the PE+CPE25 blend though these were modified. These features owing to crystalline morphology were significantly lost in the case of PE+CPE35 blend indicating that the two CPE polymers had noticeable differences in their effect on polymer morphology. In an earlier study, the morphology of these systems was also studied using differential scanning calorimetry and it was observed that the melt enthalpy in the case of PE+CPE35 blend was 130 J/g as compared to 151 J/g for pure polymer, whereas the PE+CPE25 blend had a melt enthalpy of 140 J/g, which further confirmed the findings from TEM micrographs [34]. Insights into the morphology of blends and nanocomposites samples were also achieved using atomic force microscopy performed in tapping mode, as shown in Figure 3.3. The scales in the TEM and AFM images were kept constant in order to achieve comparison of the morphological features observed in the samples. Similar to TEM, significant differences in the sample morphology were observed. The PE sample was uniform with bright areas (in the phase image) corresponding to crystal structure and dark areas related to the amorphous content. The PE+CPE25 sample also had uniform distribution of bright phase inside the dark areas indicating that the CPE25 polymer still maintained the semi-crystalline nature of the pure polymer. However, its morphology was quite different from pure PE as in this case the chains seemed to be shorter in length. Also, the addition of CPE25 generated polycrystallinity in the sample. PE+CPE35 blend had the most significant effect on the polymer microstructure, which was also confirmed earlier by significant reduction in polymer crystallinity on its addition [22]. The crystalline phases were still present but the order or periodicity was disturbed. Also, the crystalline phases were randomly distributed due to the hindrance of CPE35 phase to

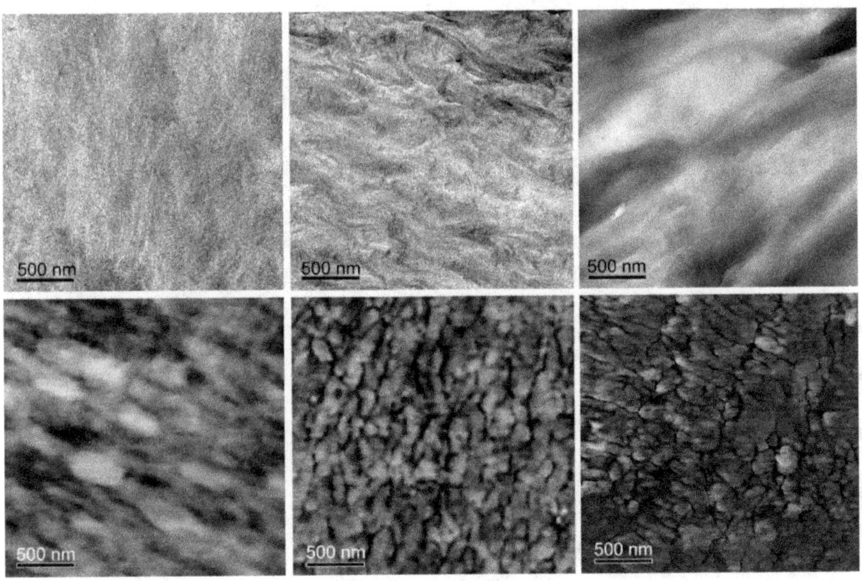

Figure 3.3 TEM and AFM images of (left) PE, (middle) PE+CPE25, and (right) PE+CPE35 samples.

PE crystallization. However, the system was still miscible as confirmed by the rheological analysis. It is difficult to comment on the distribution of CPE chains in the matrix, as the CPE25 is semi-crystalline in nature, so it is expected to align with the crystalline phase as well as amorphous phase of PE. However, in the case of CPE35 polymer, it would be expected to be in the dark regions due to its amorphous nature. It is also possible that the chlorination leads to local stiffness/hardness in CPE polymer chains, in this case, their concentration could be higher in brighter areas of phase images. This was also confirmed by the fact that CPE polymers in the blends generated similar torque as graphene oxide platelets in the composite without CPE.

Figure 3.4 also depicts the TEM characterization of nanocomposite samples. The graphene oxide platelets in the powder form in an earlier study were shown to stack with each other and were also observed to be flexible [22]. The incorporation of graphene oxide to PE without any additional compatibilizer did not lead to optimal filler dispersion and graphene oxide particles with high stack thickness were observed in the micrograph (Figure 3.4a). The PE+GO+CPE25

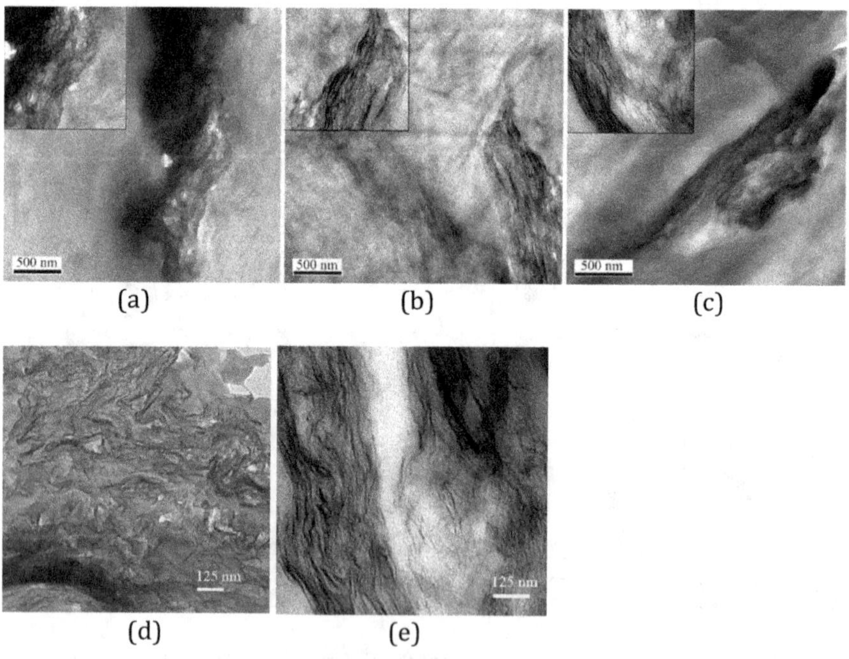

Figure 3.4 Transmission electron micrographs of a) PE+GO hybrid, b) PE+GO+CPE25 composite, and c) PE+GO+CPE35 composite. Figures d) and e), respectively, show the high-magnification micrographs of PE+GO+CPE25 and PE+GO+CPE35 nanocomposites. The black lines represent the cross-section of graphene platelets.

improved the dispersion of graphene oxide in the polymer matrix, but varying degrees of stack thicknesses could still be observed in the micrographs indicating less optimal filler dispersion (Figure 3.4b). In the other case, PE+GO+CPE35 nanocomposite showed significant extent of polymer intercalation inside the stacks of graphene oxide platelets and in the process also led to exfoliation of large extent of these stacks (Figure 3.4c). This, thus, confirmed that the higher extent of chlorination in the system resulted in better interactions between the CPE compatibilizer and graphene oxide leading to much better filler dispersion in the composite. Also, the blend as well as nanocomposite with CPE35 were observed to become extremely flexible and could not be sectioned even at -140 °C indicating the significant reduction of the T_g of polymer along with degree of crystallinity.

Figure 3.5 demonstrates the WAXRD analysis of the blends and nanocomposites. The diffraction peak for pure graphene oxide was observed at 26.4° 2Θ. However, this peak was not visible in the nanocomposites due its very low concentration in the polymer. The polymer crystalline structure was also observed to change in the presence of CPE and graphene. The 110 (21.8° 2Θ) and 200 (24.2° 2Θ) diffraction peaks of pure polymer were observed to have shoulders at higher angle indicating the presence of different crystal structures in the polymer. The shoulder at 110 peak was observed to develop into well-defined peaks in the nanocomposites, indicating possible nucleating effect of graphene oxide. This was also confirmed by the absence of such a phenomenon in PE+CPE blends. In the earlier study [22], the impact of better filler dispersion due to the addition of CPE35 in the nanocomposites was reported to enhance the mechanical properties, but the plasticization effect of the CPE35 reduced them.

In order to confirm the interactions between the CPE compatibilizer and graphene oxide, EDX studies were also performed on PE+GO+CPE25 and PE+GO+CPE35 nanocomposites. As shown in Figure 3.6 for PE+GO+CPE25 composite, the EDX analysis of matrix area (marked as EDX b in TEM image, i.e. area without graphene oxide) did not show any signal for Cl, whereas the area with graphene oxide platelets (marked as EDX a in TEM image) showed only a weak signal of Cl. It has to be noted that amount of CPE was only 5 wt% in the composites and, thus, effective amount of Cl atoms was further very low due to only 25-35% chlorination levels, thus, resulting in the weak signal. It is also possible that some amount of the

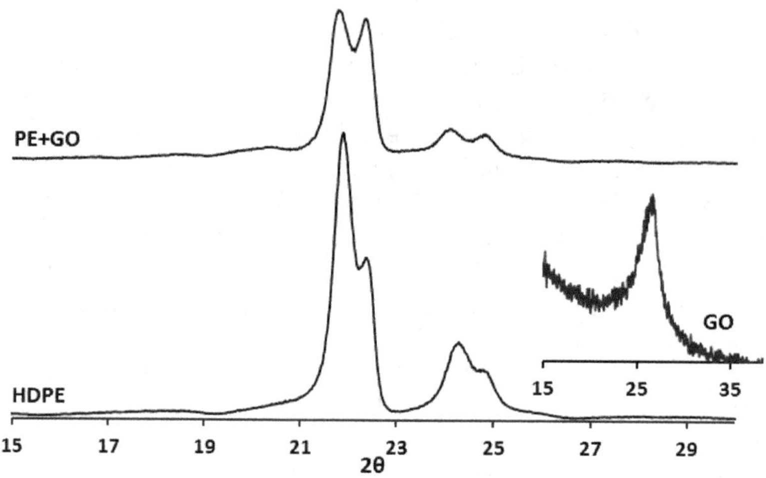

(a)

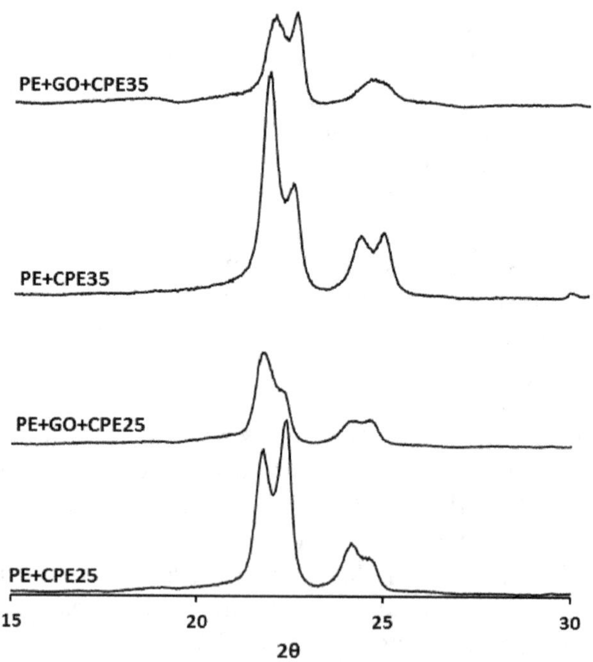

(b)

Figure 3.5 X-ray diffraction analysis of a) PE and PE+GO composite, b) PE+CPE blends and nanocomposites.

CPE phase was dispersed in the polymer, but its concentration was very low to observe its signal in EDX spectrum. However, these findings did not eliminate the possibility of a chemical interaction between the graphene and CPE phases thus leading to depletion of Cl from the CPE. Thermal degradation of CPE during melt compounding could also result in the loss of Cl atoms from the system. Figure 3.7 also shows the EDX results for PE+GO+CPE35 nanocomposite. In the selected area representing the matrix phase (marked as EDX_matr in TEM image), no signal for Cl was observed, whereas in the selected area on polymer-graphene surface (marked as EDX_gr in TEM image), a weak Cl signal (though stronger than the signal observed in PE+GO+CPE25) was observed. These studies thus necessitated further analysis to understand the interactions between the graphene and CPE phases.

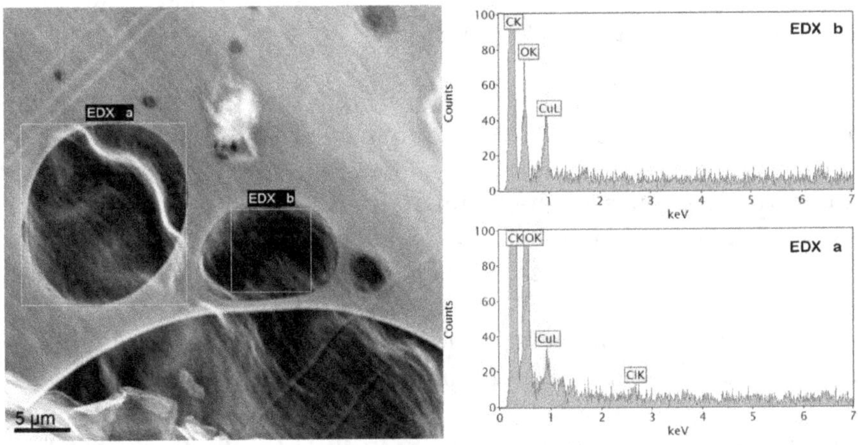

Figure 3.6 TEM image of PE+GO+CPE25 sample; EDX signal from matrix (area marked as EDX b in TEM image) and polymer-graphene surface (area marked as EDX a in TEM image).

Figure 3.8 shows higher magnification height and phase images of the nanocomposite samples PE+GO, PE+GO+CPE25 and PE+GO+CPE35 nanocomposites. The graphene oxide platelets or clusters were observed to be surrounded (or covered) by bright phase indicating hardness. The distribution of graphene oxide in PE was very poor as indicated by a large cluster of graphene oxide particles. The matrix in the background was similarly semi- crystalline in nature as observed in Figure 3.3. The distribution of graphene oxide improved on adding the compatibilizer as smaller cluster sizes

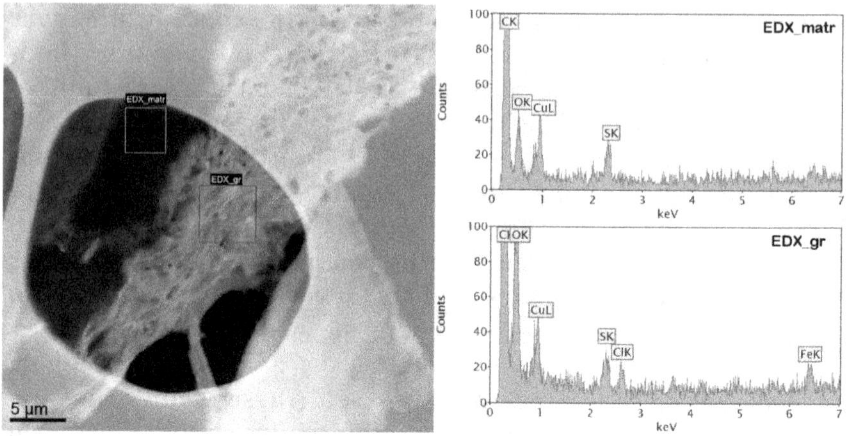

Figure 3.7 TEM image of PE+GO+CPE35 sample; EDX signal from matrix (area marked as EDX_matr in TEM image) and polymer-graphene surface (area marked as EDX_gr in TEM image).

of graphene oxide particles in both CPE25 and CPE35 containing composites were observed. The graphene platelets were occasionally also observed as dark phase (e.g. PE+GO+CPE25 composite in the middle row) because of their embedding below the surface of the sample or due to their flexibility. The distribution of harder areas was also uniform and indicated the semi-crystalline polymer phase. In the case of PE+GO+CPE35 composite, the distribution of the harder areas was significantly affected in the matrix. It should also be noted that the melt enthalpy in the case of nanocomposites was higher than that of blends [22]. In the case of PE+GO, it was 142 J/g as compared to 151 J/g for pure PE. In PE+GO+CPE25, it was 151 J/g, whereas in the case of PE+GO+CPE35 composite, a marginal decrease to 147 J/g in the melt enthalpy was observed. This may indicate that the graphene platelets act as nucleating agents and this effect also enhanced as the dispersion of graphene oxide platelets was improved in PE. The overall crystallinity is also thus accordingly affected; for PE+GO due to very poor distribution of graphene in PE, the effect was not observed. For the composite with semi-crystalline CPE25 compatibilizer, the crystallinity of pure polymer was retained due to much better nucleating effect of the filler as in the PE+CPE25 blend system, the melt enthalpy was only 140 J/g. Similarly, in the composite with amorphous CPE35 polymer, the overall crystallinity observed was result of competing factors of nucleating effect of graphene as well as decrease in crystallinity of the polymer because of CPE35 addition. As areas around the graphene oxide platelets are bright depicting the harder or crystalline phase, it would also confirm the notion of graphene oxide acting as nucleating agent. Moreover, as it was earlier observed that these areas also have higher concentration of CPE chains, it would thus indicate that the harder regions may also form due to ordering/alignment of CPE chains around the particles. However, as such regions are also

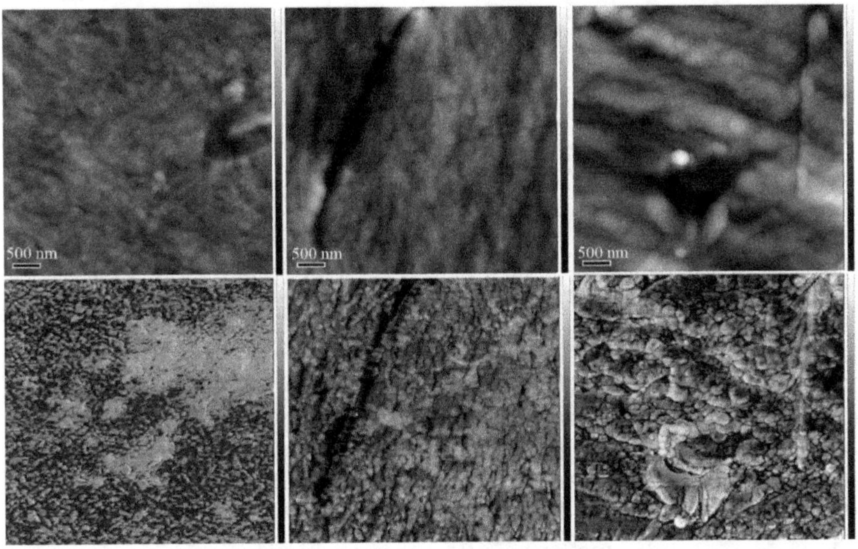

Figure 3.8 High-magnification AFM height (top) and phase (bottom) images of the samples; (left) PE+GO, (middle) PE+GO+CPE25, and (right) PE+GO+CPE35.

formed in the composite without CPE polymers, therefore, it can be concluded that both the factors would have contributed in the formation of these hard phases around the graphene platelets. Figure 3.9 also shows lower magnification height and phase images of the nanocomposites. The concentration of bright phase around the graphene oxide platelets was further confirmed. Also, Figure 3.9c demonstrates much better dispersion of the graphene oxide filler for the PE+GO+CPE35 nanocomposite.

Interactions between CPE compatibilizer and graphene oxide in PE+GO+CPE nanocomposites were also studied by IR as shown in Figures 3.10 and 3.11. In addition to the characteristic $V_{C\text{-}Cl}$ vibrations of CPE35 at 654 cm^{-1} and 607 cm^{-1}, IR spectrum of CPE35+GO showed a new peak at 1735 cm^{-1} which was attributed to $V_{C=O}$ vibrations of ester linkages (Figure 3.10 and Table 3.1). This new peak along with the reduced intensity of $V_{C\text{-}Cl}$ vibrations indicated that C-Cl groups on CPE were involved in a chemical reaction with polar functional groups on graphene oxide (-COOH, -OH) under the mild mixing conditions (100°C, 30 min sonication). Melt mixing of this composite with PE at 170 °C under shear (60rpm) facilitated more reaction between CPE and graphene oxide, that resulting in an intense $V_{C=O}$ vibration peak of ester linkages at 1715 cm^{-1} with a shoulder at 1740 cm^{-1} and $V_{C\text{-}O}$ vibration peak of ester and ether linkages at 1156 cm^{-1} (Figure 10). Chances of presence of carboxylic acid in the final composite (carboxylic acid also gives signal at 1715 cm^{-1}) could be safely ignored as the IR spectrum of PE+GO+CPE35 didn't show any hydroxyl contribution at ~3500 cm^{-1}. Utilization of chlorine for chemical bonding (ester, ether etc.) between CPE and graphene oxide in the PE+GO+CPE35 nanocomposite could be further

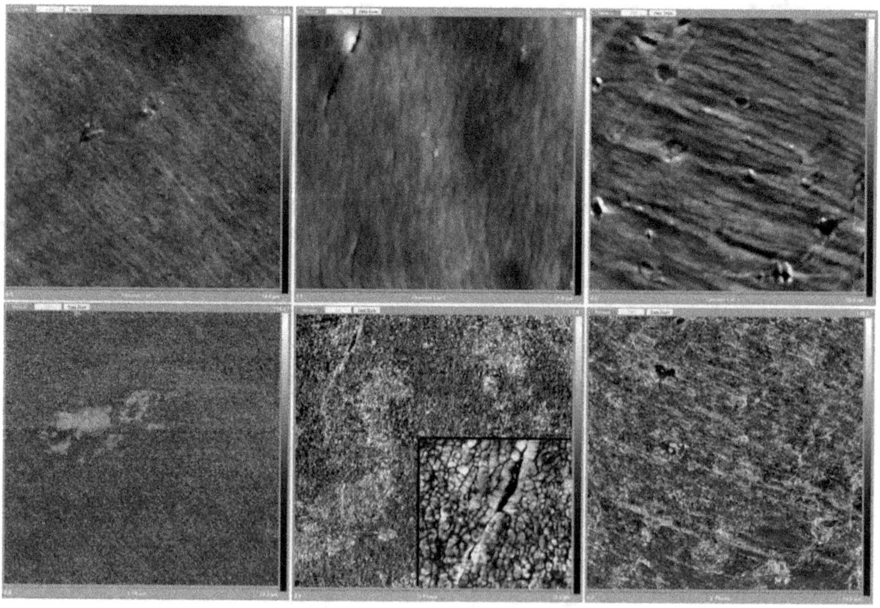

Figure 3.9 Low-resolution AFM height (top) and phase (bottom) images of the samples; (left) PE+GO, (middle) PE+GO+CPE25, and (right) PE+GO+CPE35.

confirmed from the absence of $V_{C\text{-}Cl}$ vibrations at 654 cm^{-1} and 607 cm^{-1}. CPE25 exhibited exactly similar behavior to that of CPE35 in its composite with graphene oxide and PE (Figure 3.11). However, the extent of reaction between the functional groups on graphene oxide and CPE25 was less due to the low percentage of chlorine content that could be observed in the IR spectrum of PE+GO+CPE25 as a low intense $V_{C=O}$ vibration at 1737 cm^{-1} (Figure 3.11). Interestingly, IR spectra of PE+CPE35 and PE+CPE25 didn't show any features from CPE part (Figure 3.12) which indicated that C-Cl bonds would have been destroyed at higher temperature used during compounding. Overall lower concentration of CPE in the blend might also partly contribute to this effect. Since the contribution of graphene oxide in nanocomposites was difficult to detect in IR, Raman spectra of

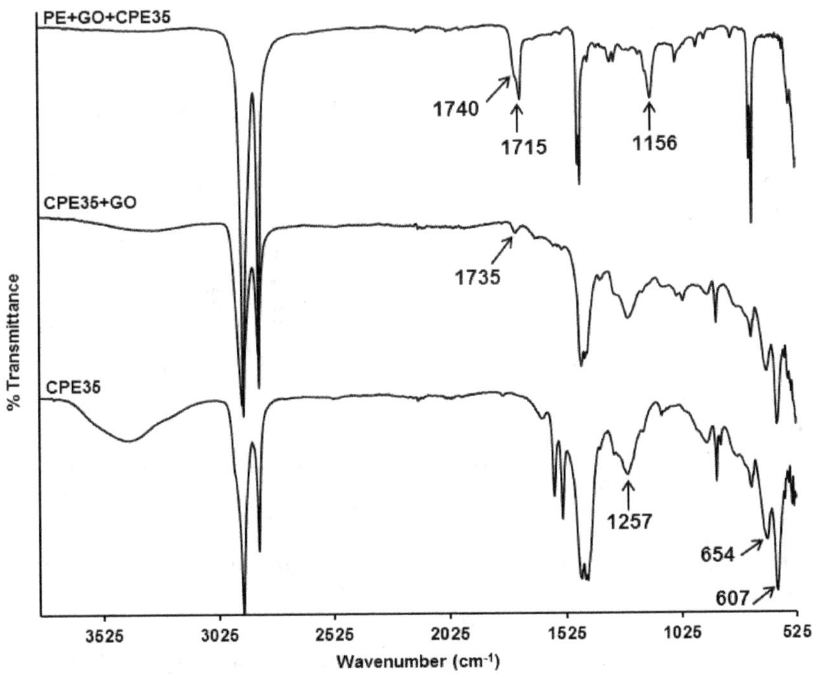

Figure 3.10 IR spectra of CPE35 and its composites with graphene oxide and PE.

PE+GO, PE+GO+CPE35 & PE+GO+CPE25 are given in Figure 3.13. Characteristic G band of graphene oxide was observed at ~1580 cm^{-1} and the D band was merged with PE signals in the 1250 cm^{-1} to 1350 cm^{-1} region. It could be concluded from above IR and Raman results that CPE compatibilizer helped graphene oxide to be dispersed well in the PE matrix through the formation of strong chemical bonds like ester and ether between CPE and graphene oxide layers, and the effect increased with the increase of chlorine content in CPE. Moreover, since the Cl atoms were sacrificed from CPE due to chemical reaction and possible thermal degradation, the amount of chlorine content in the final mixture would be negligible. The schematic representation of these findings is shown in Scheme 3.1.

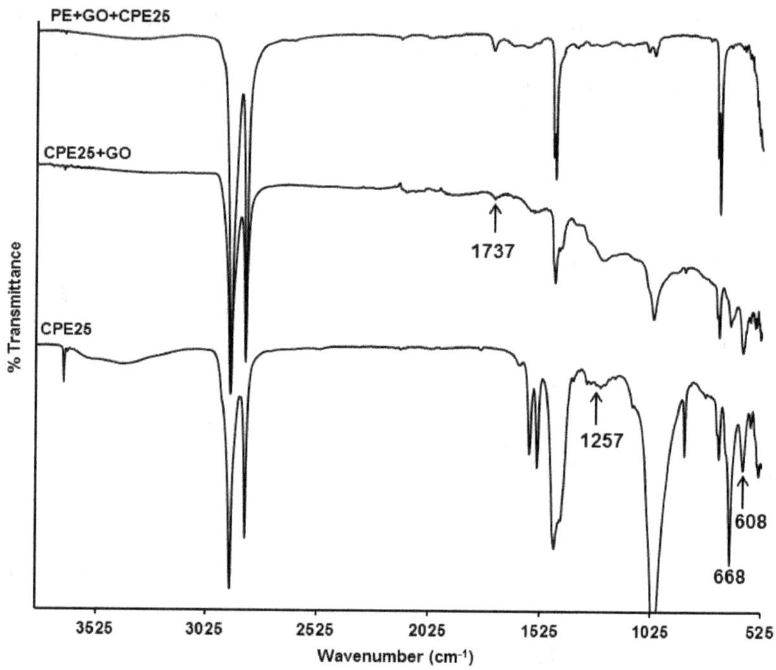

Figure 3.11 IR spectra of CPE25 and its composites with graphene oxide and PE.

Confirmation into the content of chlorine in the final product could also be obtained from the TGA /DTA/MS analysis of CPE35 and its nanocomposite with PE and graphene oxide. Since the nanocomposite contained only 5 wt% CPE35, weight loss analysis of chlorine from TGA plots was within the range of instrumental error. However, the elimination of chlorine could be clearly monitored from DTA curves when compared with corresponding mass fragments in mass spectrometry analysis. DTA plot of CPE35 given in Figure 3.14 showed three endothermic peaks corresponding to the first thermal degradation step in TGA (150 °C-360 °C). The first endothermic peak at T_{max} at 211 °C was due to sample melting. Melting

Table 3.1 FTIR band assignment of PE, CPE and its composites with graphene oxide [36]

Sample	Wavenumber (cm^{-1})	assignment
HDPE	2914	V_a CH_2
	2847	V_s CH_2
	1472, 1462	δ CH_2
	730, 719	γ_r CH_2
CPE25	2916,2849	V_{C-H} of methylene (-C**H**$_2$-CH_2-)
&	1464	δ_{C-H} of methylene (-C**H**$_2$-CH_2-)
CPE35	1430	δ_{C-H} of methylene with adjacent chlorine atom (-C**H**$_2$-CHCl-)
	1257	δ $_{C-H}$ of (-CH_2-C**H**Cl-)
	719	γ_r CH_2
	668 (654 in the case of CPE35), 608	V_{C-Cl} of CPE
	1576, 1540,1013, 874	carbontetrachloride contamination [37]
	3406, 1620	H_2O
CPE25+GO	1737	V $_{C=O}$ of (-COO-) all other peaks are same as in CPE25
CPE35+GO	1735	V $_{C=O}$ of (-COO-)
	1156	V_{C-O} of (-COO-), (C-O-C) all other peaks are same as in CPE35
PE+GO+CPE25	~3400	$C=O_{str}$ overtone of (-COO-)
	2915, 2847	V_a CH_2 & V_s CH_2
	1737	δ $_{C=O}$ vibrations of (-COO-)
	1472,1462	δ CH_2
	730,719	γ_r CH_2
PE+GO+CPE35	~3400	$C=O_{str}$ overtone of (-COO-)
	2915, 2847	V_a CH_2 & V_s CH_2
	1740	$C=O_{str}$ of saturated esters
	1715	$C=O_{str}$ of aryl esters or α,β-unsaturated esters
	1472,1462	δ CH_2
	1329	δ $_{C-H}$ of (-CH_2-C**H**OCO-)
	1156	V_{C-O} of (-COO-)
	730,719	γ_r CH_2

(V=stretching, δ=bending, γ_r= rocking)

phenomena seemed to result in the loss of water and some amorphous carbon as it showed small humps at 211 °C in the mass spectrometer profiles of H_2O (m/z=18) and CO_2 (m/z=44) given in Figure 3.15. The second endothermic peak (T_{max}=292 °C) in the DTA plot of CPE35 could be assigned to the loss of chlorine because the

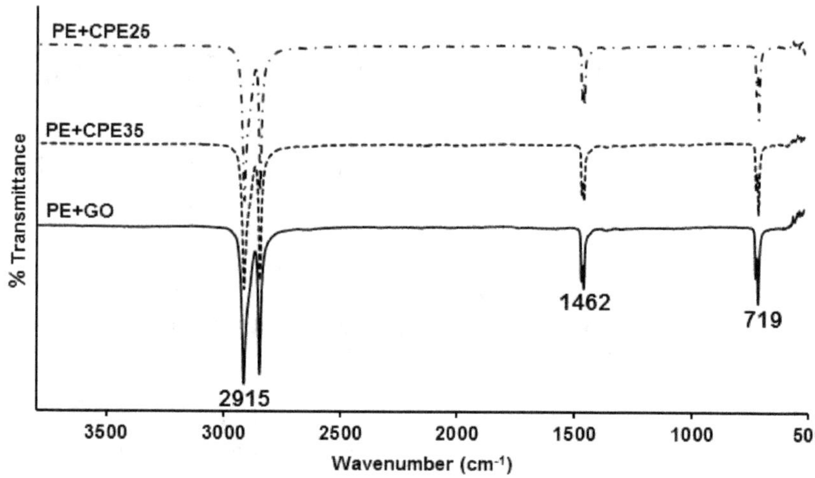

Figure 3.12 IR spectra of PE+GO composite and PE blends with CPE35 and CPE25.

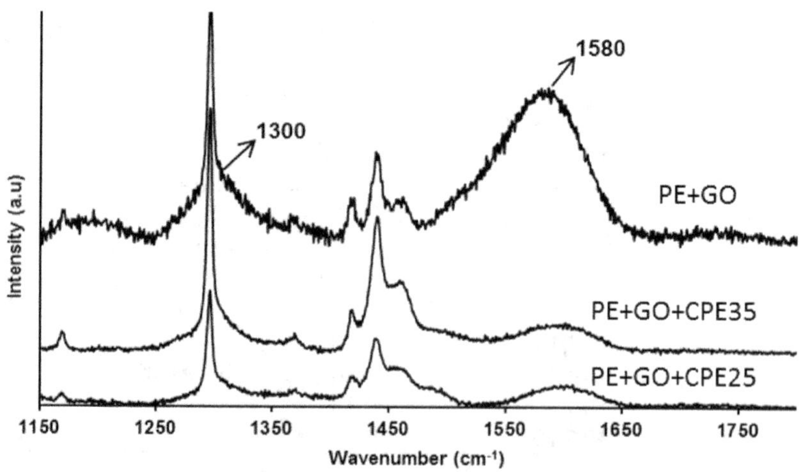

Figure 3.13 Raman spectra of PE+GO, PE+GO+CPE35, and PE+GO+CPE25 composites.

mass spectra contains HCl (m/z=36) profile in the temperature range of 250 °C to 350 °C. Comparison of DTA plot of CPE35 with those of PE as well as PE+CPE35 revealed that all plots had first endothermic peak, but for PE, the second endothermic peak was absent (Figure 3.14). It, thus, confirmed that the second endothermic peak in the DTA plot of PE+CPE35 was from the loss of HCl which appeared as smaller but clearly visible contribution at 273 °C (T_{max}). Even though the amount of CPE in PE+CPE35 was only 5 wt%, the

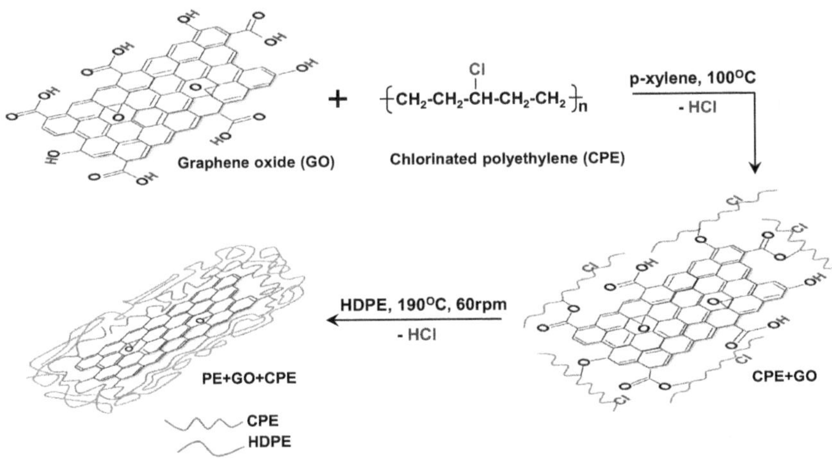

Scheme 3.1 Scheme representing the chemical interaction between the CPE and graphene oxide.

evolution of HCl was clearly detected in the DTA plot. The final composite PE+GO+CPE35 showed a significant decrease in the intensity of second endothermic peak and/or it appeared as a small shoulder around 300 °C. The presence of chlorine contamination in the final composite could not be eliminated as the IR spectra contained weak signals at 1576 cm^{-1}, 1540 cm^{-1} and 874 cm^{-1} (Figure 3.10, Table 3.1). Thus, it could be concluded from TGA/DTA /MS analysis that the amount of chlorine in PE+CPE35 and PE+GO+CPE35 decreased significantly due to the destruction of C-Cl bond during melt compounding. The amount of remaining Cl in the PE+GO+CPE35 was even lower than the PE+CPE35 blend due to significant depletion of Cl atoms during their reactions with functional groups on graphene oxide which resulted in strong chemical bonds such as ester, ether etc. between CPE chains and graphene oxide as confirmed from FTIR.

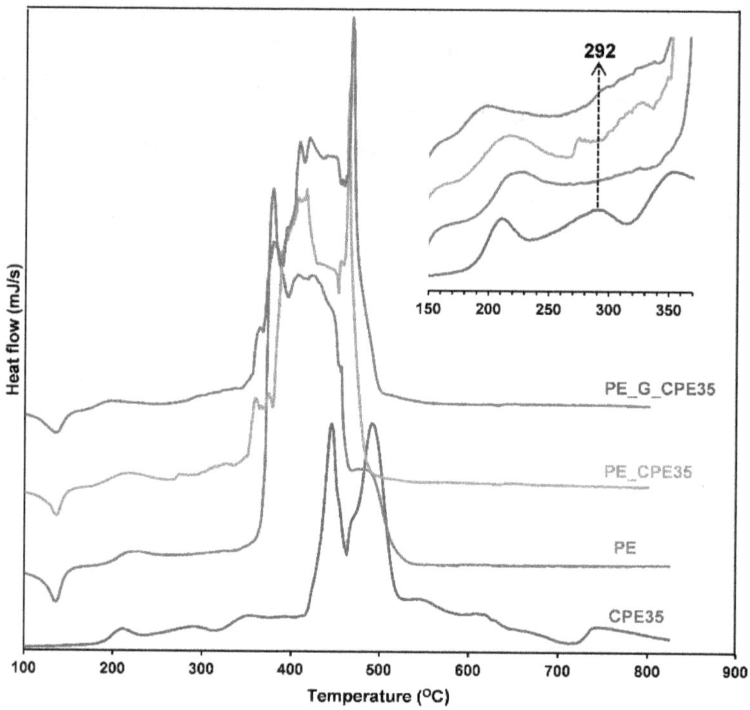

Figure 3.14 DTA plots of PE, CPE35, PE+CPE35, and PE+GO+CPE35.

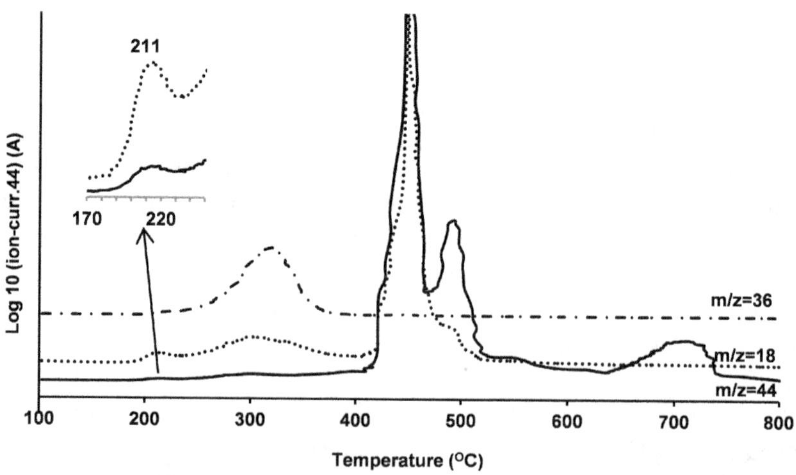

Figure 3.15 Mass spectrometer profile of H_2O (m/z = 18), CO_2 (m/z = 44), and HCl (m/z = 36) released from CPE35.

3.4 Use of Raman Mapping for Component Distribution

Luckachan and Mittal [38] developed further the use of Raman mapping technique to characterize the phase distribution of individual components in polymer blends and polymer nanocomposites systems (0.5 wt% filler and 5 wt% CPE), as shown in Figures 3.16 and 3.17.

Two visualizations shown in Figure 3.16 A and B for PE+GO nanocomposite system were obtained by selecting G band of graphene at 1580 cm^{-1} and V_{as}C-C band of PE at 1060 cm^{-1} respectively. In both cases, colour signal varying from black to yellow in the intensity bar indicated increasing intensity of peak chosen as contrast parameter. Three representative spectra of the different colour zone from the map in Figure 3.16A are shown in Figures 3.16(C)-(E) to demonstrate the variation of graphene content in the mapping area. Spectra showed maximum concentration of graphene in the red region. The concentration decreased in the green region, whereas the presence of graphene was almost nil in the black region. Moreover, the red and green coloured regions were observed to exist in the mapping image (Figure 3.16A) as bounded in discrete domains which indicated the immiscibility of graphene phase in the PE matrix. High stack thickness of graphene platelets observed in TEM micrograph of the PE+G nanocomposite (Figure 3.16F) further supported the absence of nano-scale graphene dispersion in PE+GO nanocomposite.

Similarly, mapping of PE+GO+CPE35 nanocomposite given in Figure 3.17 also showed two representative visualizations corresponding to the intensity of G band of graphene and V_{as}C-C band of PE. For instance, in Figure 3.17A, majority of the mapping areas exhibited red and green colour signal, though a small yellow domain was also observed to be confined in the red region. The representative spectra of the different colour domains exhibited the majority of the yellow domain to contain graphene, however, a significant amount of graphene was also observed in the red and green colour domains. Thus, it indicated that the distribution of graphene signal in various colour domains was an indicative of uniform graphene dispersion in the PE matrix. A significant extent of polymer intercalation inside the stacks of graphene platelets observed in the TEM micrograph of PE+GO+CPE35 nanocomposite (Figure 3.17E) also supported the findings of uniform dispersion of graphene in the PE matrix with the addition of CPE35 compatibilizer.

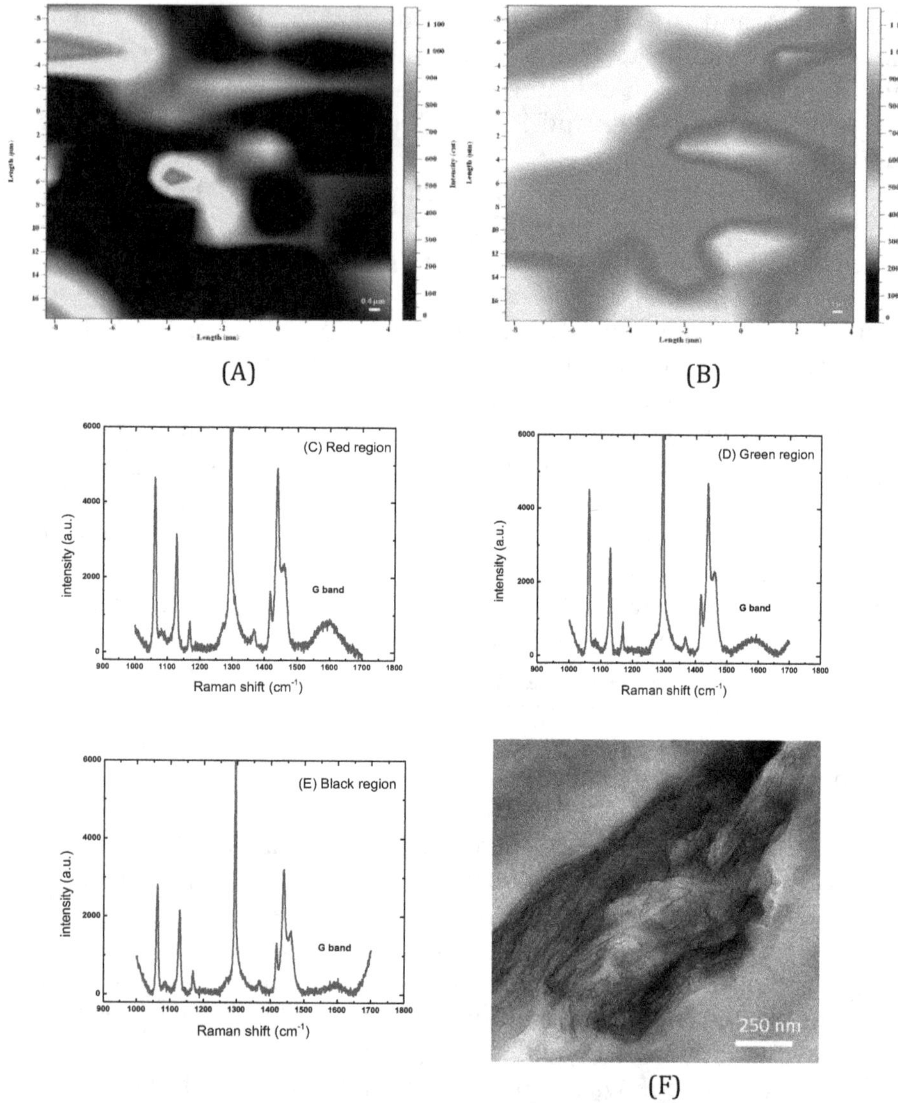

Figure 3.16 Raman mapping of PE+GO nanocomposite: (A) and (B) represent the maps of graphene and PE separately. Spectra were collected from (C) red, (D) green and (E) black regions of the map in (A). Figure 3.16F shows the TEM micrograph of PE+GO nanocomposite. Reproduced from Reference 38 with permission from Springer.

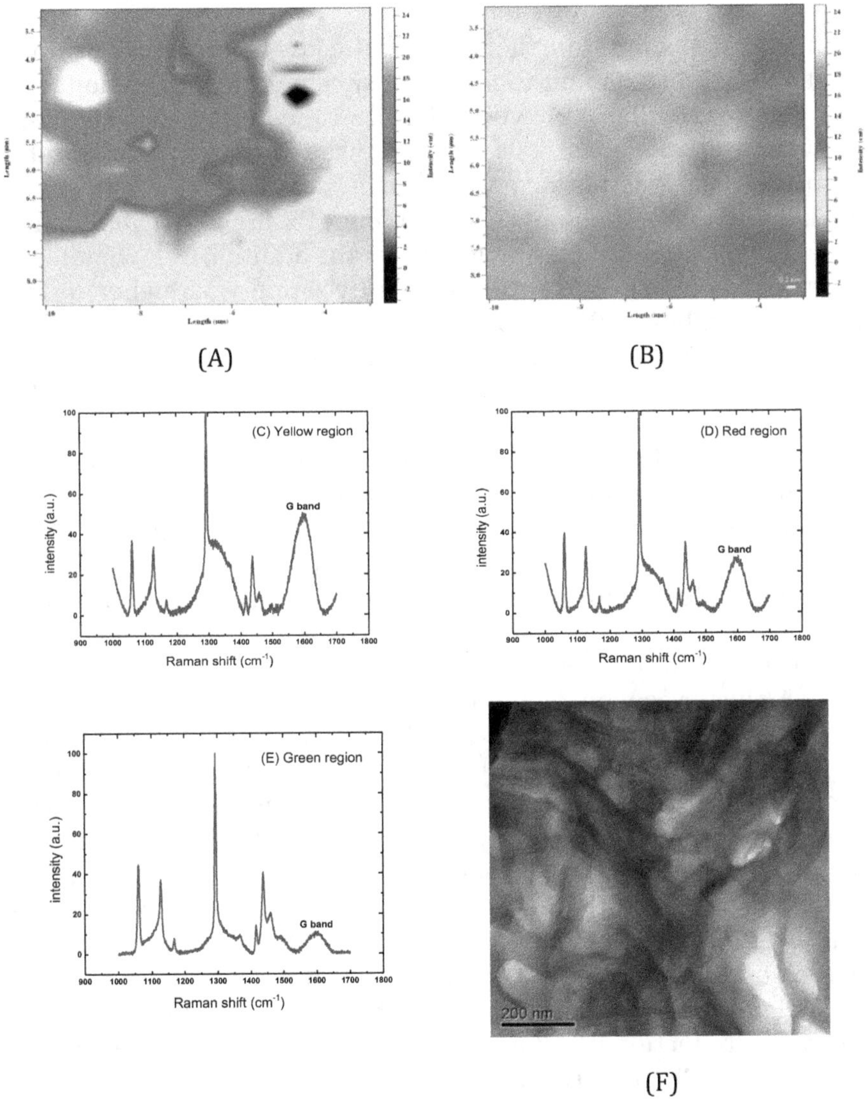

Figure 3.17 Raman mapping of PE+GO+CPE35 nanocomposite: (A) and (B) represent maps of graphene and PE separately. Spectra were collected from (C) yellow, (D) red and (E) green regions of (A). Figure 3.17F shows the TEM micrograph of PE+GO+CPE35 nanocomposite. Reproduced from Reference 38 with permission from Springer.

3.5 Reduced Low Temperature Brittleness

Mittal and Matsko [39] further studied the low temperature flexibility of PE+CPE blends and their composites with a small amount of graphene oxide filler (0.1 wt%). Surface topography analysis (surface height or roughness variation) at three different places on the sample surface was performed thorough atomic force microscopy. The height variation enhanced on adding CPE to polyethylene indicating the plasticization of the matrix on the addition of low molecular weight polymer. The extent of plasticization was higher in the case of amorphous CPE35 polymer where the height variation was significantly large as compared to PE blend with CPE25. The composites followed similar trend as the polymer blends, however, the height variation was even higher in the composite samples (Figure 3.18). For example, for PE+GO sample, a value of 113 nm for surface variation was calculated which was enhanced markedly to 198 nm for PE+GO+CPE25 nanocomposite and 564 nm for PE+GO+CPE35 nanocomposites. It indicated that the addition of small amount of graphene oxide did not negatively impact the flexibility of the samples.

Figure 3.19 also demonstrates the fracture analysis performed on the polymer as well as nanocomposite samples at two temperatures: -140 °C and -195 °C. The PE as well as PE+CPE25 and PE+CPE35 blends (with 5% CPE content) were flexible at -140 °C, whereas only PE+CPE35 sample was still flexible at -195 °C. The other two samples exhibited a brittle failure at -195 °C. It confirmed the findings from the AFM analysis about the flexibility of CPE35 containing samples at -180 °C and indicated that the samples were still flexible at -195 °C. Similar to the PE+CPE35 blend, the nanocomposite PE+GO+CPE35 was also observed to be fully flexible at -195 °C indicating that addition of graphene oxide to this system did not lead to reduction in flexibility. The other two composites PE+GO and PE+GO+CPE25 were flexible only near to the glass transition temperature of PE, whereas failed in a brittle manner at -195 °C. These findings are of substantial importance as these can ensure the application of polyethylene based materials at very low temperature conditions, which otherwise is not feasible due to low temperature brittleness. In addition, though these findings relate specifically to polyethylene based materials, however, other commercially important polyolefins such as polypropylene can also benefit from these insights.

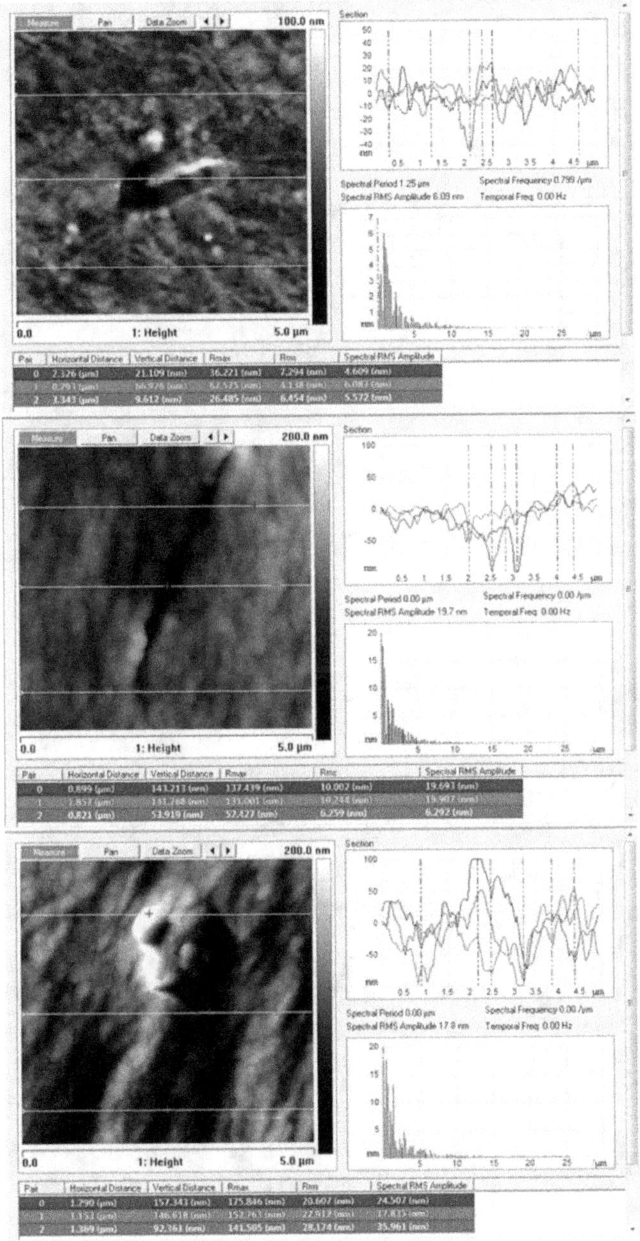

Figure 3.18 AFM height images and corresponding height corrugation analysis for PE+GO (top image), PE+GO+5%CPE25 (middle row) and PE+GO+5%CPE35 (bottom row) when sectioned at -140 °C. Reproduced from Reference 39 with permission from Springer.

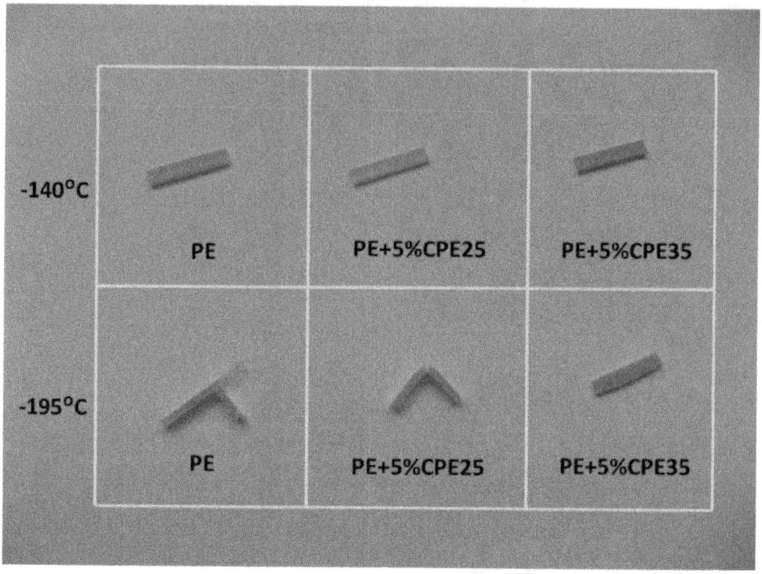

(a)

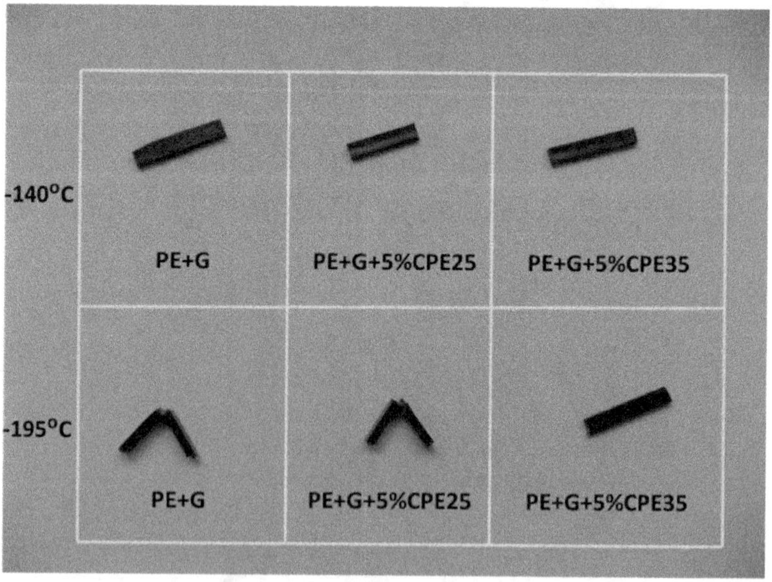

(b)

Figure 3.19 Fracture analysis of samples; (a) PE and PE+CPE blends, (b) PE+GO+CPE nanocomposites. Reproduced from Reference 39 with permission from Springer.

3.6 Conclusions

In the current study, morphological characterization of the PE+CPE blends and PE+GO+CPE nanocomposites, generated using CPE of two different chlorination contents, was studied. The rheological analysis using Cole-Cole, van Gurp and Han Chuang plots showed miscibility for the PE+GO, PE+CPE25 and PE+CPE35 samples, whereas the nanocomposite materials PE+GO+CPE25 and PE+GO+CPE35 were concluded to have immiscible phase. The extent of immiscibility was observed to be highest in PE+GO+CPE35 nanocomposite. The TEM analysis of the samples confirmed the best distribution of graphene oxide platelets in the composite containing CPE35 compatibilizer, which also resulted in least value of oscillatory torque for this composite. Apart from that, the polymer microstructure was also affected by the incorporation of CPE polymers, the most significant change was observed for CPE35 system due to its amorphous nature. The atomic force microscopy analysis also complemented the TEM findings that the crystalline microstructure of the PE polymer was drastically affected by CPE35 addition (as confirmed with lowest extent of crystallinity measured for this sample using DSC). The filler surface was also observed to be surrounded with hard phase which could have been resulted from the enhanced nucleation action of the graphene platelets as well as ordering of CPE chains near the graphene surface due to their positive interactions. This was also confirmed by higher extent of overall crystallinity in the composite samples as compared to PE+CPE blends. IR and Raman analysis confirmed a chemical reaction between the CPE and graphene oxide platelets. The blends of PE and CPE did not show C-Cl peaks thus indicating that the Cl was depleted not only due to the chemical reaction, but also by thermal degradation, which was also confirmed by DTA-MS analysis.

Role of compatibilizer in the dispersion of graphene filler in the polyethylene matrix was also successfully visualized using Raman mapping technique in PE+GO and PE+GO+CPE35 nanocomposites. The dispersion of graphene was confirmed to be less optimum in the PE+GO composites, whereas in the PE+GO+CPE35 composites, the addition of comapatibilizer resulted in improved degree of filler dispersion.

The blend and composite samples with CPE35 content could both be sectioned at -140 °C and -180 °C without brittle fracture, thus, confirming the flexibility at lower temperature, whereas the other systems with CPE25 or composite without CPE exhibited a brittle fracture at -180 °C. The fracture analysis carried out with samples further confirmed the low temperature flexibility of PE+CPE35 and PE+GO+CPE35 systems even at -195 °C.

Acknowledgements

The work (except Sections 3.4 and 3.5) presented in this study has been published earlier in the journal Macromolecular Chemistry and Physics, 2014, 215, 255-268, doi:10.1002/macp.201300613. The work has been reproduced in this chapter with permission from Wiley.

References

1. Jancar, J. (1999) *Mineral Fillers in Thermoplastics I: Raw Materials and Processing*, Springer-Verlag, Germany.
2. Mittal, V. (2012) *Functional Polymer Blends: Synthesis, Properties and Performance*, CRC Press, USA.
3. Hippi, U. (2008) *Polyolefin Composites*, John Wiley & Sons, USA.
4. Zhou, L., Wang, X., Lin, Y., Yang, J., and Wu, Q. (2003) Comparison of the toughening mechanisms of poly(vinyl chloride)/chlorinated polyethylene and poly(vinyl chloride)/acrylonitrile-butadiene-styrene copolymer blends. *Journal of Applied Polymer Science*, **90**, 916-924.
5. Eastwood, E. A., and Dadmun, M. D. (2002) Compatibilization of poly(vinyl chloride) and polyolefin elastomer blends with multi-block/blocky chlorinated polyethylenes. *Polymer*, **43**, 6707-6717
6. He, P., Huang, H., Xiao, W., Huang, S., and Cheng, S. (1997) A study on PVC/LLDPE blends with solid-state-chlorinated polyethylene as compatibilizer. *Journal of Applied Polymer Science*, **64**, 2535-2541.
7. Stoeva, S. (2006) Study of poly(vinyl chloride) blends with solid-state chlorinated polyethylene. *Journal of Applied Polymer Science*, **101**, 2602-2613.
8. Waddington, S., and Briggs, D. (1991) Adhesion mechanisms between polymer coatings and polypropylene studied by XPS and SIMS. *Polymer Communications*, **32**, 506-508.
9. Ogoe, S. A., Liou, D. W., Ainsworth, O. C., Bethea, J. R., Flores, D. R., Ellebracht, S. R., Muskopf, J. W., Werling, C. L., Wilson, E. J., Cocup, J. W., and Allison, J. L. (1996) Blends of Polycarbonate and Chlorinated Polyethylene, US patent 5,525,651.
10. Barrie, J. A., and Webb, W. D. (1989) Gas transport in miscible ble-

nds of ethylene-vinyl acetate copolymer and chlorinated polyethylene. *Polymer*, **30**, 327-332.

11. Maksimov, R. D., Ivanova, T., Zicans, J., and Kalkis, V. (2005) Mechanical properties of blends of low-density polyethylene with chlorinated polyethylene. *Mechanics of Composite Materials*, **41**, 267-276.
12. Maksimov, R. D., Ivanova, T., Kalnins, M., and Zicans, J. (2004) *Mechanics of Composite Materials*, **40**, 331-340.
13. Mukhopadhyay, P., and Gupta, R. K. (2011) Trends and frontiers in graphene-based polymer nanocomposites. *Plastics Engineering*, **67**, 32-42.
14. Kim, H., Abdala, A. A., and Macosko, C. W. (2010) Graphene/polymer nanocomposites. *Macromolecules*, **43**, 6515-6530.
15. Kuilla, T., Bhadra, S., Yao, D., Kim, N. H., Bose, S., and Lee, J. H. (2010) Recent advances in graphene based polymer composites. *Progress in Polymer Science*, **35**, 1350-1375.
16. Rafiee, M. A., Rafiee, J., Wang, Z., Song, H., Yu, Z. Z., and Koratkar, N. (2009) Enhanced mechanical properties of nanocomposites at low graphene content. *ACS Nano*, **3**, 3884-3890.
17. Kim, H., Miura, Y., and Macosko, C. W. (2010) Graphene/polyurethane nanocomposites for improved gas barrier and electrical conductivity. *Chemistry of Materials*, **22**, 3441-3450.
18. Nguyen, D. A., Lee, Y. R., Raghu, A. V., Jeong, H. M., Shin, C. M., and Kim, B. K. (2009) Morphological and physical properties of a thermoplastic polyurethane reinforced with functionalized graphene sheet. *Polymer International*, **58**, 412-417.
19. Schniepp, H. C., Li, J.-L., McAllister, M. J., Sai, H., Herrera-Alonso, M., Adamson, D. H., Prud'homme, R. K., Car, R., Saville, D. A., and Aksay, I. A. (2006) Functionalized single graphene sheets derived from splitting graphite oxide. *Journal of Physical Chemistry B*, **110**, 8535-8539.
20. Kim, H., Kobayashi, S., AbdurRahim, M. A., Zhang, M. L. J., Khusainova, A., Hillmyer, M. A., Abdala, A. A., and Macosko, C. W. (2011) Graphene/polyethylene nanocomposites: Effect of polyethylene functionalization and blending methods. *Polymer*, **52**, 1837-1846.
21. Guffey, V. O., and Sabbagh, A. B. (2002) PVC/wood-flour composites compatibilized with chlorinated polyethylene. *Journal of Vinyl & Additive Technology*, **8**, 259-263.
22. Chaudhry, A. U., and Mittal, V. (2013) High-density polyethylene nanocomposites using masterbatches of chlorinated polyethylene/graphene oxide. *Polymer Engineering and Science*, **53**, 78-88.
23. Donskoi, A. A., M. A. Shashkina, and G. E. Zaikov. (2003) *Fire Resistant and Thermally Stable Materials Derived from Chlorinated Polyethylene*, VSP, The Netherlands.

24. McAllister, M. J., Li, J. L., Adamson, D. H., Schniepp, H. C., Abdala, A. A., Liu, J., Herrera-Alonso, M., Milius, D. L., Car, R., Prud'homme, R. K., and Aksay, I. A. (2007) Single sheet functionalized graphene by oxidation and thermal expansion of graphite. *Chemistry of Materials*, **19**, 4396-4404.
25. Hummers W. S., and Offeman, R. E. (1958) Preparation of Graphitic Oxide. *Journal of American Chemical Society*, **80**, 1339-1339.
26. Cho, K., Lee, B. H., Hwang, K. M., Lee, H., and Choe, S. (1998) Rheological and mechanical properties in polyethylene blends. *Polymer Engineering and Science*, **38**(12), 1969-1975.
27. Kim, H. K., Rana, D., Kwag, H., and Choe, S. (2001) Melt rheology of ethylene 1-octene copolymer blends synthesized by Ziegler-Natta and metallocene catalysts. *Korea Polymer Journal*, **8**, 34-43.
28. Kwag, H., Rana, D., Choe, K., Rhee, J., Woo, T., Lee, B. H., and Choe, S. (2000) Binary blends of metallocene polyethylene with conventional polyolefins: Rheological and morphological properties. *Polymer Engineering and Science*, **40**(7), 1672-1681.
29. Joshi, M., Butola, B. S., Simon, G., and Kukaleva, N. (2006) Rheological and viscoelastic behavior of HDPE/octamethyl-POSS nanocomposites. *Macromolecules*, **39**(5), 1839-1849.
30. van Gurp, M., and Palmen, J. (1998) Time-temperature superposition for polymeric blends. *Rheology Bulletin*, **67**(1), 5-8.
31. Chuang, H. K., and Han, C. D. (1984) Rheological behavior of polymer blends. *Journal of Applied Polymer Science*, **29**, 2205-2229.
32. Han, C. D., and Chuang, H. K. (1985) Criteria for rheological compatibility of polymer blends. *Journal of Applied Polymer Science*, **30**(11), 4431-4454.
33. Utracki, L. A. (1989) *Rheology of Polymer Alloys and Blends*, Hanser Publishers, Germany.
34. Ko, S. W., Gupta, R. K., Bhattacharya, S. N., Choi, H. J. (2010) Rheology and physical characteristics of synthetic biodegradable aliphatic polymer blends dispersed with MWNTs. *Macromolecular Materials and Engineering*, **295**, 320-328.
35. Choi, H. J., and Ray, S. S. (2011) A review on melt-state viscoelastic properties of polymer nanocomposites. *Journal of Nanoscience and Nanotechnology*, **11**, 8421-8449.
36. Krimm, S., Liang, C. Y., and Sutherland, G. B. B. M. (1956) Infrared Spectra of High Polymers. II. Polyethylene. *Journal of Chemical Physics*, **25**(3), 549-562.
37. Smook, M. A., Pieski, E. T., and Hammer, C. F. (1953) Derivatives of chlorosulfonated polyethylene and their infrared spectra. Industrial & Engineering Chemistry, **45**(12), 2731-2737.
38. Luckachan, G. E., and Mittal, V. (2015) Evaluation of crystallinity variation and phase dispersion in polymer blends and nanocomposites by Raman mapping. *Journal of Polymer Research*, **22**:237,

doi:10.1007/s10965-015-0878-0.

39. Mittal, V., and Matsko, N. (2013) PE-CPE blends and their graphene oxide nanocomposites with reduced low temperature brittleness. *Colloid and Polymer Science*, **291**(8), 1949-1961.

4

PP/EVA-MA Blend Nanocomposites with Graphene Nano-sheets

4.1 Introduction

Polypropylene (PP) is a stereo-regular polymer, with properties depending on crystalline structure and various tacticities. Moreover, PP possesses various beneficial properties like low density, high thermal resistance, resistance to chemical attack (insoluble in almost all solvents at room temperature), ease of processing and recyclability. Interestingly, PP manifests mechanical profile closely to the engineering thermoplastics. Owing to this extraordinary set of properties, PP is commonly used thermoplastic for structural applications [1]. Unfortunately, PP also suffers from drawbacks like low temperature brittleness and low impact strength. In order to improve the impact strength, addition of elastomeric compounds has generally been employed [2-14], which leads to improved elongation, however, may negatively affect the modulus. On the other hand, incorporation of polymeric compatibilizers along with reinforcing fillers (like calcium carbonate, nano-clay, carbon nanotubes and graphene platelets, etc.) has been reported to achieve good balance of modulus and toughness enhancement, though negative impact on the elongation was observed in some studies due to compatibility issues [15-27]. Thus, use of blend composites is a functional strategy to achieve simultaneous enhancement in stiffness and toughness of the materials. Also, in order to improve compatibility of the polymer and filler phases, approaches of reactive compatibilization and chemical modification of fillers also need to be effectively implemented.

Owing to its exceptional mechanical properties (Young's modulus 1 TPa) and high thermal conductivity (3000 $W.m^{-1}.K^{-1}$), use of gra-

Shital Pawar[a,], Vikas Mittal[a,**], Nadejda Matsko[b]*
[a]The Petroleum Institute (part of Khalifa University of Science and Technology), Abu Dhabi, UAE; [b]Graz Centre of Electron Microscopy Graz, Austria
**Central address: University of Calgary, Calgary, Canada; **Central address: Bletchington, Wellington County, Australia*

phene presents opportunities for significant reinforcement of polymers [28-32]. However, due to its nano-scale dimensions and polarity mismatch with the polymer, its dispersion in the polymer matrix is challenging. In order to improve dispersion and compatibility with matrix, various functional groups can be generated/attached on graphene sheets by oxidization of graphite platelets. These groups can further be used for reactive compatibilization with either polymer matrix directly or one of the components of the polymer blend (or compatibilizer), thus, leading to advanced nanocomposites with uniform filler dispersion and selective localization. For instance, Mittal *et al.* [33] reported nanocomposites of bio-polyamide (PA) with varying amounts of thermally reduced graphene, generated by melt mixing. As shown in Figure 4.1, free amide bonds in the amorphous phase of PA were observed to undergo chemical reaction with the surface groups of graphene, thus, resulting in better interfacial interactions. Due to this reactive compatibilization, polyamide nanocomposites

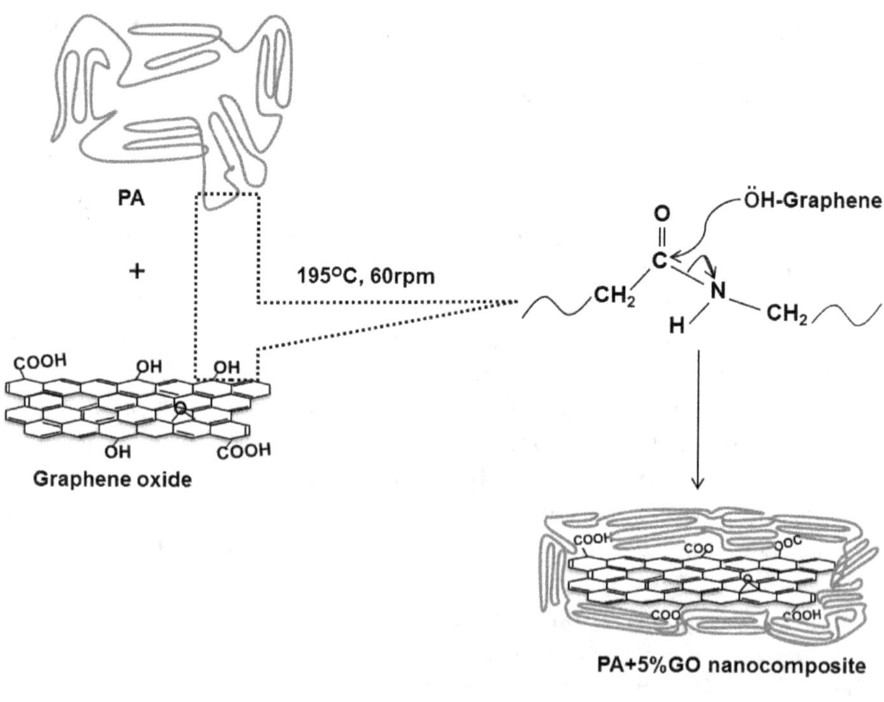

Figure 4.1 Representation of interaction of the polyamide chains with graphene surface. Reproduced from Reference 33 with permission from Elsevier.

had exfoliated morphology, and the tensile modulus of the nanocomposite with 5 wt% filler content was 1.8 times higher than the pure polymer. The polyamide nanocomposites were also observed to exhibit enhanced thermal stability owing to polymer immobilization on filler surface as well uniform filler dispersion in the polymer, as the peak degradation temperature in PA composite with 5 wt% graphene content was enhanced by 10 °C, and was also a function of the filler content. The findings, thus, confirmed the role of filler-matrix interactions and compatibilization in achieving filler dispersion and superior nanocomposite properties.

In another study reporting the compatibilization of polymer and filler through the chemical interaction of the functional groups present on the filler surface with the compatibilizer, Vengatesan *et al.* [34] developed nanocomposites using polypropylene-grafted-maleic anhydride (PP-g-MA) as compatibilizer with different weight ratios (0.5, 1.0, 1.5, 2.5, 5.0, and 10.0 wt%) of amine functionalized mullite fibers (AMUF) via solution blending method. Chemical grafting of AMUF with PP-g-MA resulted in enhanced filler dispersion in the polymer as well as effective polymer-filler interactions. Figure 4.2 shows the SEM images of the fracture surface of the PP matrix and PP-AMUF nanocomposites. In Figure 4.2(a) for pure PP, a smooth fracture surface was observed. The fracture surface in Figure 4.2(b) and (c) exhibited a plastic deformation during the Izod impact test. The voids and cracks were minimal in these 1 wt% and 5 wt% AMUF-filled PP nanocomposites. These results suggested that covalently bound AMUF along with its uniform dispersion provided dissipation of large amount of energy, thus, leading to high toughness of the PP nanocomposites. On the other hand, the fractography of 10 wt% AMUF filled PP nanocomposite exhibited a large number of voids, agglomeration, and cracking (Figure 4.2(d)). This behavior resulted from the aggregation of AMUF within the PP matrix at higher filler fractions, which was also evidenced by the SEM elemental mapping.

Vengatesan *et al.* [35] also reported the effect of amine functionalized reduced graphene oxide (r-AGO) on the properties of high density polyethylene (HDPE) matrix, using chlorinated polyethylene (CPE) as a compatibilizer (Figure 4.3). r-AGO chemically interacted with CPE, which was confirmed by Fourier-transform infra-red spectroscopy (FT-IR) analysis. For instance, in the CPE spectrum, the peak at 668 cm^{-1} corresponded to the C-Cl stretching frequency. In addition, the two peaks at 2926 cm^{-1} and 2853 cm^{-1} corresponded to the asymmetric and symmetric characteristic $-CH_2$ stretching vibrations

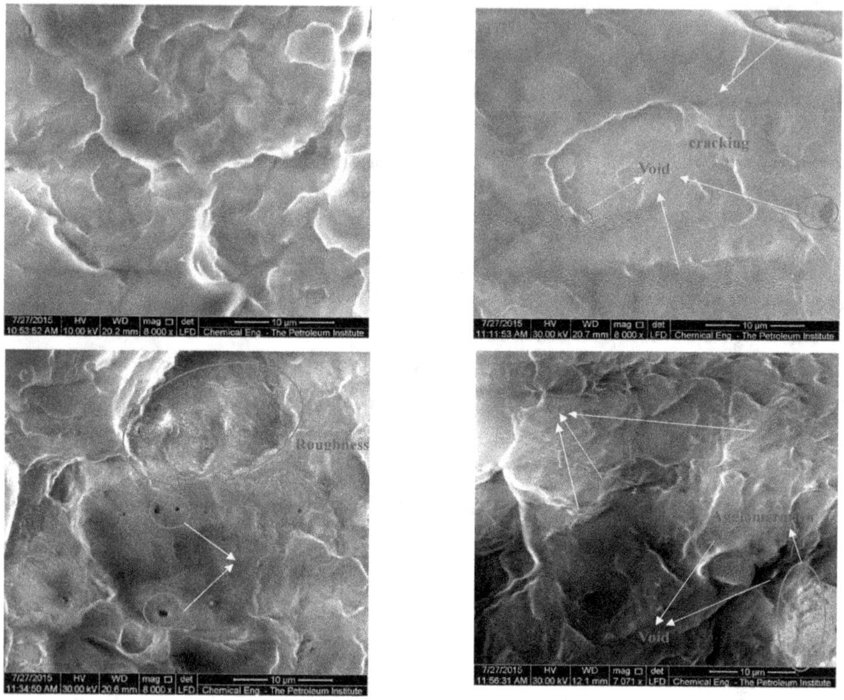

Figure 4.2 SEM images of fracture surface of (a) pure PP, (b) PP-AMUF2 (1 wt % AMUF), (c) PP-AMUF5 (5 wt % AMUF) and (d) PP-AMUF6 (10 wt % AMUF). Reproduced from Reference 34 with permission from Wiley.

of CPE. For the filler/CPE hybrid, a reduction in the intensity of peak at 668 cm^{-1} for C-Cl stretching frequency was observed, which indicated the reaction between r-AGO and CPE. Furthermore, the peak at 1029 cm^{-1} corresponding to –CNO- stretching frequency also represented the nucleophilic reaction between the amine groups in r-AGO and Cl group in CPE. The improved interfacial interactions between the polymer and filler resulted in high degree of filler dispersion in the nanocomposites. Young's modulus of the composites increased with filler content and a maximum increment was observed at 3 wt% of r-AGO content. The tensile strength of the polymer was also retained in the composites. 3 wt% of r-AGO filled HDPE nanocomposites also exhibited lowest surface resistivity indicating enhancement of conducting pathways in the composites on filler addition. Due to the similar architecture and molecular nature of HDPE and polypropylene, similar functional approaches for the filler compatibilization through surface immobilization of the polymer chains can, thus, also

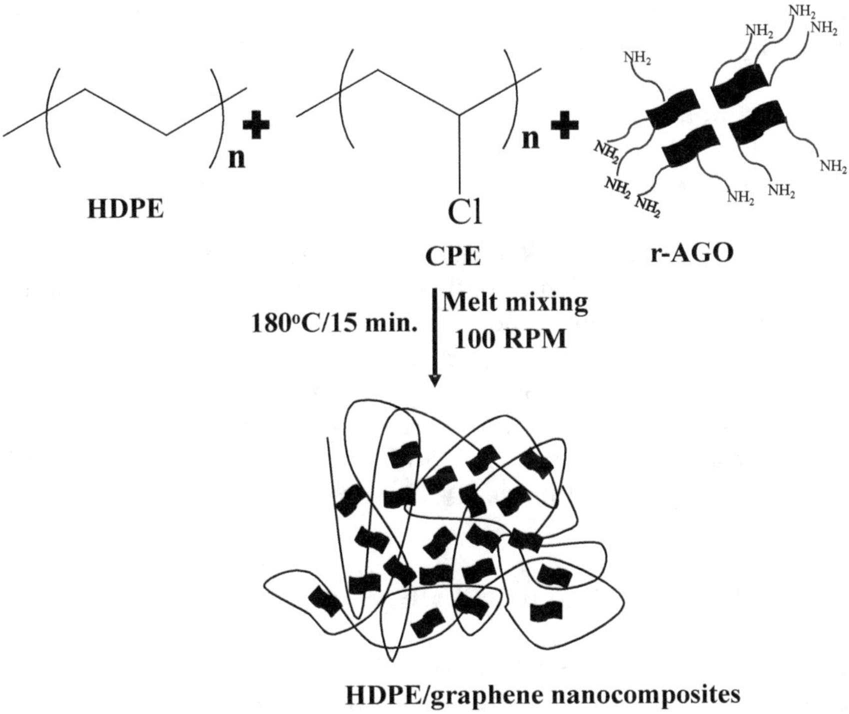

Figure 4.3 Schematic representation of the generation of polyethylene nanocomposites employing chemical interaction of compatibilizer with the filler [35].

be applied for generating high performance polypropylene composites.

In this study, the approach of generating blend nanocomposites using reactive compatibilization method has been further explored to achieve significant enhancement in the composite properties, especially at a low filler content. For this purpose, a 80:20 blend of PP with ethylene vinyl acetate-g-maleic anhydride (EVA-MA) was used. The surface of the graphite oxide platelets was functionalized with amine groups so as to achieve chemical interaction with EVA-MA chains. The chemical interaction was characterized using different structural analyses, and the resulting property enhancements were compared with composite generated without reactive compatibilization as well as pure polymer blend. The blend and nanocomposites were prepared using industrially accepted melt mixing method using optimized processing conditions so as to achieve good degree of phase mixing, but avoiding thermal degradation of the polymers at high processing temperature.

4.2 Experimental

4.2.1 Materials

Isotactic PP (HD951CF) was received from Abu Dhabi Polymers (Borouge), UAE. EVA-MA polymer with a trade name of Dupont Fusabond C190 was obtained from Dupont. It had a melt flow index (MFI, g/10 min at 2.16 kg, 190 °C) of 16, specific gravity of 0.950 and peak melting temperature of 71 °C. Solvents required for graphite oxide synthesis and surface modification were obtained from local suppliers and used without further purification. Ethylene diamine (EDA) and N,N'-dicyclohexylcarbodiimide (DCC) were procured from Sigma Aldrich.

4.2.2 Synthesis of GrO and GrO-NH_2

Graphite oxide (GrO) was synthesized through acid treatment, where graphite flakes were mixed with concentrated mixture of H_2SO_4 and H_3PO_4 (9:1 ratio). After homogenization of the solution, $KMnO_4$ was added slowly. Subsequently, the reaction mixture was transferred to the closed reactor and reaction was carried out for 24 h at room temperature. After completion of the reaction, mixture was diluted with deionised (DI) water and stirred vigorously for 2 h. Hydrogen peroxide (35%) solution was added dropwise till the reaction mixture turned bright yellow in color. The reaction product was filtered to obtain solid mass, which was washed several times with DI water, HCl and ethanol. The obtained graphite oxide was finally vacuum dried at room temperature.

Graphite oxide was modified with amine functional groups using ethylene diamine. Briefly, 100 mg GrO was dispersed in 50 mL THF using bath sonication for 1 h. *N,N'*-Dicyclohexylcarbodiimide was used as catalyst for the functionalization reaction. DCC and EDA were added in GrO dispersion and stirred for 1 h at room temperature. Concentration of reaction components (GrO:DCC:EDA) employed for the reaction was 1:20:20. Finally, the reaction mixture was refluxed at 80 °C for 24 h. The obtained surface modified GrO (designated as GrO-NH_2) was washed with DI water and vacuum dried at room temperature.

4.2.3 Generation of PP/EVA-MA Blend and Blend Nanocomposites

PP/EVA-MA (80/20) blend was generated by melt mixing using mini twin conical screw extruder (MiniLab HAAKE Rheomex CTW5, Germany). A mixing temperature of 190 °C for 10 min at 60 rpm under inert atmosphere with batch size of 5 g was used. The screw length and screw diameter were 109.5 mm and 5/14 mm conical, respectively. Both polymers were vacuum dried overnight at 60 °C before melt processing. Similarly, the blend nanocomposites with GrO and GrO-NH_2 were generated using the same processing conditions with 0.1, 0.5 and 1 wt% filler content. All three components were physically mixed prior to feeding to the compounder. The disc and dumbbell-shaped test samples were prepared by mini injection molding machine (HAAKE MiniJet, Germany) at a processing temperature of 190 °C. The injection pressure was 1000 bar for 6 s, whereas holding pressure was 400 bar for 3 s. The temperature of the mold was kept at 55 °C.

4.2.4 Material Characterization

IR spectra of the fillers, blend and nanocomposites were collected on Nicolet iS10 spectrometer equipped with SmartiTR diamond ATR accessory (angle of incidence of 45O), DTGS KBr detector and KBr beam splitter. It had a diamond ATR crystal (index of refraction 2.4 at 1000 cm^{-1}) and a depth penetration of 2 μm at 1000 cm^{-1} for sample with refractive index of 1.5. Spectra were recorded by the OMNIC software in the 4000-525 cm^{-1} region with a resolution of 4 cm^{-1} from 16 scans. Raman spectra were recorded on a Jobin Yvon Horiba LabRAM spectrometer with backscattered confocal configuration using a HeNe laser (633 nm). A long working distance objective with magnification 50x was used both to collect the scattered light and to focus the laser beam on the sample surface. CCD (charge-coupled device) detector exposure time was 10 s and an average of 2 cycles was used to increase S/N ratio. Raman shifts were calibrated using Si wafer signal at 520.7 cm^{-1}. Perkin Elmer elemental analyzer was used to quantify the elemental composition of graphite oxide before and after surface modification with amine groups.

Wide angle X-ray diffraction analysis of the fillers, blend and composites was performed on Panalytical Powder Diffractometer (X'Pert PRO) using CuKα radiation (λ = 1.5406 A°) in reflection mode. Zero-background holder was used to minimize the noise. The samples were step-scanned from 10-70° 2Θ at room temperature using a step size of 0.02° 2Θ and a step time of 10 s.

DHR-3 rheometer from TA Instruments was used to characterize the rheological properties such as storage modulus (G′) and complex viscosity elasticity (η*). Disc shaped samples of 25 mm diameter and 2 mm thickness were measured at 190 °C using a gap opening of 1.2 mm. Flow sweep experiment was carried out apriori and all experiments were performed in linear viscoelastic region (LVR) at 0.5% strain from ω = 0.1 to 100 rad/s.

Thermal properties of the blend and nanocomposites were recorded using Discovery thermogravimetric analyzer (TGA) from TA Instruments. Nitrogen was used as a carrier gas and the scans were obtained from 50 to 700 °C at a heating rate of 10 °C/min. Calorimetric properties of pure polymers, blend and nanocomposites were recorded on a Discovery differential scanning calorimeter (DSC) from TA Instruments under nitrogen atmosphere. The scans were obtained from 30 to 180 °C (first heating), 180 to -50 °C (first cooling) and -50 to 180 °C (second hating) using heating and cooling rates of 10 °C/min and 5 °C/min, respectively. Second heating runs were used for the comparison purpose.

Tensile analysis of the blend and composites was carried out using universal testing machine (Instron, USA) (ASTM D 638). The dumbbell-shaped samples were used. A loading rate of 10 mm/min was used and the tests were carried out at room temperature. The tensile strength and modulus of the composites were calculated using Win Test Analysis software. An average of five sample values is reported.

For the microscopy analysis, ultrathin sections (30-90 nm) as well as block faces of the composite samples were obtained using a Leica Ultracut E microtome (Leica, Austria) equipped with a diamond knife (Diatome, Switzerland) at -120 °C. The so-obtained sections for transmission electron microscopy (TEM) analysis were collected on 400 mesh formvar electron microscopy grids (coated with holey carbon film), and were examined in a Philips CM 20 (Philips/FEI, Germany) electron microscope at 200 kV at room temperature without additional staining. TEM image processing was performed using DigitalMicrograph software (Gatan, USA). Electron energy loss spectroscopy (EELS) of the blend and composites was performed in TEM mode and the spectra were visualized using DigitalMicrograph software (Gatan, USA). The surface of the block faces after cryo-ultramicrotomy was observed in Hitachi field emission in-lens S-900 high resolution scanning electron microscope at accelerating voltages of 10-20 kV. In addition, the block faces of specimens were also investigated at ambient conditions using ZEISS Axioplan light microscope (Zeiss, Germany) equipped with Zeiss Axio Cam ICc 1 CCD camera in reflected polarized light.

4.3 Results and Discussion

Chemically binding the filler phase with the polymer can lead to uniform filler dispersion as well as enhancement of composite properties at lower filler fraction. It, thus, helps to retain the polymer density, which is an important parameter for many structural applications of polymer nanocomposites. In addition, the requirement of lower filler fraction also helps to retain the polymer flexibility, which has been reported in the literature to significantly reduce due to large extent of filler in the conventional composites. In the case of blend based composites, such chemical binding of the filler with one of the polymer phases can also lead to better phase intermixing, and reduction in domain sizes through interfacial compatibilization, thereby leading to synergistic combination of component properties. Thus, in the current study, the EVA-MA interaction with functionalized graphite oxide was explored in order to generate polypropylene based functional nanocomposites with enhanced properties in comparison with the composite systems where such interactions are absent.

Presence of oxygen functional groups on graphite surface after oxidation was confirmed using FT-IR and elemental analysis. Figure 4.4a depicts FT-IR spectra for oxidized graphite sheets (GrO) where newly emerged peak at 1736 cm^{-1} was characteristic stretching of carbonyl functional group. Similarly, two distinct peaks evident at 850 and 1213 cm^{-1} were attributed to epoxy groups. These observations clearly supported the generation of oxygen functional groups on graphite surface for carrying out the amine functionalization reaction. Furthermore, functionalization was confirmed using elemental analysis (Table 4.1), where graphite oxide powder exhibited significant extent of oxygen. Amine functionalization of graphene sheets was also confirmed using FT-IR and elemental analysis. As evident in Figure 4.4a, the FT-IR spectrum of GrO-NH_2 exhibited the emergence of two new peaks evident at 1346 and 1166 cm^{-1} which were signature stretching characteristic of C-N bonds. The peak for amide bond at 1608 cm^{-1} also confirmed the successful tethering of amine molecules on the graphite oxide surface by chemical reaction. Further, the

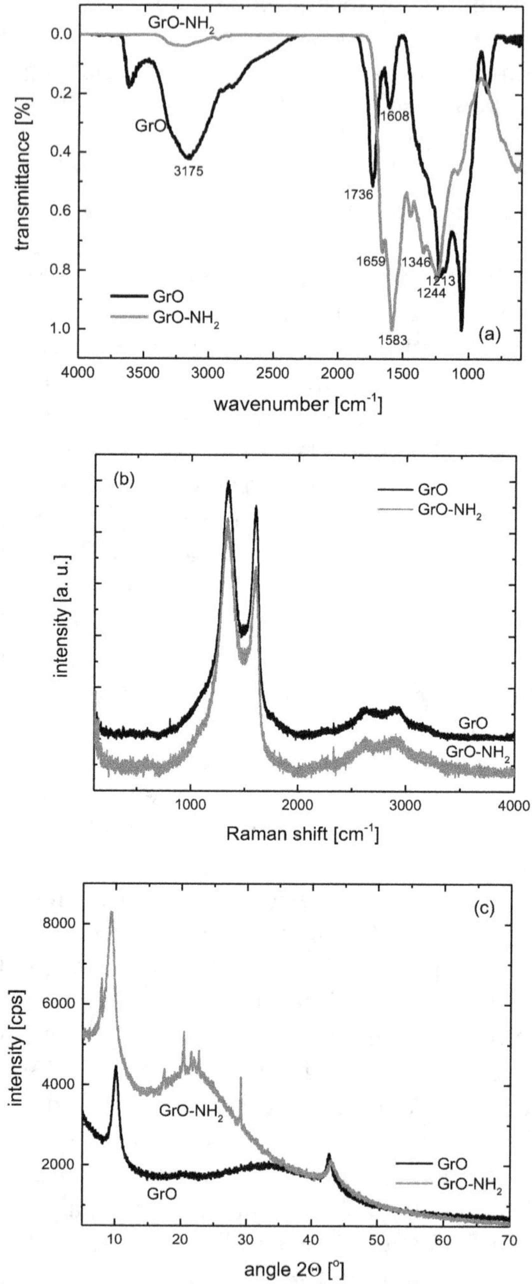

Figure 4.4 (a) FTIR, (b) Raman spectroscopy and (c) X-ray diffraction analysis of GrO and GrO-NH$_2$.

Table 4.1 Elemental analysis for graphite oxide and amine functionalized graphite oxide

Material	**N%**	**C%**	**O%**
GrO	0.56	40.05	53.89
GrO-NH_2	9.36	47.84	37.77

elemental analysis confirmed the presence of nitrogen in the GrO-NH_2 to be 9.4%, thus, confirming the presence of nitrogen containing moieties on graphite oxide surface. Figure 4.4b demonstrates the Raman spectra of GrO and GrO-NH_2. GrO exhibited the D band at 1335 cm^{-1}, whereas a weaker G band was observed at 1592 cm^{-1}. In the case of GrO-NH_2, the characteristic D and G bands were also observed to appear at the same Raman shift. The intensity ratio of the D/G bands (I_D/I_G), regarded as degree of measurement of the disorder, was slightly increased in GO-NH_2 probably due to the restoration of the sp^2 network in GrO after amine functionalization as well as due to the presence of edge oriented structural defects. The successful attachment of amine groups on GrO surface was also confirmed through X-ray diffraction. As shown in Figure 4.4c, the characteristic 2Θ diffraction peak for GrO was observed at 10.9°, which was shifted to a lower diffraction angle at 9.9°, thus, indicting the intercalation of amine groups inside the interlayers of GrO. It should be noted that the interlayer spacing of GrO did not increase to the large extent, however, it was expected as the molecular size of the amine attached to the GrO surface was also small. Enhancement of even a small amount of interlayer distance can be expected to provide better opportunities for the functional groups to interact with the polymer during composite generation. Overall, the structural characterization along with elemental analysis confirmed the modification of GrO surface.

In the IR spectra of PP+EVA-MA and its composites, the peaks between 2837 and 2948 cm^{-1} were attributed to the C–H asymmetric and symmetric stretching frequencies for both PP and EVA-MA. The peak at 1457 cm^{-1} represented the asymmetric C-H bending vibration of –CH_2 and –CH_3 groups of PP and EVA-MA. The peak at 1374 cm^{-1} indicated the symmetric C-H bending vibration of –CH_2 and –CH_3 groups and the peak at 1166 cm^{-1} represented the C-C stretching of C-CH_3 groups of PP and EVA-MA. The band at 1735 cm^{-1} represented the –C=O stretching of MA in PP/EVA-MA. The bands at 1019 and 1236 cm^{-1} indicated the ester C-O stretching of EVA-MA. The amide linkage in the region 1600-1650 cm^{-1} was observed in Figure 4.5(a)

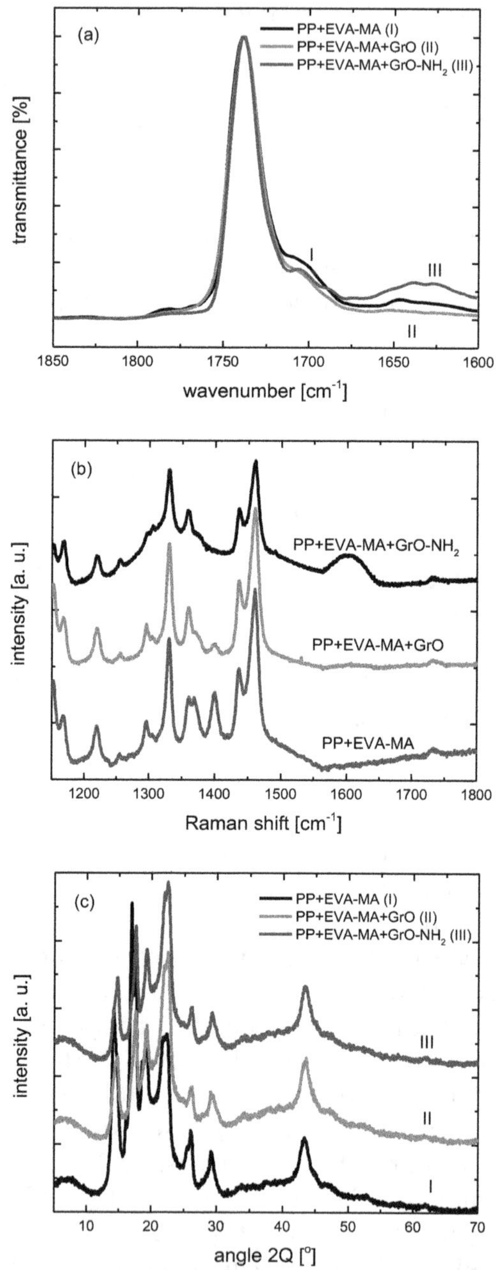

Figure 4.5 (a) FTIR, (b) Raman spectroscopy and (c) X-ray diffraction analysis of PP+EVA-MA, PP+EVA-MA+1%GrO and PP+EVA-MA+1%GrO-NH_2.

for the PE+EVA-MA+1%GrO composite, which was attributed to the filler, as observed earlier. This band was observed to increase in intensity in the case of PE+EVA-MA+1%GrO-NH_2 composite owing to the reaction of NH_2 groups on the graphene surface with the MA groups in EVA-MA. Thus, it indicated that some degree of chemical linking of filler platelets took place with the EVA-MA phase. Even though the observed signal was not sharp, however, a small extent of reactive compatibilization can lead to high degree of interfacial compatibilization of filler and EVA-MA phases, thus, leading to enhanced dispersion levels, which would subsequently affect the composite properties. Since the contribution of graphite oxide in nanocomposites was difficult to detect in IR, Raman spectra of PE+EVA-MA, PP+EVA-MA+1%GrO and PP+EVA-MA+1%GrO-NH_2 were also recorded and are provided in Figure 4.5(b). Characteristic G band of graphite oxide was observed at ~1590 cm^{-1} for PP+EVA-MA+GrO-NH_2, however, it was completely missing in the case of PP+EVA-MA+GrO composite. The D band was observed to be merged with PP signals in the 1325 cm^{-1} to 1350 cm^{-1} region. PP+EVA-MA+1%GrO-NH_2 also exhibited the absence of Raman shifts at 1300 cm^{-1} and 1400 cm^{-1}, in comparison with PE+EVA-MA and PP+EVA-MA+1%GrO, probably owing to the chemical interactions between GO-NH_2 and EVA-MA. As observed in Figure 4.5(c), the X-ray diffraction of the PP+EVA-MA blend and its composites with 1% filler content did not exhibit the presence of characteristic diffraction peak of GO or GO-NH_2, due to the low filler content. Good degree of filler intermixing with the polymers may have also resulted in the loss of diffraction peak of the filler platelets.

Further insights into the morphology of the PP+EVA-MA blend and composites were gained through microscopy. Figure 4.6 demonstrates the TEM micrographs of the PP+EVA-MA blend. Though EVA-MA was not molecularly miscible with PP under the processing conditions used in the study, however, its dispersion in PP was observed to be uniform (observed as darker phase in the TEM micrographs). Specifically, EVA-MA was dispersed as smaller domains inside the PP matrix due to its 20% fraction in the blend. Thus, it also indicated that due to its uniform dispersion in PP, EVA-MA can automatically lead to an enhanced degree of dispersion of GrO-NH_2 platelets owing to the chemical interaction with them. The optical analysis of the blend also represented the similar morphology, as observed in TEM. Figure 4.7 represents the microscopy analysis of the PP+EVA-MA+GrO and PP+EVA-MA+GrO-NH_2 with 1% filler content. Composites with other

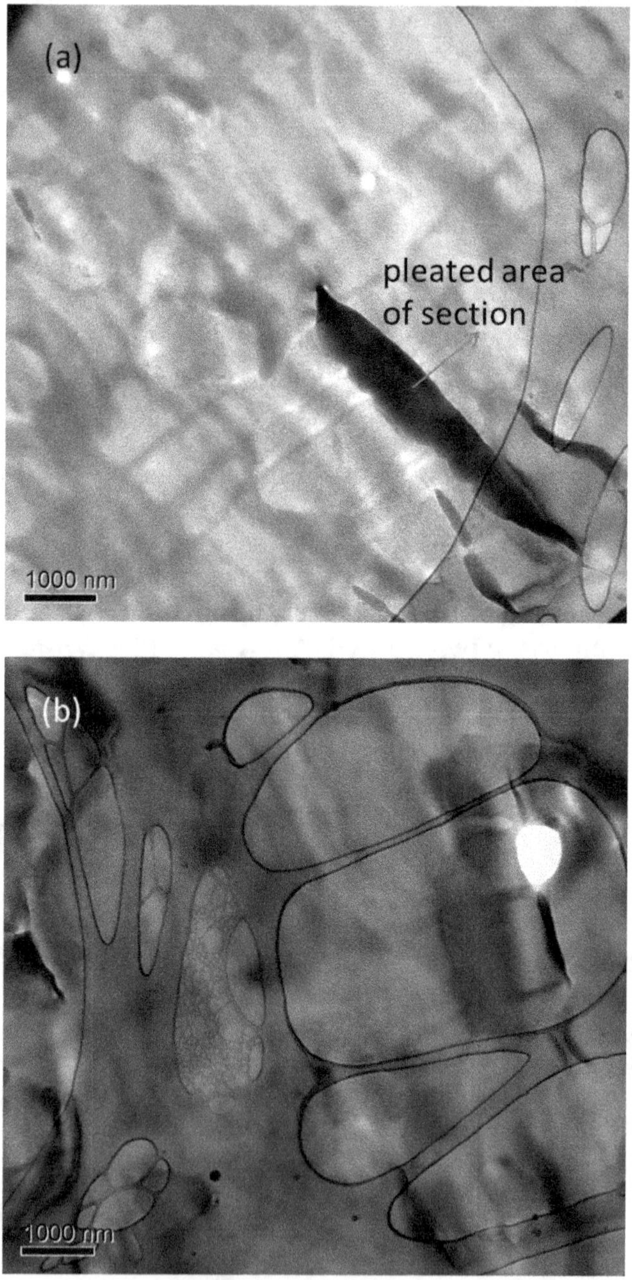

Figure 4.6 TEM micrographs of PP+EVA-MA blend.

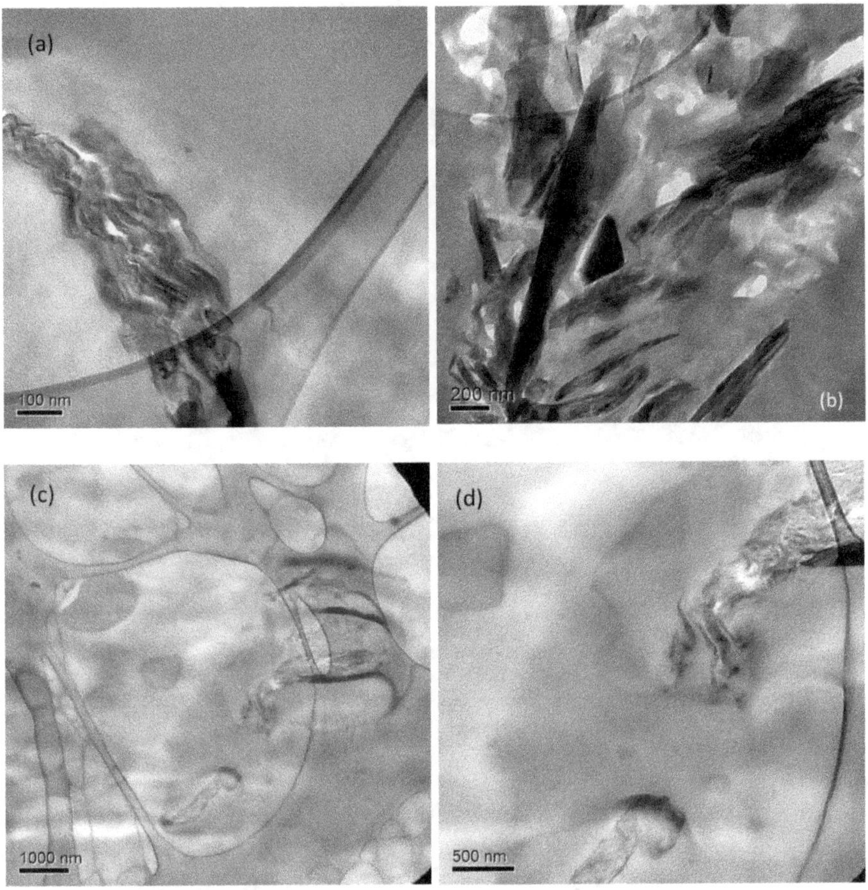

Figure 4.7 TEM micrographs of (a,b) PP+EVA-MA+1%GrO and (c,d) PP+EVA-MA+1%GrO-NH_2.

filler fractions also exhibited similar morphology. In the case of PP+EVA-MA+GrO composite, the filler was largely observed in the form of thick stacks mostly localized at the interface of PP and EVA-MA phases, though localization of these platelets in the PP phase was also observed. On the other hand, the filler in the case of PP+EVA-MA+GrO-NH_2 nanocomposite was observed to be largely embedded in the EVA-MA (darker dispersed phase in the TEM micrographs). In addition, the stack thickness was observed to be lesser than observed in the earlier case of PP+EVA-MA+GrO composite. Similar finding was also observed from the SEM analysis presented in Figure 4.8. The filler was largely in the PP phase as well as at PP/EVA-MA interface in the case of PP+EVA-MA+GrO composite, whereas the filler platelets

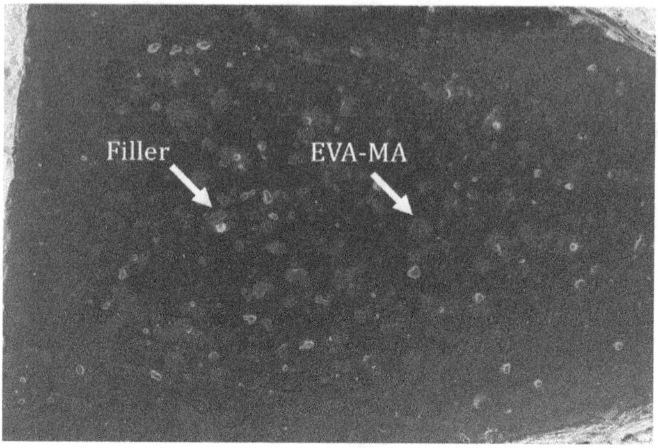

(a)

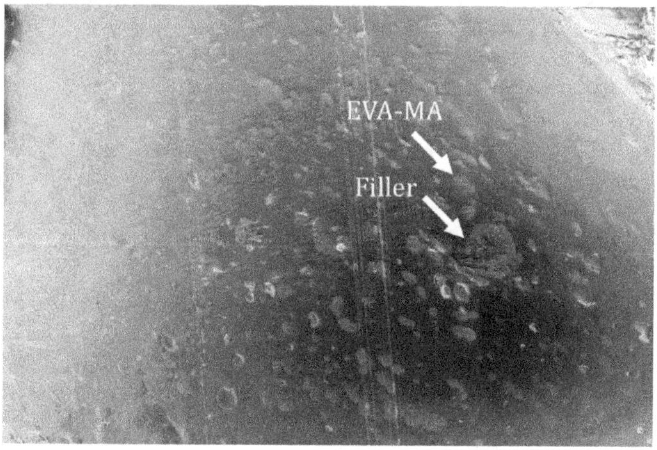

(b)

Figure 4.8 SEM images of (a,b) PP+EVA-MA+1%GrO and (c,d) PP+EVA-MA+1%GrO-NH_2.

were largely localized in the EVA-MA phase in the case of PP+EVA-MA+GrO-NH_2 nanocomposite. Such a morphology would only result due to the strong interaction of EVA-MA phase with the filler platelets. As also observed in the TEM micrographs, the SEM micrographs also revealed uniform distribution of EVA-MA phase as spherical domains inside PP, which due to chemical interaction with GrO-NH_2 filler resulted in better filler distribution as well as delamination inside the polymer. Thus, the component distribution was altered depending on the filler-polymer interactions. Figure 4.9 also compares

PP+EVA-MA+GrO and PP+EVA-MA+GrO-NH_2 nanocomposites at different magnifications. As clearly observed, the PP+EVA-MA+GrO composite did not exhibit any exfoliation of the GrO platelets. On the other hand, PP+EVA-MA+GrO-NH_2 nanocomposite exhibited a good degree of filler delamination in the matrix, as observed earlier. Thus, the microscopy findings confirmed the earlier findings from FTIR about the interaction of EVA-MA with GrO-NH_2 in the composite. The observed enhancement in composite morphology in the case of PP+EVA-MA+GrO-NH_2 nanocomposite is, thus, also expected to lead to superior composite performance.

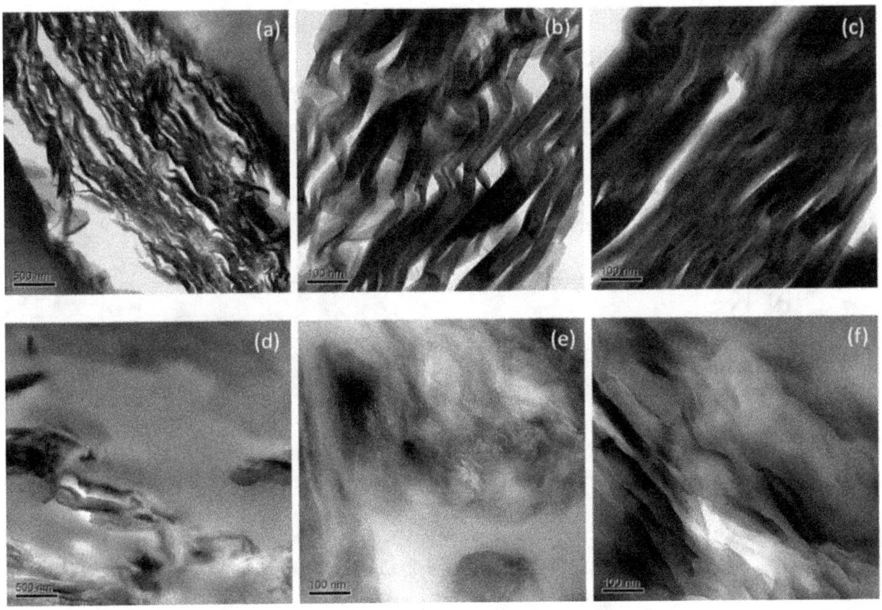

Figure 4.9 Comparison of the morphology of PP+EVA-MA+1%GrO (a-c) with PP+EVA-MA+1%GrO-NH_2 (d-f) at different magnifications.

Electron energy loss spectroscopy in the TEM mode was also carried out on PP, PP+EVA-MA blend and nanocomposites with 1 wt% filler content to gain further insights about the polymer filler interactions, as demonstrated in Figure 4.10. PP and EVA-MA exhibited signals for both nitrogen and oxygen in very low concentration, though the presence of N in these materials was unexpected. The presence of some additive added occasionally to commercial polymers may have also resulted in such weak signal. The PP+EVA-MA+GO composite exhibited increased concentration of O due to the contribution from

GrO platelets along with EVA-MA. On the other hand, no specific increase in the N signal was observed. The concentration of N in the PP+EVA-MA+GrO-NH_2 composite was much higher than the corresponding GO composite due to the functionalization of GrO with amine groups. In addition, the shape of the electron energy-loss near-edge structure (ELNES) N K edge indicated that N was most probably present as mixture of NH_2 and some other forms. This also possibly confirmed the change of NH_2 form to amide form, as observed earlier in IR analysis. However, it was not possible to identify if this signal indicated change of NH_2 form during GrO functionalization with diamine or due to the reaction of NH_2 groups on graphene surface with EVA-MA, as both reactions formed an amide linkage. Similar to the PP+EVA-MA+GO composite, the PP+EVA-MA+GrO-NH_2 composite also exhibited O signal due to the EVA-MA and filler contributions. Thus, though it was not possible to absolutely confirm the reaction between amine functionalized graphite oxide and EVA-MA from the EELS studies, however, these findings largely corroborated the earlier findings from structural analysis and microscopy.

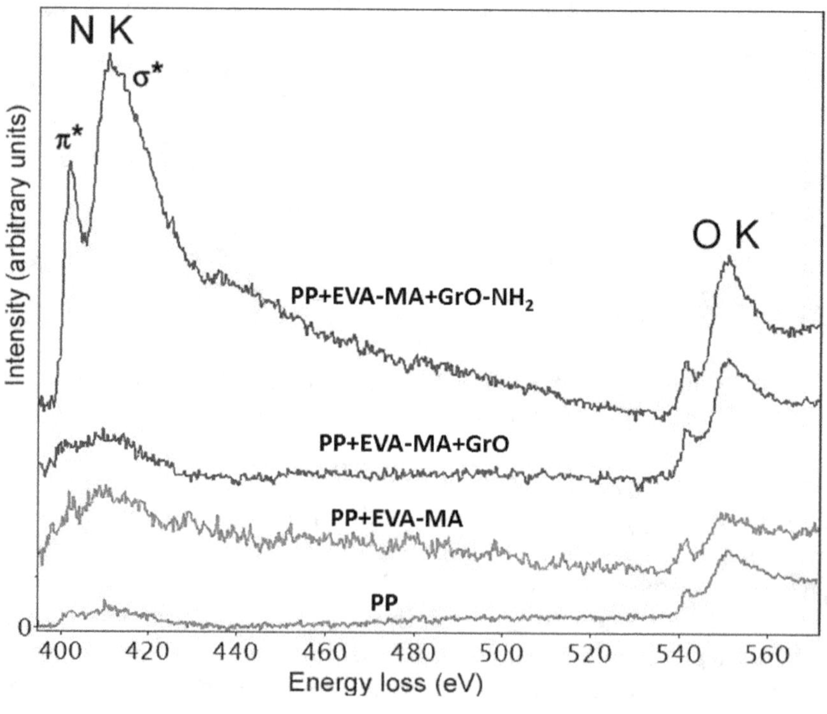

Figure 4.10 Energy loss spectra of the materials.

Rheological, thermal and mechanical properties of the composites were studied further to analyze the effect of polymer-filler interactions in the case of PP+EVA-MA+GrO-NH_2 composites, in comparison with PP+EVA-MA+GrO composites as well as PP+EVA-MA blend. In the case of polymer nanocomposites, the viscoelastic properties in the melt state are an effective tool to analyze the dispersion of nanoparticles in polymer matrices. Thus, in the current study too, the dispersion and network formation of nano-filler in PP+EVA-MA matrix was studied using frequency sweep, where all rheological parameters were recorded as a function of oscillation frequency in the range of 0.1 to 100 rad/s. Figure 4.11 shows the storage modulus of the blend and nanocomposite materials as a function of frequency. The storage modulus of the materials increased with angular frequency, however, the rate of increase slowed down at higher frequencies due to shear thinning effect. The PP+EVA-MA blend exhibited lower storage modulus than the nanocomposites over the whole frequency range. The composites with GrO enhanced the storage modulus gradually as a function of filler content, thus, indicating matrix reinforcement on the

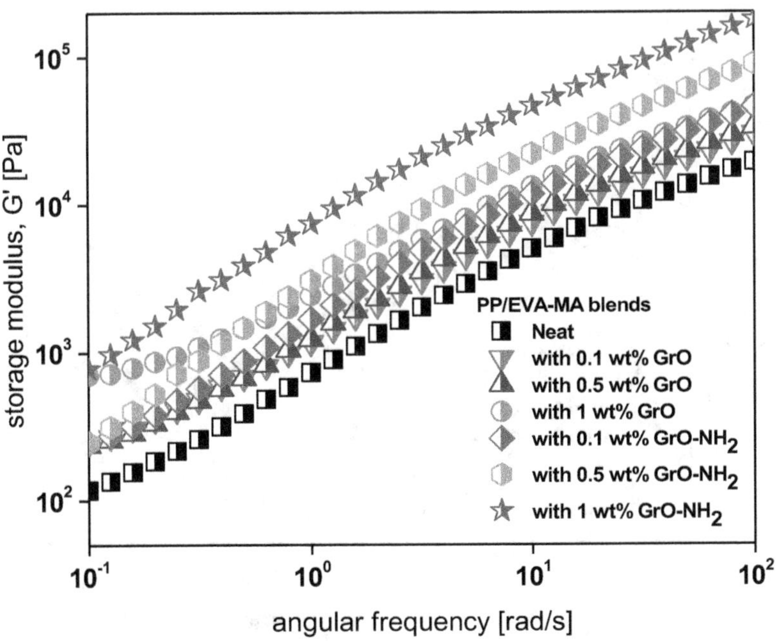

Figure 4.11 Storage modulus of the blend and composites as a function of angular frequency.

addition of GrO. In comparison, the incorporation of GrO-NH_2 filler in PP+EVA-MA led to even higher extent of storage modulus increment. For instance, the storage modulus of PP+EVA-MA at angular frequencies of 1 and 10 rad/s was measured to be 733 and 5035 Pa, which was enhanced to 7427 and 46403 Pa respectively for the composite with 1 wt% GrO-NH_2. Comparing the storage modulus of PP+EVA-MA+GrO and PP+EVA-MA+GrO-NH_2 composites, the composite with 0.5 wt% GO-NH_2 filler was observed to have higher storage modulus than the composite with even 1 wt% GrO. It indicated that better polymer-filler interfacial interactions achieved due to chemical reaction between EVA-MA and GO-NH_2 filler resulted in effective reinforcement of the polymer matrix due to efficient stress transfer. The loss modulus of the composites was also lower than storage modulus over the complete frequency range due to the dominance of elastic nature over the viscous nature.

Figure 4.12 shows the complex viscosity of the blend and nanocomposites as a function of angular frequency. As expected, the PP+EVA-MA blend and blend nanocomposites exhibited shear

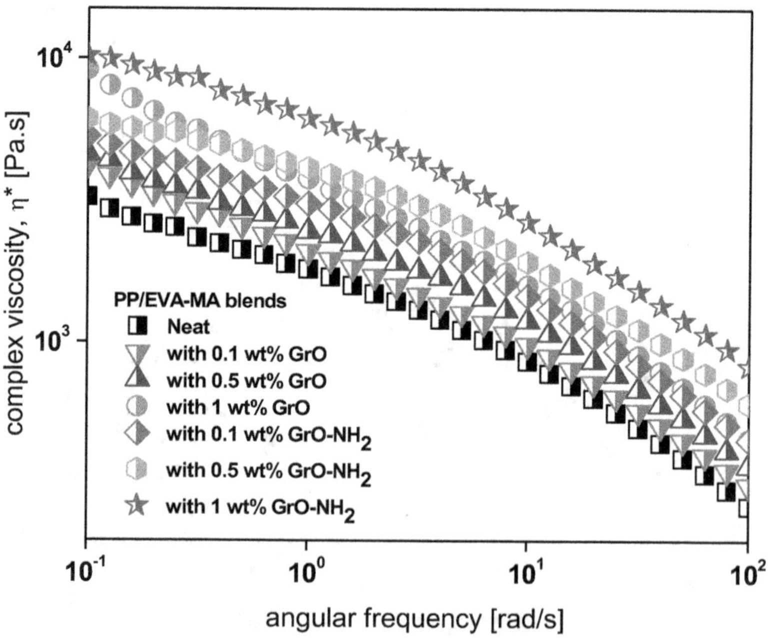

Figure 4.12 Complex viscosity of the blend and composites as a function of angular frequency.

thinning behavior with increasing angular frequency. As observed in the case of storage modulus, the complex viscosity of the nanocomposites increased with the addition of filler in the polymer matrix. The increment in viscosity for both GO and GO-NH_2 filler systems was a function of filler content in the composites. Comparing the GrO and GrO-NH_2 filled nanocomposites systems, the GrO-NH_2 composites were observed to exhibit larger extent of increase in viscosity owing to effective delamination of filler platelets and immobilization of the polymer chains on the surface of filler platelets. Though the viscosity of the composites increased significantly in comparison with PP+EVA-MA blend, however, the composites were still processable using the same processing conditions as the blend.

Figure 4.13(a) depicts the DSC melting thermograms of the pure polymers, polymer blend as well as composites. Crystalline melting of PP was observed at 167 °C, whereas EVA-MA exhibited the endothermic crystalline melting peak at 67.5 °C. The thermograms of PP+EVA-MA blend as well as nanocomposites with GrO and GrO-NH_2 exhibited a small hump at 67.5 °C corresponding to the melting transition of EVA-MA, along with the transition corresponding to the melting of PP crystals. These findings supported the earlier findings of absence of molecular mixing between PP and EVA-MA. Also, the melt transition corresponding to PP did not shift in the blend as well as blend nanocomposites, thus, indicating that the EVA-MA phase as well as nanoparticles had no effect on the melting behavior of PP. The melt enthalpy of polypropylene was observed to reduce slightly in the blend as compared to pure polypropylene due to the hindrance to the crystallization of PP chains. The melt enthalpy of PP remained constant on the addition of the filler platelets to the blend, irrespective of the type and amount of filler. Figure 4.13(b) presents the crystallization exotherms of the materials. Neat PP and EVA-MA exhibited peak crystallization temperatures of 126.5 °C and 51 °C respectively. Similar to the melting behavior, the peak crystallization temperature of PP remained unchanged in the blend as well as nanocomposites, thus, indicating that the filler particles did not exhibit any nucleating effect on PP crystallization. In addition, the thermograms of blend and nanocomposites exhibited a small hump at 51 °C corresponding to the crystallization of EVA-MA. Overall, the DSC findings confirmed that the filler particles did not change the melting and crystallization behavior of the polymer blend, even though the nanocomposites with GrO and GrO-NH_2 fillers exhibited different phase morphologies in microscopy analysis.

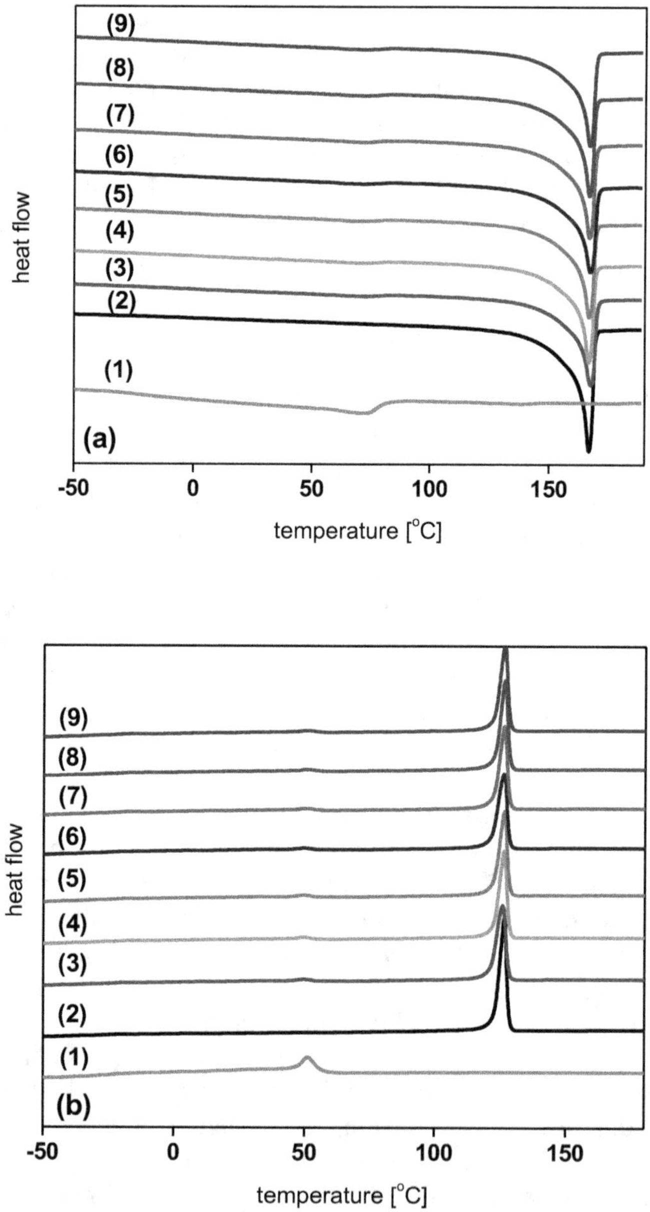

Figure 4.13 (a) Melting and (b) crystallization thermograms of the materials. Samples denoted as (1) neat EVA-MA, (2) neat PP, (3) PP+EVA-MA 80/20 blend, and blends with (4, 5 and 6) 0.1, 0.5, 1 wt% GrO respectively and blends with (7, 8 and 9) 0.1, 0.5, 1 wt% GrO-NH_2 respectively.

The thermal degradation behavior of the PP+EVA-MA+GrO and PP+EVA-MA+GrO-NH_2 nanocomposites with 1 wt% filler content is also compared with PP+EVA-MA in Figure 4.14. The thermogram of the PP+EVA-MA blend exhibited two step degradation pattern with a low temperature degradation step having onset of degradation at 350 °C attributed to EVA-MA, whereas the higher temperature degradation step with an onset of degradation at 425 °C was attributed to pure polypropylene. The incorporation of GrO and GrO-NH_2 fillers in the blend led to an improvement in the thermal stability as the lower temperature degradation step due to EVA-MA was absent in the thermograms, thus, indicating better heat transfer between the polymer chains and filler platelets. There was no difference in the behavior of the composites with GrO and GrO-NH_2 fillers. In addition, similar improvement in the thermal performance was exhibited by all the composites, indicating that even 0.1 wt% filler was able to impart significant thermal resistance. Dynamic TGA of the nanocomposites was also carried out by analyzing the degradation as a function of time at 190 °C (compounding temperature) and the composites were observed to have insignificant weight loss for the test duration equivalent to the compounding time, thus, confirming the durability of the

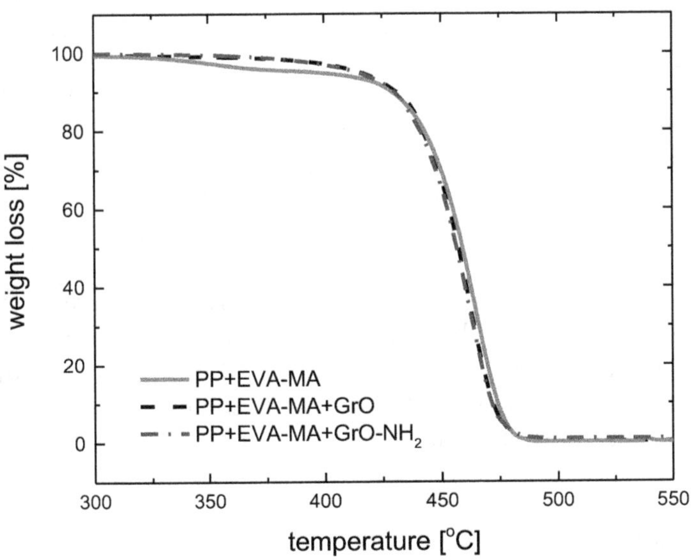

Figure 4.14 TGA thermograms of PP+EVA-MA, PP+EVA-MA+1%GrO and PP+EVA-MA+1%GrO-NH_2.

composites for high temperature processing and applications.

The tensile properties of the nanocomposites are compared with the PP+EVA-MA blend as a function of filler content in Figure 4.15. The tensile modulus of the blend was measured as 916 MPa, which was enhanced to 1202 MPa for the PP+EVA-MA-GrO composite with 1 wt% filler content, thus exhibiting an increase of 30% in the modulus (Figure 4.15(a)). On the other hand, the corresponding composite with 1 wt% GrO-NH_2 filler had the tensile modulus of 1343 MPa, which was an increment of >45%. It, thus, confirmed the earlier findings from rheological analysis that the chemical interaction between the EVA-MA and GrO-NH_2 resulted in better stress transfer between the polymer chains and filler platelets. As no change in the extent of polymer crystallinity was observed in the blend and composites, thus, the observed increment in tensile modulus was solely attributed to the polymer-filler interactions and filler dispersion. Similar behavior was also observed in the case of tensile strength, as shown in Figure 4.15(b). In comparison to 29 MPa for the PP+EVA-MA blend, the tensile strength of the nanocomposites with 1% GrO and GrO-NH_2 fillers was observed to be 30.9 and 33.4 MPa respectively. It again underlined the impact of reactive compatibilization on the composite properties. As observed in Figure 4.15(c), the elongation of the composites decreased with the addition of filler. In the case of PP+EVA-MA+GrO nanocomposites, the reduction in the elongation was considerable. Even the composite with 0.1 wt% filler had a significant reduction in elongation to 7.7%, in comparison with 41% for the polymer blend. The elongation decreased gradually in the case of PP+EVA-MA+GrO-NH_2 composites, and still remained appreciable at 21% even in the composite with 1 wt% filler. It indicated that similar to other tensile properties, the toughness of the nanocomposites with GO-NH_2 filler was much superior than the corresponding GrO containing composites due to interfacial interactions between EVA-MA and filler platelets leading to better filler dispersion. This subsequently reduced the strain hardening effect in these composites. On the other hand, the composites with GrO had higher extent of stain hardening due to the presence of large filler stacks, which negatively impacted the toughness of the polymer.

4.4 Conclusions

In the current study, blend nanocomposites and reactive compatibilization approaches have been successfully demonstrated to generate

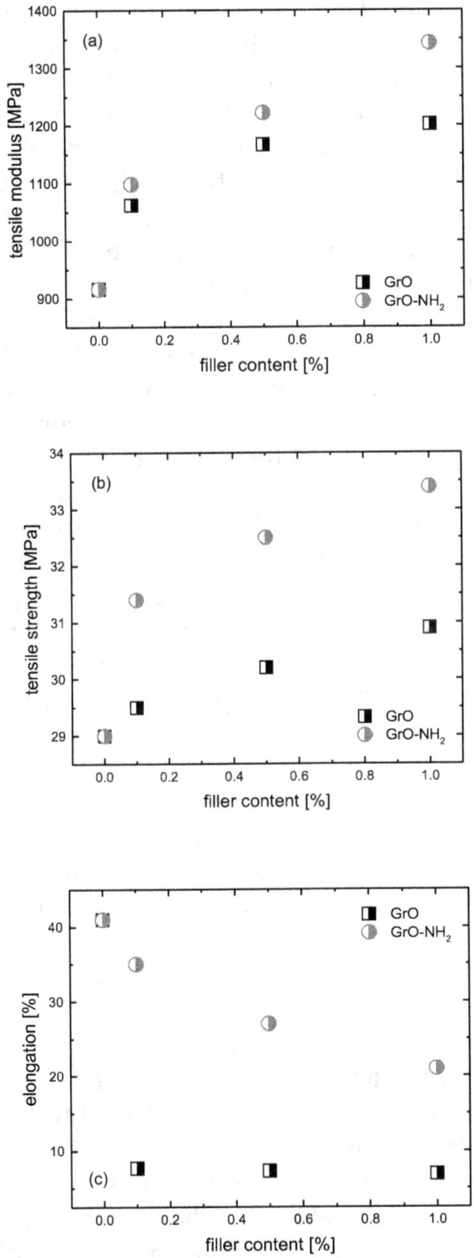

Figure 4.15 Tensile properties of the composites as a function of filler content. The relative probable error for tensile modulus and tensile strength was ±2%, whereas it was ±5% for elongation.

functional PP/EVA-MA nanocomposites with amine functionalized graphite oxide, GrO-NH_2. The graphite oxide could be modified using ethylene diamine to generate amine functionalities on the surface. The structural characterization as well as elemental analysis confirmed the tethering of the amine groups on the surface of graphite oxide platelets. The interlayer spacing was though increased by a small degree, however, it was expected due to the small size of the ethylene diamine molecules. The nanocomposites generated with GrO-NH_2 also exhibited an increase in the intensity of amide linkage in the FTIR analysis in comparison with the composites with GrO, thus, confirming the successful chemical binding of EVA-MA chains with the filler platelets. The morphology of PP+EVA-MA was observed to be uniformly mixed with small sized spherical domains of EVA-MA in the PP phase. The morphology of the composites was significantly affected depending on the type of the filler. The composites with GrO filler had the presence of thick stacks of GrO platelets largely localized at the PP/EVA-MA interface or in PP itself. In the case of composites containing GrO-NH_2 filler, the filler stacks were only a few layers thick and were largely present in the EVA-MA phase owing to the chemical interactions between the two. Thus, due to uniform dispersion of EVA-MA in PP, the GrO-NH_2 platelets were also observed to be uniformly dispersed in the polymer matrix. The observed differences in the morphology of the composites also affected the composites properties significantly. The storage modulus of the GrO-NH_2 composites was observed to be the maximum owing to the polymer-filler interactions and effective filler dispersion leading to higher stress transfer from the polymer chains to the filler platelets. For instance, the composite with 0.5 wt% GrO-NH_2 had higher storage modulus than the composite with 1 wt% GrO, thus, indicating effective reinforcement in the case of GrO-NH_2 composites. Similarly, the viscosity of the composites was also enhanced on the addition of filler, with GrO-NH_2 based composites exhibiting higher resistance to flow. Even though the viscosity of the composites was significantly enhanced, the composites still remained processable using the same processing conditions as the PP+EVA-MA blend. The thermal degradation properties of the composites were also enhanced with filler, though both GrO and GrO-NH_2 exhibited similar behavior. The composites with even 0.1 wt% filler were observed to impart significant thermal stability to the polymer blend. The DSC thermograms also confirmed the absence of molecular mixing between PP and EVA-MA. In addition, the peak melting and crystallization temperatures of the polymer were not affected in the blend and composites indicating that EVA-MA and fillers did not impact the melting and crystallization behavior of PP. However, a small reduction in the melt enthalpy of PP was observed in the blend and corresponding blend composites due to the hindrance to crystallization of PP chains in the presence of EVA-MA. The specific polymer-filler interactions and subsequent enhanced degree of filler dispersion were also observed to result in considerable enhancement in the mechanical properties of GrO-NH_2 based composites. For instance, the composite with 1 wt% GrO-NH_2 filler led to an increment of >45% in the tensile modulus, in comparison to 30% for the composite with 1 wt% GrO. The GrO-NH_2 based composites also exhibited larger tensile strength than the corresponding GrO based composites. The elongation of the composites was significantly reduced in GrO based composites to around 7% even with 0.1 wt% filler content. On the other hand, the GrO-NH_2 based composites had much higher elongation with 1 wt% composite exhibiting the elongation of 21%. The GrO-NH_2 based composites, thus, represent advanced composites with optimal properties achieved at low weight fractions of filler and have the potential to compete with many engineering plastics.

Acknowledgement

The authors are indebted to Dr. Gisha E. Luckachan for support in structural characterization (X-ray, Raman, and IR) of the materials.

References

1. Mittal, V. (2007) Polypropylene-layered silicate nanocomposites: Filler matrix interactions and mechanical properties. *Journal of Thermoplastic Composite Materials*, **20**, 575-599.
2. Zhang, M., Liu, Y., Zhang, X., Gao, J., Huang, F., Song, Z., Wei, G., and Qiao, J. (2002) The effect of elastomeric nano-particles on the mechanical properties and crystallization behavior of polypropylene. *Polymer*, **43**, 5133-5138.
3. van der Wal, A., Verheul, A., and Gaymans, R. (1999) Polypropylene–rubber blends: 4. The effect of the rubber particle size on the fracture behaviour at low and high test speed. *Polymer*, **40**, 6057-6065.
4. Wahit, M., Hassan, A., Mohd Ishak, Z., and Czigany, T. (2009) Ethylene-octene copolymer (POE) toughened polyamide 6/polypropylene nanocomposites: Effect of POE maleation. *EXPRESS Polymer*

Letters, **3**, 309-319.

5. Martins, C., Larocca, N., Paul, D., and Pessan, L. (2009) Nanocomposites formed from polypropylene/EVA blends. *Polymer*, **50**, 1743-1754.
6. Bassani, A., and Pessan, L. (2002) Toughening of polypropylene with styrene/ethylene-butylene/styrene tri-block copolymer: Effects of reactive and nonreactive compatibilization. *Journal of Applied Polymer Science*, **86**, 3466-3479.
7. Jafari, S., and Gupta, A. (2000) Impact strength and dynamic mechanical properties correlation in elastomer-modified polypropylene. *Journal of Applied Polymer Science*, **78**, 962-971.
8. Zebarjad, S., Lazzeri, A., Bagheri, R., Seyed Reihani, S., and Frounchi, M. (2003) Fracture mechanism under dynamic loading of elastomer-modified polypropylene. *Materials Letters*, **57**, 2733-2741.
9. Wang, Z. (1996) Toughening and reinforcing of polypropylene. *Journal of Applied Polymer Science*, **60**, 2239-2243.
10. Wei, G. X., and Sue, H. J. (2000) Fracture behavior of styrene-ethylene-propylene rubber-toughened polypropylene. *Polymer Engineering & Science*, **40**, 1979-1988.
11. Bai, H., Wang, Y., Song, B., and Han, L. (2008) Synergistic toughening effects of nucleating agent and ethylene–octene copolymer on polypropylene. *Journal of Applied Polymer Science*, **108**, 3270-3280.
12. Lim, J., Hassan, A., Rahmat, A., and Wahit, M. (2006) Rubber-toughened polypropylene nanocomposite: Effect of polyethylene octene copolymer on mechanical properties and phase morphology. *Journal of Applied Polymer Science*, **99**, 3441-3450.
13. Abreu, F., Forte, M., and Liberman, S. (2005) SBS and SEBS block copolymers as impact modifiers for polypropylene compounds. *Journal of Applied Polymer Science*, **95**, 254-263.
14. Bassani, A., Pessan, L., and Hage, E. (2001) Toughening of polypropylene with styrene/ethylene-butylene/styrene tri-block copolymer: Effects of mixing condition and elastomer content. *Journal of Applied Polymer Science*, **82**, 2185-2193.
15. Ma, Y., Wu, D., Liu, Y., Li, X., Qiao, H., and Yu, Z.-Z. (2014) Electrically conductive and super-tough polypropylene/carbon nanotube nanocomposites prepared by melt compounding. *Composites Part B: Engineering*, **56**, 384-391.
16. Song, P., Cao, Z., Cai, Y., Zhao, L., Fang, Z., and Fu, S. (2011) Fabrication of exfoliated graphene-based polypropylene nanocomposites with enhanced mechanical and thermal properties. *Polymer*, **52**, 4001-4010.
17. Yuan, B., Bao, C., Song, L., Hong, N., Liew, K. M., and Hu, Y. (2014) Preparation of functionalized graphene oxide/polypropylene nanocomposite with significantly improved thermal stability and studies

on the crystallization behavior and mechanical properties. *Chemical Engineering Journal*, **237**, 411-420.

18. Song P., Liu, L., Huang, G., Fu, S., Yu, Y., and Guo, Q. (2013) Facile Fabrication of Polyolefin/Carbon Nanotube Composites via in Situ Friedel–Crafts Polyalkylation: Structure and Properties. *Industrial & Engineering Chemistry Research*, **52**, 14384-14395.
19. Kuila, T., Bose, S., Hong, C. E., Uddin, M. E., Khanra, P., Kim, N. H., and Lee, J. H. (2011) Preparation of functionalized graphene/linear low density polyethylene composites by a solution mixing method. *Carbon*, **49**, 1033-1037.
20. Shin, K.-Y., Hong, J.-Y., Lee, S., and Jang, J. (2012) Evaluation of anti-scratch properties of graphene oxide/polypropylene nanocomposites. *Journal of Materials Chemistry*, **22**, 7871-7879.
21. Hsiao, M.-C., Liao, S.-H., Lin, Y.-F., Wang, C.-A., Pu, N.-W., Tsai, H.-M., and Ma, C.-C. M. (2011) Preparation and characterization of polypropylene-graft-thermally reduced graphite oxide with an improved compatibility with polypropylene-based nanocomposite. *Nanoscale*, **3**, 1516-1522.
22. Page, D. J., and Gopakumar, T. G. (2006) Properties and crystallization of maleated polypropylene/graphite flake nanocomposites. *Polymer Journal*, **38**, 920-929.
23. Clemons, C. (2010) Elastomer modified polypropylene–polyethylene blends as matrices for wood flour–plastic composites. *Composites Part A: Applied Science and Manufacturing*, **41**, 1559-1569.
24. Souza, A., and Demarquette, N. (2002) Influence of coalescence and interfacial tension on the morphology of PP/HDPE compatibilized blends. *Polymer*, **43**, 3959-3967.
25. Modesti, M., Lorenzetti, A., Bon, D., and Besco, S. (2005) Effect of processing conditions on morphology and mechanical properties of compatibilized polypropylene nanocomposites. *Polymer*, **46**, 10237-10245.
26. Liu, S.-P., Ying, J.-R., Zhou, X.-P., Xie, X.-L., and Mai, Y.-W. (2009) Dispersion, thermal and mechanical properties of polypropylene/magnesium hydroxide nanocomposites compatibilized by SEBS-g-MA. *Composites Science and Technology*, **69**, 1873-1879.
27. Lee, H.-s., Fasulo, P. D., Rodgers, W. R., and Paul, D. (2005) TPO based nanocomposites. Part 1. Morphology and mechanical properties. *Polymer*, **46**, 11673-11689.
28. Mukhopadhyay, P., and Gupta, R. K. (2011) Trends and frontiers in graphene-based polymer nanocomposites. *Plastics Engineering*, **67**, 32-42.
29. Mittal, V. (2014) Functional polymer nanocomposites with graphene: A review. *Macromolecular Materials and Engineering*, **299**, 906-931.

30. Kuilla, T., Bhadra, S., Yao, D., Kim, N. H., Bose, S., and Lee, J. H. (2010) Recent advances in graphene based polymer composites. *Progress in Polymer Science*, **35**, 1350-1375.
31. Cai, D., and Song, M. (2010) Recent advance in functionalized graphene/polymer nanocomposites. *Journal of Materials Chemistry*, **20**, 7906-7915.
32. Rafiee, M. A., Rafiee, J., Wang, Z., Song, H., Yu, Z. Z., and Koratkar, N. (2009) Enhanced me- chanical properties of nanocomposites at low graphene content. *ACS Nano*, **3**, 3884-3890.
33. Mittal, V., Chaudhry, A. U., and Luckachan, G. E. (2014) Biopolymer - thermally reduced graphene nanocomposites: Structural characterization and properties. *Materials Chemistry and Physics*, **147**, 319-332.
34. Vengatesan, M. R., Singh, S., Pillai, V. V., and Mittal, V. (2016) Crystallization, mechanical, and fracture behavior of mullite fiber-reinforced polypropylene nanocomposites. *Journal of Applied Polymer Science*, doi:10.1002/APP.43725
35. Vengatesan, M. R., Al Asafen, H. Y., Varghese, A. M., and Mittal, V. (2017) Effect of amine functionalized reduced graphene oxide on the thermal, mechanical, rheological and electrical properties of CPE compatibilizer HDPE. *In preparation*.

5

Polypropylene/Polystyrene Blend Nanocomposites with Graphene

5.1 Introduction

The physical polymer-reinforcement means such as blending or filler incorporation have gained favorable position for acquiring new polymeric materials with altered properties [1-3]. The properties of the resultant materials depend on features like interaction between the phases, compatibilization, processing conditions, composition, etc. The generation of polymer blends specially offers advantages such as property profile combinations, aid in processing, low capital costs as compared with the development of new monomers and polymers of required properties [4-7]. The advances in nanofillers and nanofibers have also made it possible to produce light weight nanocomposites with better physical and mechanical properties at a low filler concentration than conventional composites. This is owing to the much higher number of interfacial contacts of inorganic particles with the polymer chains in such nanocomposites, which generate a completely different interfacial morphology as compared with the bulk polymer [8,9]. In the class of polymer nanocomposites, further development of blend composites, combining the useful features of polymer blending as well as filler incorporation has also resulted in many advanced materials recently. For instance, Mittal and Matsko [10] recently reported poly(3-hydroxybutyrate-co-3-hydroxyvalerate) (PHBV) and poly(butylene adipate-co-terephthalate) (PBAT) based blend nanocomposites with vermiculite and graphene as reinforcing fillers. The authors observed co-continuous blend morphology in the blend composites. The filler (graphene and vermiculite platelets) was observed to be localized in the tougher PBAT phase due to its early melting, as shown in Figure 5.1, which helped to retain the toughness

V. Mittal[a,] and M. Pannirselvam[b] and Gregory Griffin[b]*
[a]The Petroleum Institute (part of Khalifa University of Science and Technology), Abu Dhabi, UAE; [b]Royal Melbourne Institute of Technology, Melbourne, Australia
**Current address: Bletchington, Wellington County, Australia*

of the blend nanocomposites. Overall, owing to the generated composite morphology, simultaneous enhancement in both stiffness and toughness in the blend nanocomposites was achieved, which otherwise is quite challenging to attain.

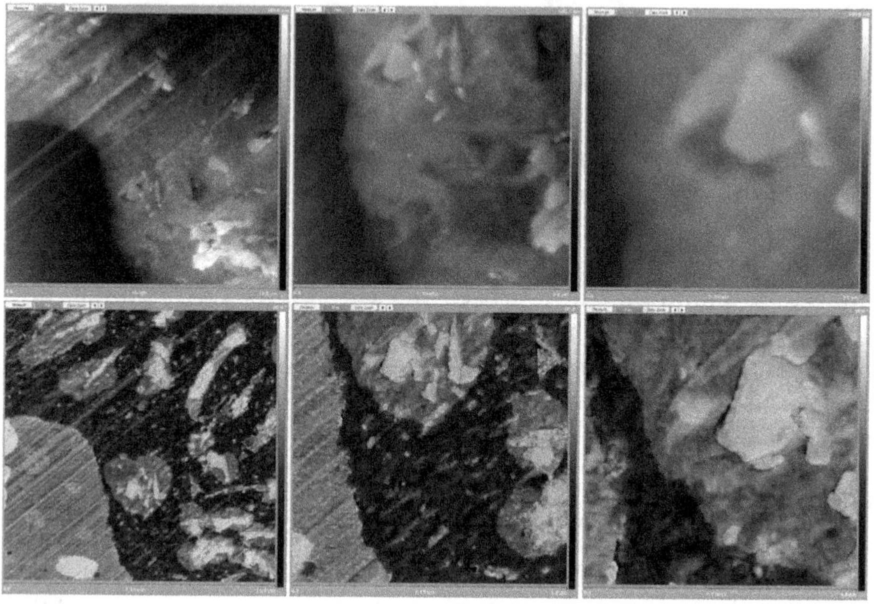

Figure 5.1 AFM height (top row) and phase (bottom row) images of the PHBV/PBAT/5% graphene composite [10]. In the images, the light phase corresponded to PHBV, whereas the dark phase contained PBAT. The filler platelets were observed as the bright platy inclusions.

Polyolefins are the largest group of industrial thermoplastics in terms of production and have superior rank among commodity plastics owing to their use in a large number of applications. The number of individual polyolefin members is limited and the number of applications based on such materials is constantly increasing. Various polyolefin-based blends and composites are, thus, required to meet the increasing demand of materials with required properties and processing suitable for specific applications [1,3]. A large variety of reinforcing fillers such as silica, clay, nanotubes and graphene have been explored as fillers in polyolefin matrices. Graphene, which consists of one atomic thick sheets of covalently sp^2-bonded carbon atoms in a hexagonal arrangement, has already received attention of researchers for the generation of polymer nanocomposites [11,12]. Its choice as a filler is due to its excellent electrical and mechanical properties, which are significantly better than other inorganic filler materials. A single defect-free graphene layer has Young's modulus of ~1.0 TPa, intrinsic strength ~42 N/m, thermal conductivity ~4840-5300 W/(m.K), electron mobility exceeding 25,000 cm^2/V.s, excellent gas impermeability and specific surface area of ~2630 m^2/g [11]. All these properties make this material superior to carbon nanotubes and layered silicates, which have been commonly used in polymer nanocomposites.

Among various polyolefins, polypropylene (PP) and polyethylene (PE) have outstanding properties including low cost, recyclability, good processability, high thermal and chemical resistance, non-toxicity, etc., and compete with many engineering plastics over a wide spectrum of application areas. However, these polymers suffer from limitations such as high oxygen permeation, shrinkage and relatively lower stiffness as compared to many engineering polymers [13-18]. To overcome these limitations, blends of these polyolefins with stiff polymers like polystyrene (PS) represent effective alternatives. For instance, Diaz *et al.* [19] and Xu *et al.* [20] reported the polyethylene/polystyrene blends and stabilized these using reactive compatibilization and incorporation of copolymer based compatibilizers. Other alternative is to combine these blends additionally with a functional filler, as mentioned earlier. For instance, Inberg *et al.* [21] reported the long glass fiber reinforcement of PP/PS blends. The authors observed that on the addition of 30 wt% of fibers, the modulus increased by a factor of 3.5, whereas the fracture stress was observed to increase by a factor of 2.5. In addition, the unnotched Izod impact strength in the blend composites increased by a factor of 10. In a recent study, Mittal *et al.* [22] also studied functional PE/PS blend nanocomposites with different fillers like silica, alumino-silicate and graphene. One step melt mixing process could be used for generating the composites, which resulted in the kinetic embedding of the filler particles in the PE phase due to its earlier melting during compounding. A fraction of the filler was also observed at the interface between PE and PS phases, as shown in Figure 5.2. The tensile modulus in the blend composites increased with filler content with maximum extent of increase provided by graphene platelets. High aspect ratio platelets of graphene were observed to be beneficial in generating good degree of stress transfer. In another study, Parameswaranpillai *et al.* [23] also reported polypropylene/polystyrene/carbon nanofiber nanocomposites and tailored the interface in the blend composite through

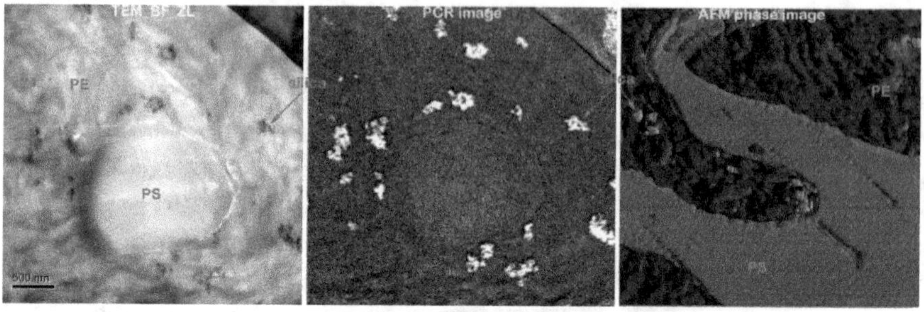

(a)

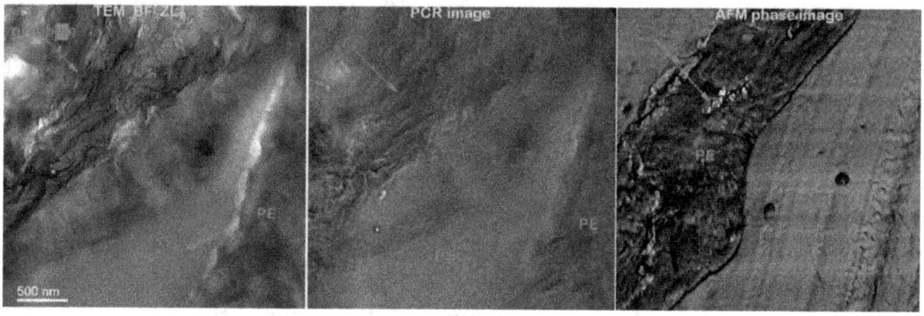

(b)

Figure 5.2 Morphological analysis of (a) $PE_{75}PS_{25}$/Silica and (b) $PE_{75}PS_{25}$/Graphene composites [22].

the addition of polystyrene-block-poly(ethylene-ran-butylene)-block-polystyrene.

In the current study, the blend nanocomposites of polypropylene and polystyrene have been further explored by using graphene as filler. Two different blend compositions were employed to study the effect on composite morphology and properties. In addition, in order to analyze the effect of additional compatibilizer on filler dispersion and subsequent composite performance, a small amount of ethylene-co-acrylic acid copolymer was also incorporated in the blend composites. As many earlier studies have reported the localization of the filler in the tougher polymer phase in the blend nanocomposites to be beneficial in retaining the overall toughness of the composites [10,22], therefore, processing conditions were also tuned in the current study so as to achieve similar morphology. In addition, due to the risk of thermal degradation of polymers above 200 °C, the compounding of the materials was carried out at 200 °C for optimal duration, so as to achieve good degree of phase intermixing, however, avoiding the thermal degradation of the polymers.

5.2 Experimental

5.2.1 Materials

Injection-molding PP grade RA130E-1498 was received from Abu Dhabi Polymers Company Limited. PS with average molecular weight (M_w) of 400,000 was procured from Sigma. Ethylene-co-acrylic acid (EAA) with a trade name of ESCOR 5050 was a product of Exxon Mobil. It has an acrylic acid content of 9 wt% and melt flow index (MFI, 190 °C and 2.16 kg) of 8.4 g/10 min. The EAA copolymer had a specific gravity of 0.936 and peak melting temperature of 97 °C. Graphene sample with trade name Grafen-iGP was kindly provided by Grafen Chemical Industries, Turkey. The polymers and filler were used as received, without further modification.

5.2.2 Generation of PP/PS Blends and Nanocomposites

Mini twin conical screw extruder (MiniLab HAAKE Rheomex CTW5, Germany) was used for the melt mixing of PP, PS and graphene to generate PP/PS (70/30 and 30/70) blends and their composites with 2 wt% graphene. In the composites incorporating EAA, 5% of the copolymer was added along with PP/PS (70/30 and 30/70) blends and 2% graphene. One step melt mixing process was used for the generation of blends and nanocomposites by pre-mixing the components physically followed by feeding into the compounder. The materials were dried overnight at 60 °C before performing the melt mixing. The screw length and screw diameter were 109.5 mm and 5/14 mm conical respectively. A mixing temperature of 200 °C and rotational speed of 70 rpm were used. Approximately 20 g of each sample was dry mixed prior to extrusion, and an initial mass of 6 g was added to the compounder. After 5 min, 3 g was flushed from the machine so that another 3 g could be added. This process was repeated until 20 g material was blended. This method ensured that the compounder would not be overloaded and also provided small batches to ensure more intense mixing without any significant thermal degradation of the polymer.

For the subsequent characterization, the disc, dumbbell and bar-shaped test samples were prepared by hot pressing the compounds at 200 °C for 5 min. The sample were cooled down to 50 °C by using flowing water through the plates by keeping a controlled cooling rate of 5 °C.min^{-1}.

5.2.3 Material Characterization

For the bright field transmission electron microscopy (TEM) analysis of the nanocomposite samples, Philips CM 20 (Philips/FEI, Eindhoven) electron microscope at 120 kV and 200 kV accelerating voltages was used. Thin sections of 70-90 nm thickness were microtomed from the sample blocks using a Leica Ultracut E microtome (Leica, Austria) equipped with a diamond knife (Diatome, Switzerland) at -120 °C. The so-obtained sections were supported on 100 mesh grids sputter coated with a 3 nm thick carbon layer. In addition, the block faces of the specimens were also investigated at ambient conditions using ZEISS Axioplan light microscope (Zeiss, Germany) equipped with Zeiss Axio Cam ICc 1 CCD camera in reflected polarized light.

X-ray diffraction analysis was conducted using a Bruker AXS D4 Endeavour system using Cu-Kα radiation operated at 40 kV and 40 mA and a Lynxeye linear strip detector. All polymer samples (hot pressed) were loaded onto the sample holder. Microscope slides were used to flatten the surface of all samples before loading. All samples were tested between 6 and 90° 2Θ with variable slit v20 with step size 0.02°/s. The instrument was calibrated using a NIST standard of Al_2O_3 (corundum ceramic block). Endeavour is equipped with a large sample magazine that allows a wide range of samples to be docked in a single stream (manual). In the current study, a search/match technique was used to compare the constituents present in the samples (ICDD Database, Bruker DIFFRAC.EVA software version 3.0). It should be noted that background was removed from all spectra before further processing. Diffraction peaks were not smoothened for any spectra as smoothing might vary area of the peaks of interest. Raman spectra were recorded on a Jobin Yvon Horiba LabRAM spectrometer (in the range 200-3200 cm^{-1}, data interval 2 cm^{-1}) with backscattered confocal configuration using a HeNe laser (633 nm). A long working distance objective with magnification 50x was used both to collect the scattered light and to focus the laser beam on the sample surface. CCD (charge-coupled device) detector exposure time was 200 s and an average of 2 cycles was used to increase S/N ratio. Raman shifts were calibrated using Si wafer signal at 520.7 cm^{-1}.

Thermal degradation properties of the samples were analyzed using TA Discovery thermogravimetric analyzer (TGA). Nitrogen was used as carrier gas and the scans were obtained from 50 to 750 °C at a heating rate of 10 $°C.min^{-1}$. Differential scanning calorimetric (DSC) measurements on the polymer blends and nanocomposites were performed using TA Discovery DSC under nitrogen atmosphere. The second heating scans were obtained by heating the samples from -20 °C to 200 °C at a heating rate of 10 °C.min^{-1}, followed by cooling from 200 °C to -20 °C at a cooling rate of 10 °C.min^{-1}.

Tensile analysis of the samples was carried out using universal testing machine (Instron, USA) (ASTM D638). The dumbbell-shaped samples were used. A loading rate of 10 mm.min^{-1} was used and the tests were carried out at room temperature. The tensile strength and modulus of the composites were calculated using Win Test Analysis software. An average of five sample values was reported. ARES rheometer was used to characterize the rheological properties such as storage modulus (G′), loss modulus (G″), as well as complex viscosity (η*). Disc-shaped samples of 8 mm diameter and 2 mm thickness were measured using a gap opening of 1.2 mm. Strain sweep scans were recorded at 180 °C with ω = 1 rad.s^{-1} and 0.1% to 100% strain. For the comparison, temperature and frequency sweep scans (dynamic testing) of all the samples were recorded at 1% strain from ω = 0.1 to 100 rad.s^{-1} in the temperature range of 180-200 °C. DMA (dynamic mechanical analyzer) RSA3 of TA Instruments was used to study the change in modulus with temperature (-30 to 120 °C). Dynamic temperature ramp test was performed at a frequency of 5 rad.s^{-1}, ramp rate 3 °C.min^{-1} and 1% strain for all samples.

5.3 Results and Discussion

In the current study, blend nanocomposites of polypropylene and polystyrene with two different blend compositions were generated to study the effect of graphene filler on the composite morphology and properties. In addition, the blend composites were also generated using a small amount of polar copolymer as compatibilizer.

Figure 5.3 demonstrates the TEM micrographs of the PP/PS 70/30 and 30/70 blends. There was a difference in the phase morphology of the two blends. In the case of PP/PS 70/30 blend, though PP formed the continuous phase surrounding the large sized PS domains, however, it was also observed to be dispersed as spherical domains inside the PS phase. The size of these PP domains was 1-2 μm in diameter, indicating some phase intermixing between the PP and PS phases. As there is no thermodynamic interaction between the two polymers, therefore, the observed morphology would have solely resulted from the kinetic mixing in the compounder. In the case of PP/PS 30/70 blend, the polymer phases were more elongated in nature, though some degree of mixing of PP within the PS phase was still visible.

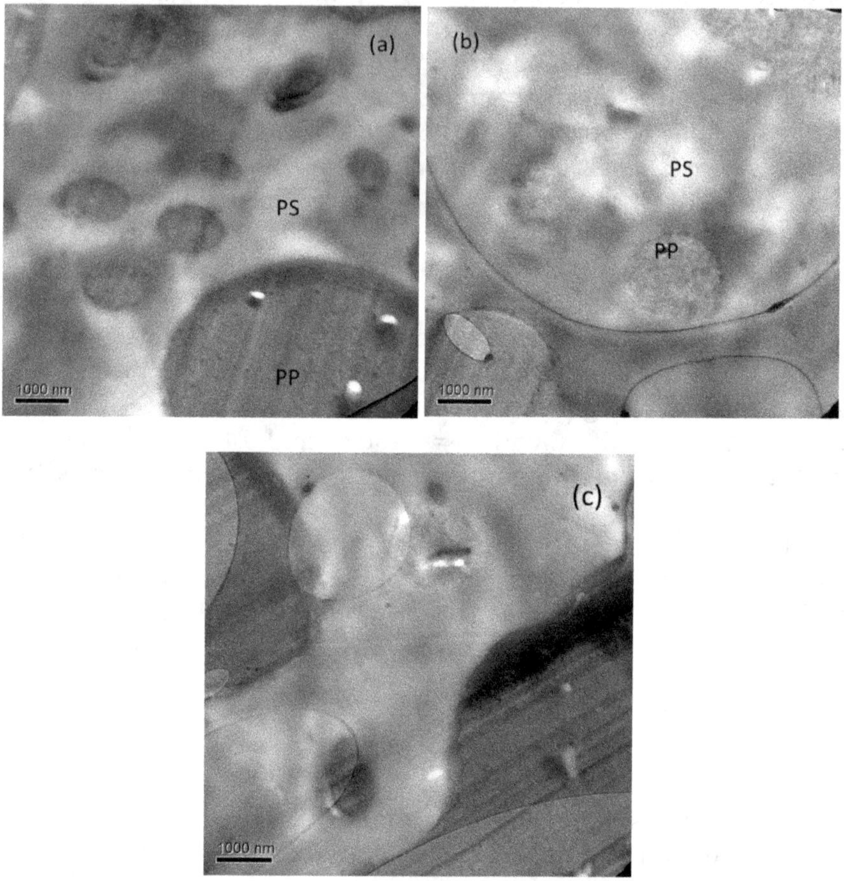

Figure 5.3 TEM images of the (a,b) PP/PS (70/30) and (c) PP/PS (30/70) blend.

Figures 5.4-5.5 present the optical images of the PP/PS (70/30)+2%G+5%EAA and PP/PS (30/70)+2%G+5%EAA blend composites. In these images, the dark phase was semi-crystalline PP, whereas amorphous PS was observed as the light phase. As observed in the TEM images of the PP/PS 70/30 blend, the composite exhibited the dispersion of the PS phase as large domains inside PP and a small amount of PP was also dispersed inside the PS domains. The filler particles (observed as white signal in the optical images) were observed to be largely localized inside the PP phase. Similarly, the PP/PS 30/70

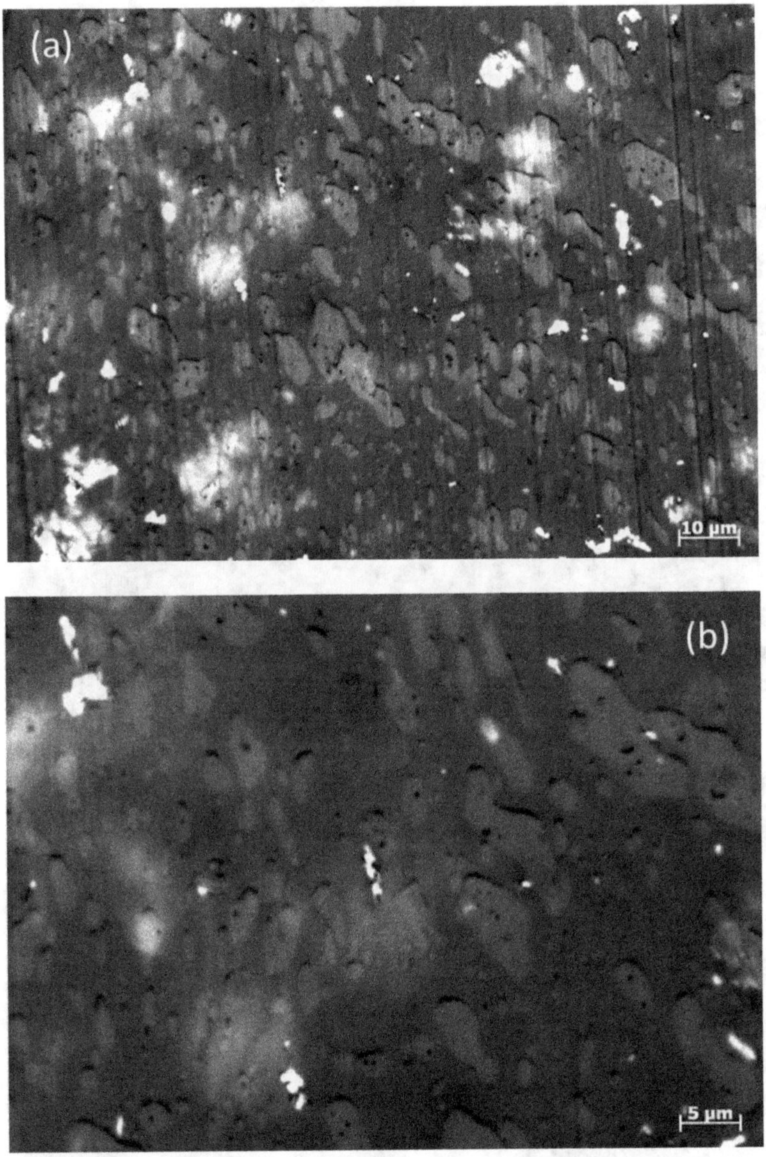

Figure 5.4 Optical images of the PP/PS (70/30)+2%G+5%EAA blend nanocomposite.

+2%G+5%EAA blend nanocomposite exhibited elongated domains with filler particles dispersed inside the PP phase. The observed composite morphology resulted from the one-step melt mixing approach used to generate the composites. As PP melted before PS inside the compounder, thus, the filler particles would have been mixed with the PP melt before the PS phase formed the melt. Therefore, the filler particles could be localized largely inside the PP phase. Occasionally, the filler platelets were also observed at the interface between the

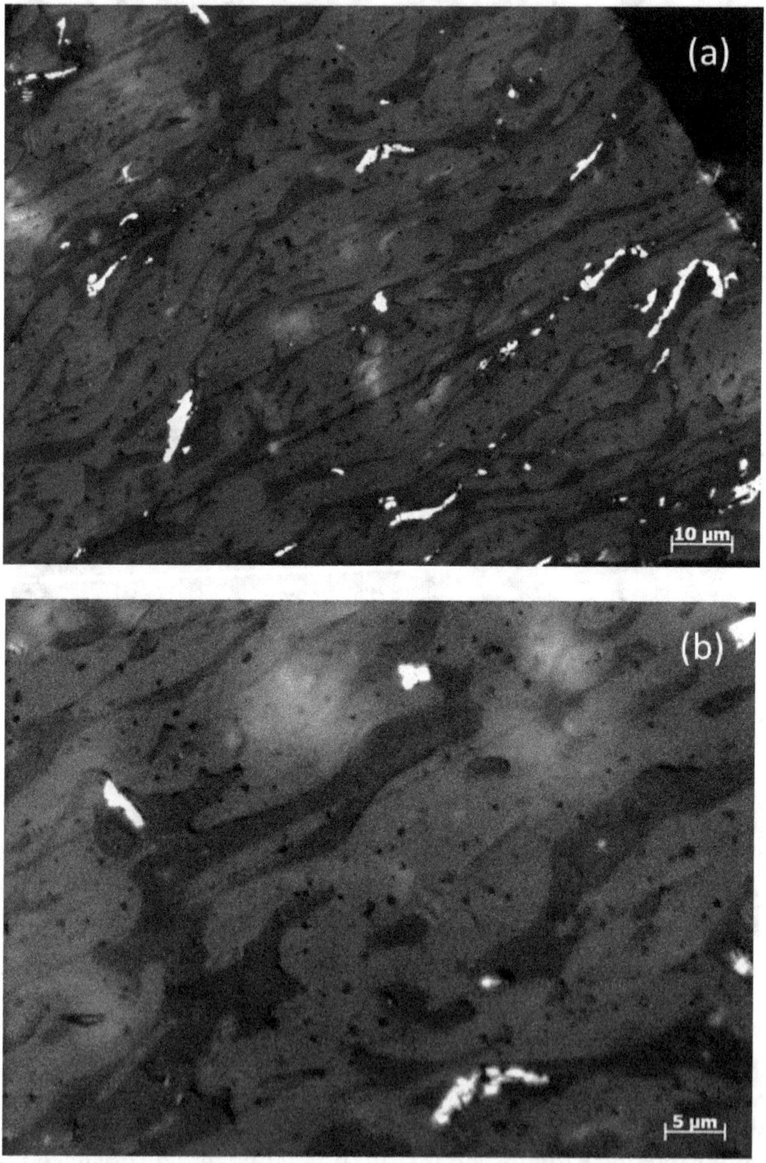

Figure 5.5 Optical images of the PP/PS (30/70)+2%G+5%EAA blend nanocomposite.

PP and PS phases, thus, providing some degree of interface compatibilization, which has been reported in the literature to be beneficial in enhancing phase intermixing by reducing the domain sizes. EAA could not be specifically observed in the composite morphology. As EAA had much lower melting point, thus, it is expected to have melt mixed with filler particles, followed by its mixing in the PP phase. As mentioned earlier, localization of the filler particles inside the PP phase is expected to be beneficial in retaining the overall flexibility of the composites. The morphology of the composites was also evaluated further using TEM, as shown in Figure 5.6. Similar to the findings from optical microscopy, the filler was localized in the PP phase and

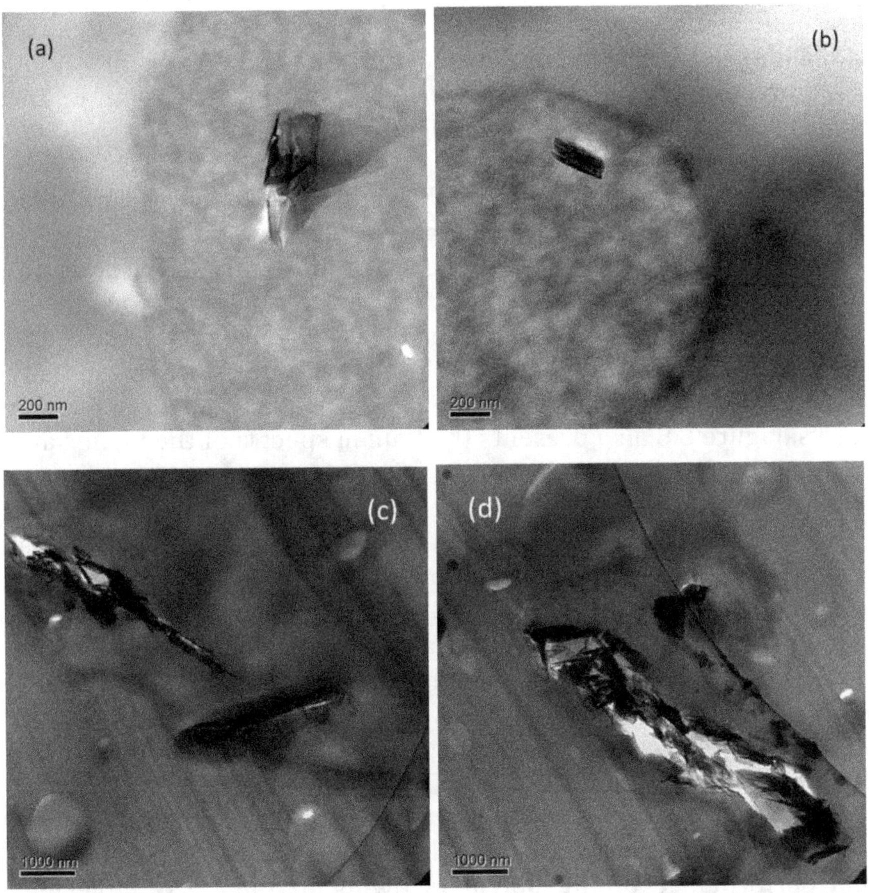

Figure 5.6 Transmission electron micrographs of the (a,b) PP/PS (70/30)+2%G+5%EAA and (c,d) PP/PS (30/70)+2%G+5%EAA blend nanocomposite.

the filler particles were also observed at the interface between PP and PS on many occasions (Figure 5.6 c and d). In addition, the filler was observed to be present mostly in the form of stacks rather than single layers, thus, indicating the need to further delaminate the filler platelets. Probably higher content of EAA is required to enhance the compatibilization of filler and the polymer phases. Another alternative would be to generate a masterbatch of EAA and graphene, which can then be added to PP and PS for effective mixing [24]. However, it should be noted that observed composite morphology can still be expected to result in enhanced composite properties owing to overall good degree of filer dispersion and large aspect ratio graphene platelets. The phase distribution in the composites was also retained even after multiple processing cycles, indicating stable morphology even though the components were mixed only kinetically.

Figure 5.7 also demonstrates the X-ray diffractograms of the blends and blend composites. No significant difference was observed between the diffraction patterns of blends and their composites indicating that the crystal structure of PP was not affected in the presence of PS, filler and EAA, even though the degree of crystallinity and crystal size may have been affected due to the hindrance to the crystallization of PP chains in the presence of other components. The diffraction peak corresponding to graphene was also observed in the composites as a sharp signal at $2\Theta \sim 27°$, indicating that the periodicity was retained in the filler due to the presence of multilayered platelet stacks. Figure 5.8 also presents the Raman spectra of the blends and composites. The characteristic D and G bands of graphene overlapped with the signals originating from the polymer phases. In addition, no new Raman shift was observed, which was expected due to the physical mixing of the composite components.

Insights into the thermal degradation of the blends and nanocomposites were gained through TGA, as demonstrated in Figure 5.9 for the PP/PS 70/30 and 30/70 blends and their composites. In PP/PS 70/30 blend, the onset of degradation was observed around 350 °C, which was observed to be unchanged in the composite of the 70/30 blend with 2% graphene. However, an increment in the onset of degradation temperature was observed in the case of composite with 5 wt% EAA. Generally, the compatibilizers tend to deteriorate the thermal performance of the polymers owing to their low molecular weight. However, the observed increment in the thermal degradation performance indicated that the filler dispersion must have been enhanced in comparison with the composite without EAA addition, thus

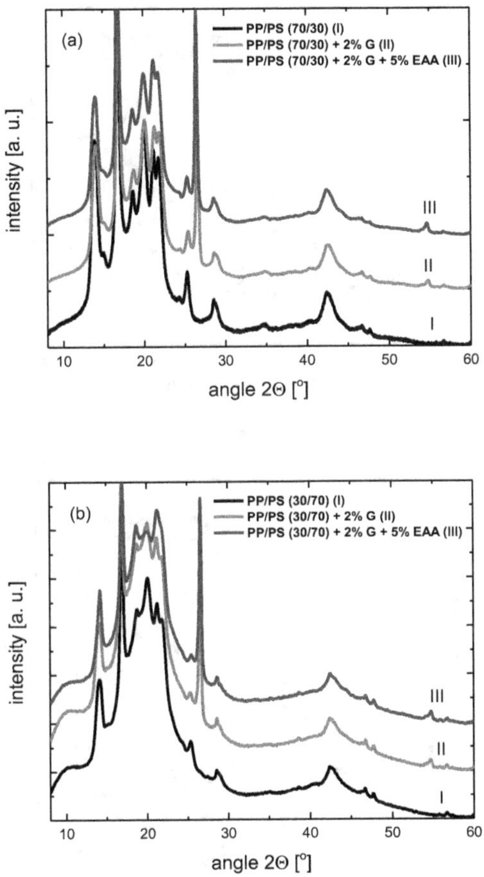

Figure 5.7 X-ray diffractograms of (a) PP/PS (70/30) blend and its composites; (b) PP/PS (30/70) blend and its composites.

enabling the polymer chains to have effective heat transfer with the high aspect ratio graphene platelets through increased extent of interfacial contacts. Though the enhancement in the onset degradation temperature in the composite was about 15 °C, however such an increase is also significant for generating effective thermal resistance. Similar enhancement was also observed in the case of PP/PS 30/70 blend composite system. In fact, the composite without EAA was also observed to enhance the thermal stability of the polymer in this case, though the composite with EAA still exhibited higher extent of thermal resistance enhancement. This confirmed that the addition of EAA was beneficial in achieving better thermal resistance owing to higher

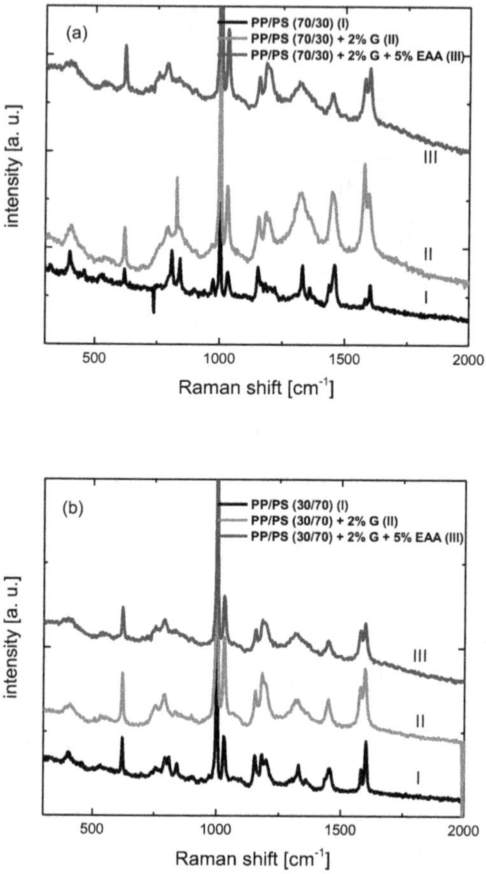

Figure 5.8 Raman spectra of (a) PP/PS (70/30) blend and its composites; (b) PP/PS (30/70) blend and its composites.

extent of filler dispersion. In addition, enhancement of thermal stability in the composite without EAA could be attributed to the blend composition. Lower extent of PP in the blend would have resulted in higher extent of its reinforcement due to filler dispersion solely in PP phase, in comparison with the blend with 70 wt% PP, thereby, leading to an enhancement in the overall thermal performance. Thus, the blend nanocomposites were suitable for high temperature processing and applications. Dynamic TGA carried out the compounding temperature of 200 °C as a function of time also revealed the stability of the composites, as no significant weight loss occurred for the test duration equivalent to the compounding time.

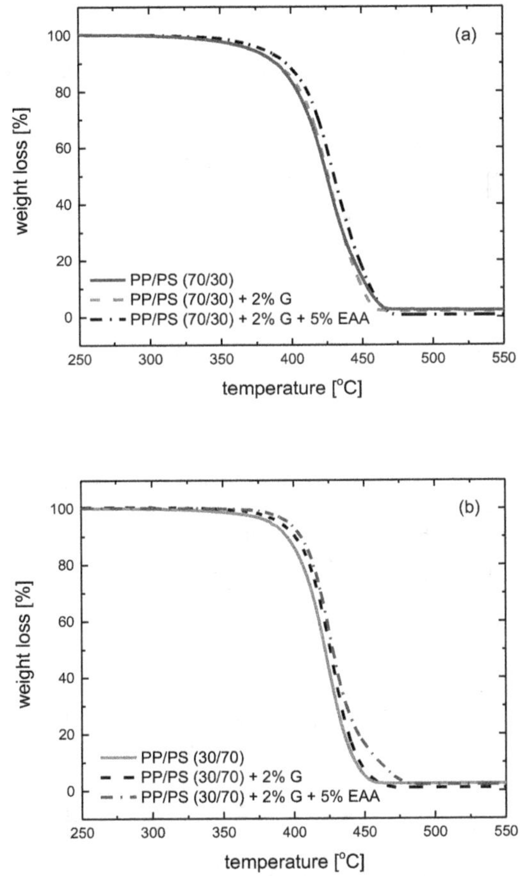

Figure 5.9 TGA thermograms of (a) PP/PS (70/30) blend and its composites; (b) PP/PS (30/70) blend and its composites.

Figures 5.10-5.11 also present the melting and crystallization thermograms of PP/PS blends and their composites. As observed in Figure 5.10(a), the melting behavior of the PP phase in the PP/PS 70/30 blend remained unaffected in the PP/PS (70/30)+2%G composite. However, the melting behavior was observed to be affected on the addition of EAA in the composites, as indicated by the decreased peak melting temperature. The melt enthalpy of PP (normalized to PP weight in the composite) was also observed to decrease in this case. It, thus, confirmed that even though the crystal structure of PP was not affected in the composite with EAA (as observed from X-ray), the degree of crystallinity was affected probably due to the hindrance to

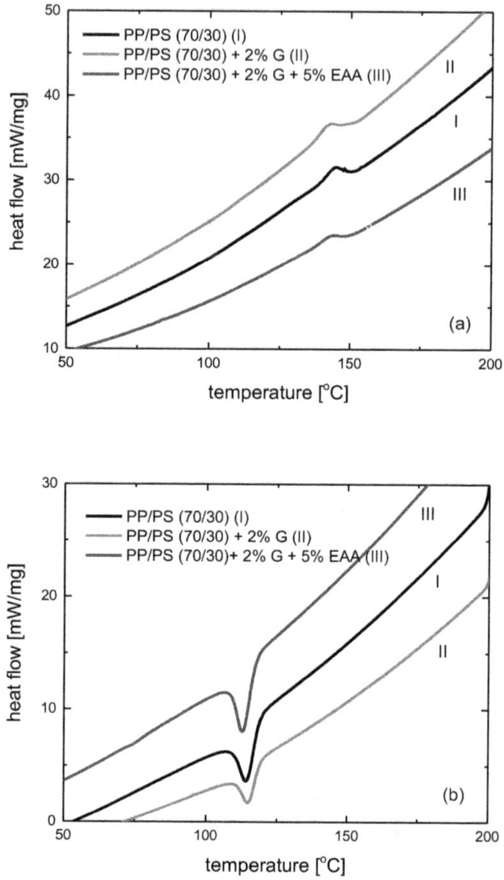

Figure 5.10 DSC (a) melting and (b) crystallization thermograms of PP/PS (70/30) blend and its composites.

crystallization in the presence of EAA. It, thus, also indicated the EAA was present in the PP phase, as opined earlier through findings from microscopy. Similar to the melting thermograms, the crystallization thermogram of the PP/PS (70/30)+2%G+%EAA composite also revealed slight reduction in the peak crystallization temperature (Figure 5.10(b)), thus, confirming again the hindrance to polymer crystallization. No change in the peak crystallization temperature was observed in the case of PP/PS (70/30)+2%G composite in comparison with the PP/PS (70/30) blend. In the case of PP/PS (30/70) blend, the endothermic transition corresponding to the crystalline melting of PP phase was significantly suppressed, indicating the higher extent

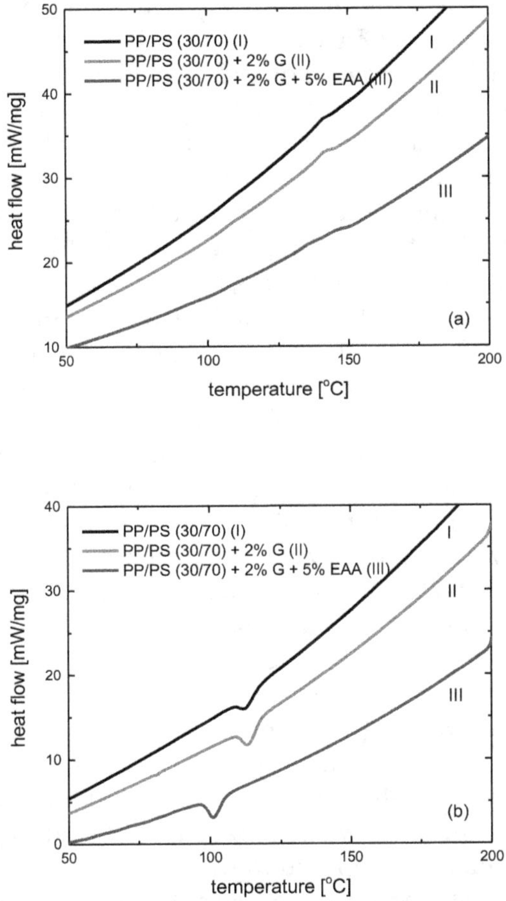

Figure 5.11 DSC (a) melting and (b) crystallization thermograms of PP/PS (30/70) blend and its composites.

of hindrance to PP crystallization in the blend. The composite without EAA also exhibited similar transition at the same peak melting temperature. However, the composite with EAA exhibited further suppression of the peak melting temperature as well as melt transition in the composite. Similarly, the crystallization behavior was also impacted in the case of PP/PS (30/70)+2%G+5%EAA composite as a large decrease in the peak crystallization temperature of PP was observed in this case. In comparison, the peak crystallization temperature in the composite without EAA was not affected and was observed to be the same as the PP/PS 30/70 blend. It should also be noted that the extent of reduction in the peak crystallization temperature of the PP/PS (30/70)+2%G+5%EAA composite was much higher than the PP/PS (70/30)+2%G+5%EAA composite. The effect of enhanced filler dispersion in the presence of EAA may have affected the PP/PS 30/70 composite more due to the higher concentration of filler in the PP phase in this case, thereby resulting in much higher hindrance to the PP crystallization. Overall, the presence of EAA was observed to significantly affect the melting and crystallization behavior of PP, therefore, confirming that the EAA was present in the PP phase, which resulted due to the melt mixing of filler with EAA during initial stages of melt compounding process, followed by melt mixing with PP and PS. Such an effect on the polymer crystallinity is also expected to impact the mechanical performance of the composites.

Figures 5.12-5.14 elaborate the rheological performance of the composites in comparison with the pure blends. The measurements were performed over the temperature range of 180-200 °C. As expected, the storage and loss modulus of the materials decreased with temperature, however, the curves were observed to merge at higher angular frequencies. The storage and loss modulus of the composites also increased with increasing angular frequency, however, the rate of increase slowed down at higher frequency values due to shear thinning effect. In all blend and nanocomposite materials, the loss modulus was observed to be higher than storage modulus over the complete angular frequency range, thus, indicating the dominance of the viscous behavior. In the case of PP/PS 70/30 blend nanocomposites system, the composite without EAA did not exhibit increment in the storage modulus as compared to the PP/PS 70/30 blend. On the other hand, an increment in the storage modulus was observed in the composite with EAA, thus, confirming the earlier findings of enhanced filler dispersion in the presence of EAA, thereby leading to effective stress transfer from the polymer chains to the platelets. In the case of PP/PS 30/70 blend nanocomposites system, a large increase in the modulus was observed in the composites with and without EAA, therefore, confirming the findings from TGA about the enhancement in the performance due to the localization of filler in the lower fraction PP phase, which together with higher modulus PS phase resulted in overall much higher degree of increment. As observed in Figure 5.14, the tan δ of the composites decreased in comparison with the pure blends, thus, indicating the increment in the elastic contribution. The decrease in tan δ was more significant in the PP/PS 70/30 composite with EAA as well as both PP/PS 30/70 composites due to

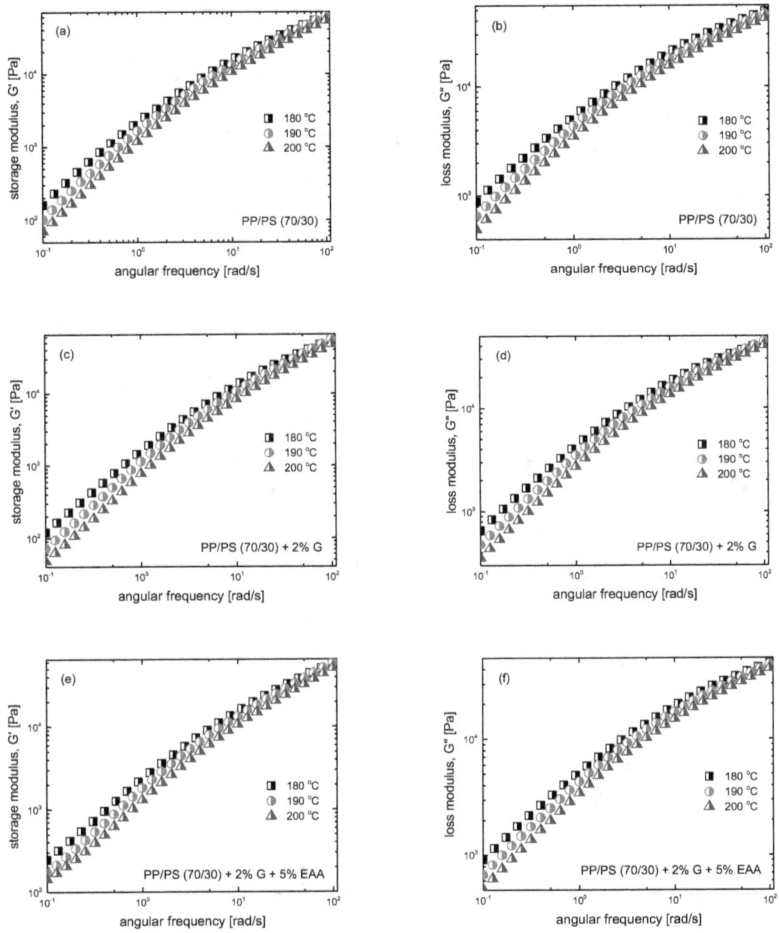

Figure 5.12 (a,c,e) Storage and (b,d,f) loss modulus of PP/PS (70/30), PP/PS (70/30)+2%G and PP/PS (70/30)+2%G+5%EAA respectively, as a function of angular frequency in the temperature range of 180-200 °C.

the same reasons observed earlier.

Dynamic mechanical analysis of the PP/PS blends and their nanocomposites is also presented in Figure 5.15 as a function of temperature (-30 to 120 °C). In the PP/PS 70/30 composite system, the composites with EAA exhibited an increment in the storage modulus at lower temperatures, whereas the composite without EAA did not exhibit much enhancement in the modulus. These findings corroborated with earlier observations from the TGA and rheological analysis

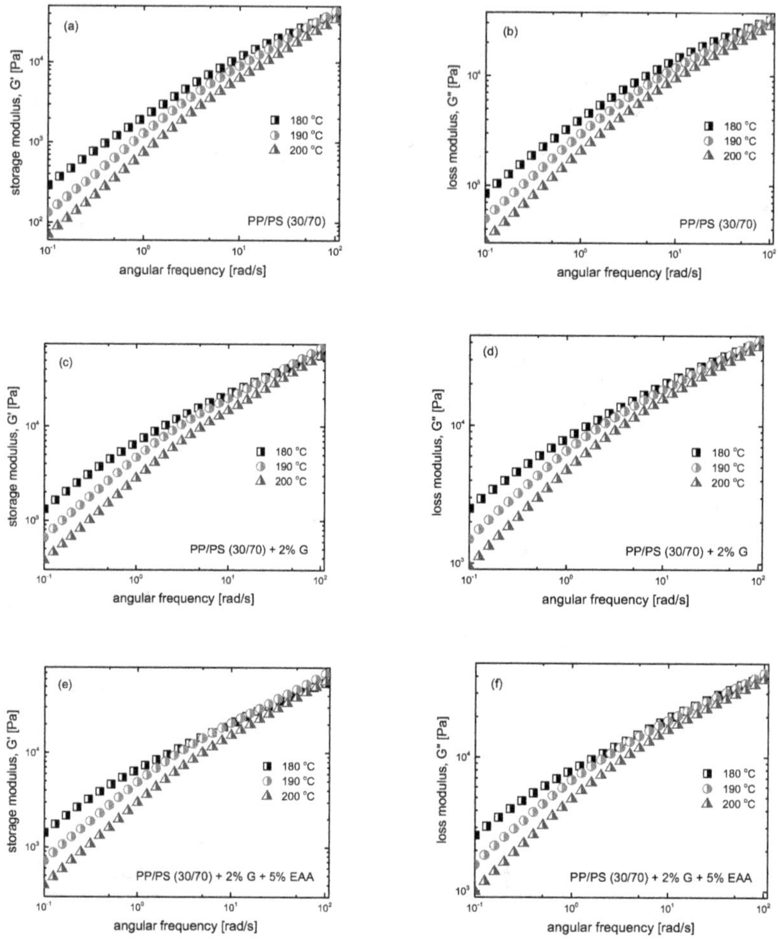

Figure 5.13 (a,c,e) Storage and (b,d,f) loss modulus of PP/PS (30/70), PP/PS (30/70)+2%G and PP/PS (30/70)+2%G+5%EAA respectively, as a function of angular frequency in the temperature range of 180-200 °C.

where higher extent of filler dispersion in the presence of EAA was observed to lead to effective heat and stress transfer. It should be noted that the modulus of the composite increased even though the crystallinity of the composite with EAA was observed to decrease due to the hindrance to crystallization caused by EAA. Thus, the observed increment in modulus in the EAA composite resulted from the combined action of decrease in properties due to reduced crystallinity and increased modulus due to enhanced extent of filler dispersion. In

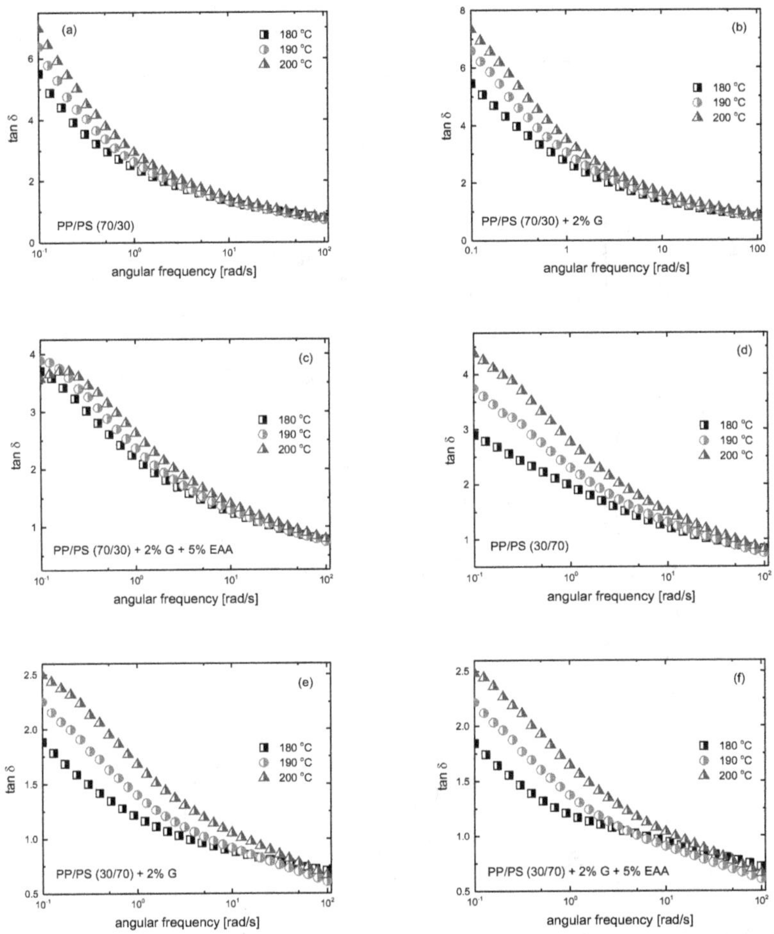

Figure 5.14 tan δ of the materials as a function of angular frequency in the temperature range of 180-200 °C: (a) PP/PS (70/30), (b) PP/PS (70/30)+2%G, (c) PP/PS (70/30)+2%G+5%EAA, (d) PP/PS (30/70), (e) PP/PS (30/70)+2%G and (f) PP/PS (30/70)+2%G+5%EAA.

the case of PP/PS 30/70 composites, both composites (with and without EAA) resulted in an enhancement of storage modulus at lower temperatures, which also correlated with the earlier findings. Reinforcement of the lower fraction PP phase with the filler led to an overall enhancement of the modulus of the composites. The composites also exhibited sharp reduction in the modulus around 30 °C, irrespective of the blend composition. Even though the modulus decreased on

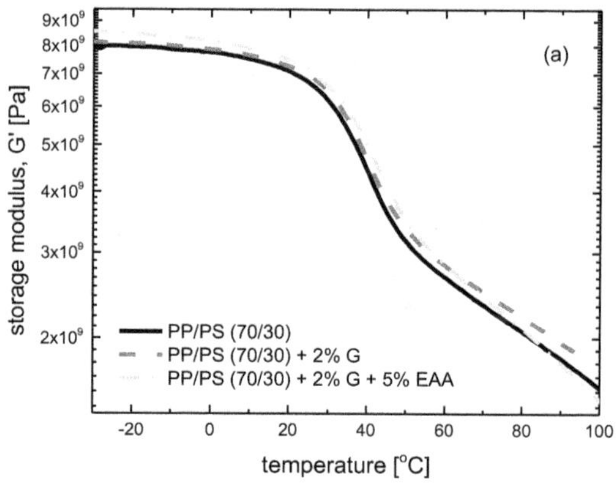

Figure 5.15 Dynamic mechanical analysis of (a) PP/PS (70/30) blend and its composites; (b) PP/PS (30/70) blend and its composites, as a function of temperature.

on increasing the temperature further, however, the modulus of the composites still remained appreciable even at 100 °C.

Figure 5.16 and Table 5.1 depict the tensile properties of the blends and nanocomposites. The PP/PS 70/30 blend exhibited a tensile modulus of 1517 MPa, which increased to 1589 MPa in the composite with 2 wt% graphene. The modulus was further enhanced marginally to 1618 MPa in the composite where EAA was also added, due to enhanced level of filler dispersion. The effect of enhanced filler dispersion on the addition of EAA copolymer was also observed on tensile strength, strain at break as well as stress at break. All of these

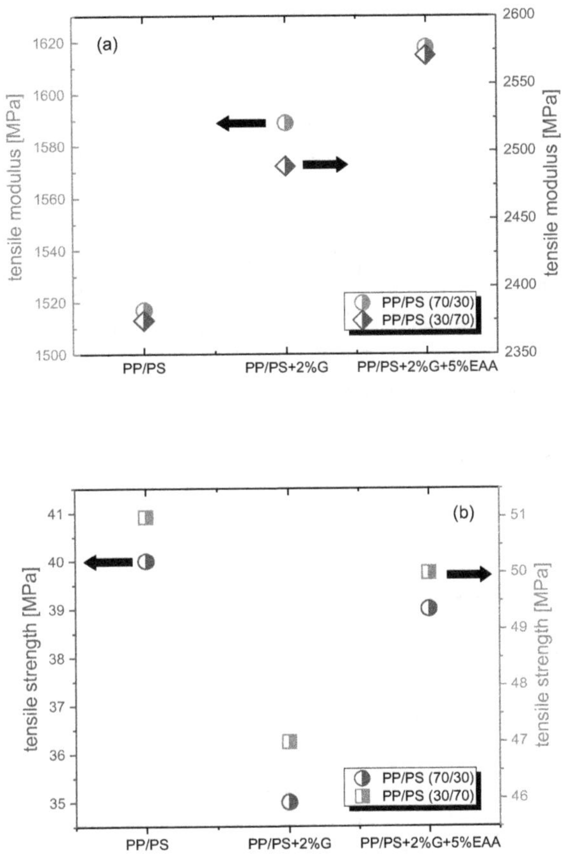

Figure 5.16 Tensile properties of (a) PP/PS (70/30) blend and its composites; (b) PP/PS (30/70) blend and its composites.

quantities were observed to be improved in the case of PP/PS (70/30)+2%G+5%EAA composite, in comparison with the composite without EAA. For instance, the strain at break of the composite with EAA was almost similar to the pure blend, which indicated that the modulus of the composites was enhanced without negatively impacting the flexibility. In the case of PP/PS 30/70 composites, the tensile modulus for the EAA containing composite exhibited higher extent of increment than the composite without EAA due to the combined effect of reinforcement of PP phase by the filler as well as higher extent of filler dispersion in the presence of EAA, which was achieved even though the crystallinity of PP phase was significantly reduced. The tensile strength, strain at break and stress at break for the EAA containing composites were also similar to the pure blend. It, thus, indicated that the specific localization of the filler in the PP phase led to largely retaining the flexibility of the composites in the case of EAA containing composites, probably also due to some degree of plasticization of the matrix caused by EAA. Overall, even though the filler did not exfoliate extensively as observed from microscopy analysis, the EAA addition still helped to achieve better composite performance by enhancing the overall extent of filler dispersion, thereby leading to the enhancement in interfacial contacts, thus, resulting in effective heat and stress transfer. In addition, the localization of filler in the PP phase also correspondingly affected the composite properties, as a function of blend composition.

Table 5.1 Tensile properties of the blends and blend nanocomposites

Material	Tensile modulus, MPa	Tensile strength, MPa	Strain at break, %	Stress at break, MPa
PP/PS (70/30)	1517	40	9.2	38
PP/PS (70/30) + 2% G	1589	35	6.5	33
PP/PS (70/30) + 2% G + 5% EAA	1618	39	8.1	37
PP/PS (30/70)	2375	51	6.3	46
PP/PS (30/70) + 2% G	2489	47	4.8	42
PP/PS (30/70) + 2% G + 5% EAA	2571	50	5.9	46

5.4 Conclusions

The current study demonstrated the successful generation of blend nanocomposites of polypropylene and polystyrene with graphene and ethylene-co-acrylic acid copolymer. PS/PP blends with two different compositions were employed. Single step melt mixing method was used for the generation of blend nanocomposites. The morphology of the blends was affected by the blend composition. The blend with PP/PS 70/30 ratio exhibited spherical domains of PS inside the

PP phase. In addition, smaller spherical domains of PP were also present inside the PP phase. The blend with PP/PS 30/70 exhibited the morphology with more elongated conformation of the components. The composites exhibited the localization of the filler particles in the PP phase due to the earlier melting of PP during the melt mixing process. Occasionally, the filler particles were also observed to be present at the interfaces of PP and PS phases. In addition, the presence of EAA was suggested to be in the PP phase owing to the much earlier melting of EAA during the compounding process. The filler did not delaminate extensively in the composites. The X-ray diffractograms of the composites also revealed the characteristic diffraction peak of graphene, indicating that the periodicity of the filler platelets was largely retained in the composites. The thermal degradation of the PP/PS (70/30)+2%G+5%EAA was observed to be enhanced in comparison with the PP/PS (70/30)+2%G composite, thereby, signaling the effect of EAA on the filler dispersion. The composites with PP/PS (30/70) blend had also enhanced thermal stability, though the addition of EAA led to higher extent of increment. The reinforcement of minority PP phase by the filler would have resulted in effective thermal resistance in the case of PP/PS (30/70)+2%G composite, whereas the composite with EAA additionally benefited from the positive interaction of EAA with filler surface thereby leading to enhanced extent of filler dispersion. The peak melting and crystallization temperatures were observed to decrease in the PP/PS (70/30)+2%G+5%EAA composite due to the additional hindrance caused by EAA chains to the PP crystallization. The melt enthalpy was also observed to decrease in this composite, thus, indicating reduced extent of crystallinity, whereas no such effect was observed for the PP/PS (70/30)+2%G composite. More significant effect on melting and crystallization of PP was observed in the case of PP/PS (30/70)+2%G+5%EAA. Probably, the effect of enhanced filler dispersion in the presence of EAA affected the PP/PS 30/70 composite more due to the higher concentration of filler in the PP phase. These findings also corroborated the earlier suggestion from microscopy analysis that EAA was present in the PP phase. The storage modulus of the PP/PS (30/70) based composites was observed to enhance significantly owing to the superior reinforcement of the minority PP phase, as mentioned earlier. In addition, the EAA containing composite of PP/PS (70/30) blend also exhibited higher storage modulus. The tan δ of the composites also decreased as compared to the pure blends due to the increment in the elastic contribution. Similar to the storage modulus observed from melt rheology, the dynamic mechanical analysis also exhibited an increment in the storage modulus of the composites due to the same reasons. It should also to be noted that the increment in the modulus of EAA containing composites was achieved even though the crystallinity was negatively impacted. Thus, the observed enhancement resulted from the competing factors of reduction in crystallinity leading to deterioration of composite performance and enhanced extent of filler dispersion resulting in increment of composite properties. Similar to rheology and dynamic mechanical analysis, the tensile analysis of the composites also revealed increment in the modulus due to the incorporation of EAA. Higher extent of increment in the modulus was observed in the case of PP/PS (30/70)+2%G+5%EAA. In addition, the composites with EAA also retained the tensile strength, elongation at break as well as stress at break due to filler localization in PP. Some degree of plasticization resulting from the addition of EAA may have also helped to enhance the flexibility of the EAA containing composites. Overall, the generated blend nanocomposites exhibited superior properties than the corresponding blends and are more useful alternatives than not only the parent materials but also other thermoplastics due to optimal properties. In addition, the morphology of the blend nanocomposites was also stable even after few high temperature processing cycles, though the components were only kinetically mixed. The composite performance can be enhanced further by employing strategies such as enhancing the content of polar copolymer, generating masterbatch of the filler and copolymer prior to melt mixing with PP and PS, and surface functionalization of graphene surface to generate chemical moieties which can chemically interact with the copolymer chains leading to reactive compatibilization.

Acknowledgement

The authors are indebted to Dr. Nadejda Matsko and Bradley Burren for support in material analysis.

References

1. Jancar, J. (1999) *Mineral Fillers in Thermoplastics I: Raw Materials and Processing*, Springer-Verlag, Germany.
2. Mittal, V. (2012) *Functional Polymer Blends: Synthesis, Properties &*

Performance, CRC Press, USA.

3. Hippi, U. (2008) *Polyolefin Composites*, John Wiley & Sons, USA.
4. Fekete, E., Foldes, E., and Pukanszky, B. (2005) Effect of molecular interactions on the miscibility and structure of polymer blends. *European Polymer Journal*, **41**, 727-736.
5. Lipatov, Y. S. (2002) Polymer blends and interpenetrating polymer networks at the interface with solids. *Progress in Polymer Science*, **27**, 1721-1801.
6. Litmanovich, A. D., Plate, N. A., and Kudryavtsev, Y. V. (2002) Reactions in polymer blends: interchain effects and theoretical problems. *Progress in Polymer Science*, **27**, 915-970.
7. Sundararaj, U., and Macosko, C. W. (1995) Drop breakup and coalescence in polymer blends: The effects of concentration and compatibilization. *Macromolecules*, **28**, 2647-2657.
8. Nwabunma, D., and Kyu, T. (2008) *Polyolefin Composites*, Wiley, USA.
9. Liu, H., and Brinson, L. C. (2008) Reinforcing efficiency of nanoparticles: A simple comparison for polymer nanocomposites. *Composites Science and Technology*, **68**(6), 1502-1512.
10. Mittal, V., Burren, B., and Matsko, N. (2017) Eco-friendly bio-nanocomposites of PHBV/PBAT blend with graphene and vermiculite. In: *Biopolymer Based Advanced Materials*, Mittal, V. (ed.), Central West Publishing, Australia, pp. 57-85.
11. Mukhopadhyay, P., and Gupta, R. K. (2011) Trends and frontiers in graphene-based polymer nanocomposites. *Plastics Engineering*, **67**, 32-42.
12. Mittal, V. (2014) Functional polymer nanocomposites with graphene: A review. *Macromolecular Materials and Engineering*, **299**, 906-931.
13. Katz, H. S., and Milewski, J. V. (1987) *Handbook of Fillers for Plastics*, Van Nostrand, USA.
14. Rothon, R. (1995) *Particulate-Filled Polymer Composites*, Longman, UK.
15. Brydson, J. A. (1975) *Plastic Materials*, Newnes-Butterworths, UK.
16. Karger-Kocsis, J. (1995) *Polypropylene Structure, Blends and Composites*, Chapman & Hall, UK.
17. Elias, H.-G. (2003) *Anwendungen von Polymeren*, Wiley-VCH, Germany.
18. Robertson, G. L. (1993) *Food Packaging - Principles and Practice*, volume 6 of Packaging and Converting Technology, Marcel Dekker, Inc., USA.
19. Diaz, M. F., Barbosa, S. E., and Capiati, N. J. (2007) Reactive compat-

ibilization of PE/PS blends. Effect of copolymer chain length on interfacial adhesion and mechanical behavior. *Polymer*, **48**, 1058-1065.

20. Xu, S.-A., and Cham, C.-M. (1998) Polystyrene/high density polyethylene blends compatibilized by a tri-block copolymer I. Properties and morphology. *Polymer Journal*, **30**, 552-558.
21. Inberg, J. P. F., Hunse, P. H., and Gaymans, R. J. (1999) Long fiber reinforcement of polypropylene/polystyrene blends. *Polymer Engineering and Science*, **39**(2), 340-346.
22. Mittal, V., Chaudhry, A. U., Patwary, F., and Matsko, N. (2017) Filler distribution and properties of polyethylene/polystyrene blend composites with 0-D and 2-D nanomaterials. In: *Functional Polymer Blends and Nanocomposites*, Mittal, V. (ed.), Central West Publishing, Australia, pp. 1-34.
23. Parameswaranpillai, J., Dubey, V. K., Sisanth, K. S., Jose, S., Zachariah, A. K., Siengchin, S., Salim, N. V., and Hameed, N. (2016) Tailoring of interface of polypropylene/polystyrene/carbon nanofibre composites by polystyrene-block-poly(ethylene-ran-butylene)-block-polystyrene. *Polymer Testing*, **51**, 131-141.
24. Chaudhry, A. U., and Mittal, V. (2013) High-density polyethylene nanocomposites using masterbatches of chlorinated polyethylene/graphene oxide. *Polymer Engineering and Science*, **53**, 78-88.

6

Polyethylene Nanocomposites with Psyllium Husk, Graphene, Sepiolite, and Their Hybrids

6.1 Introduction

Polyolefins have one of the largest tonnage in the world with respect to production. It is estimated that more than 60% of produced polyolefins (mainly polyethylene (PE) and polypropylene (PP)) is introduced to the market as compounds, while only about 23% of the volume of other thermoplastics is used for compounding. The prominent properties of the polyolefins include low cost, recyclability, good processability, non-toxicity, and biocompatibility, thus, leading to a wide range of applications including orthopedic implants, automobile parts, consumer goods, durable equipment and industrial machinery [1]. Especially in the case of polyethylene, there is a need to further improve the stiffness, wear resistance, and rigidity so as to expand its spectrum of applications [2,3].

The combination of polymers and reinforcing fillers is a fast and practical technique for the development of new materials having improved properties. Significant attempts have been made to enhance the properties of high density polyethylene (HDPE), by the incorporation of different varieties of fillers. For instance, the effect of carbon nanotubes (CNTs) on the mechanical properties such as Young's modulus, ultimate stress, and toughness has been studied in nanocomposites of HDPE and CNT [4]. It was observed that with increase in percentage of CNTs, the composite properties also exhibited improvement owing to good load transfer effect and interface linking between polymer matrix and CNTs [4]. In another study, solution and melt mixing using co-counter- and modified co-rotating screw system

Vikas Mittal[a,], Ali U. Chaudhry[a,**], M. Cole[a,***], Nadejda Matsko[b]*
[a]The Petroleum Institute (part of Khalifa University of Science and Technology), Abu Dhabi, UAE; [b]Graz Centre for Electron Microscopy, Graz, Austria
**Current address: Bletchington, Wellington County, Australia; **Current address: Texas A&M University, Qatar; ***Current address: University of Waterloo, Canada*

were employed to achieve uniform dispersion of graphene in linear low density polyethylene (LLDPE) and the effect on the mechanical properties was reported [5]. It was observed that solution mixing resulted in better reinforcement of the polymer [5]. In another study reporting the use of graphene, solution mixing of the graphene oxide filler with chlorinated polyethylene compatibilizers followed by melt mixing of the solution mixed masterbatches with high density polyethylene was reported [6]. The mechanical performance of the nanocomposites at ambient conditions was observed to be affected by the factors like filler delamination, plasticization of the matrix, compatibilizer content as well as their chlorination level and amorphous/semi-crystalline nature. Thus, the use of both 1-D and 2-D fillers has become predominant for the enhancement of polyolefin properties. Graphene has specifically evolved as the filler of interest in recent years. It consists of one atomic thick sheets of covalently sp^2-bonded carbon atoms in a hexagonal arrangement [7]. Its choice as a filler is due to its excellent electrical and mechanical properties, which are significantly better than other inorganic filler materials. Among different 1-D filler materials, sepiolite is a non-swelling natural, abundant and low-cost clay having theoretical unit cell formula $(Si_{12}Mg_8O_{30})(OH)_4(OH_2)_4{\cdot}8H_2O$. It belongs to the class of layered silicates comprising of hydrated magnesium silicate having micro-fibrous needle like morphology. As compared to phyllosilicates (BET surface area 82±1 m^2/g), it has high surface area (BET 374±7 m^2/g) and contains large amount of silanol groups present at the surface [8]. The silanol groups have been proven to be effective as sepiolite acts as reinforcing and compatibilizing agent for different polymer matrices owing to coupling reactions. In a related study, Chen *et al.* [9] prepared well dispersed sepiolite/polyurethane nanocomposite using in-situ polymerization technique. Sepiolite was observed to improve the tensile strength and modulus of polyurethane due to the strong interfacial interaction between the organic and the inorganic phases [9].

Many biofillers have also been developed for enhancement of polymer properties as well as to reduce cost. For example, Ghazanfari *et al.* [10] reported the incorporation of date seed flour in HDPE. The melt flow index of the composites was decreased, whereas the thermal conductivity of the composites increased. The advantages of using biofillers over conventional fillers are their low cost, recyclability, biodegradability, adequate physical and mechanical properties, etc. Another example of such composites reporting the use of biofillers include LDPE incorporated with oyster shell hybirds, which exhibited effective compatibility and good oxidative properties [11]. Similarly, HDPE/rice husk composites have been fabricated in order to add biodegradability to HDPE and also exhibited improved tensile and flexural properties [12]. Mechanically removed natural polysaccharides such as psyllium husk seeds of *Plantago ovata Forsk* are also known for use in many applications as binding, thickening, emulsifying, gelling and controlled release agent [13,14]. These are also known for usefulness in numerous medicinal applications. Composites of psyllium husk with polymers have also been reported in literature, however, only a few studies are available.

In recent years, the use of hybrid fillers for generating composites has also gained considerable research attention. Hybrid filler systems contain two or more different types of incorporated fillers in order to achieve synergistic effect on the properties associated with each phase. In a study reporting the use of hybrid fillers, it was observed that the incorporation of hybrid filler in PE, using different combination of aluminum nitride (AlN), wollastonite, silicon carbide whisker (SiC) and boron nitride (BN), enhanced the thermal conductivity of the composite system [15]. The enhancement of thermal conductivity was owing to the presence of fibrous filler acting as an interactive agent between the spherical shaped filler, hence facilitating conduction [15]. Similarly, using the same concept, systems of carbon nanotube/graphite nano-platelet/epoxy [16] and carbon nanotube/silica/carbon black/EPDM rubber [17] enhanced the thermal and electrical conductivity of filled composites. Further, the use of biofiller additionally within the conventional polymer composite systems can also enhance the biodegradability of the system without sacrificing its mechanical or thermal properties. For instance, Gwon *et al.* [18] reported the hybridization of mineral fillers with wood fibers in polyolefin matrices.

In the current study, further efforts have been made to gain insights into the various aspects related to the composites of polyethylene with a variety of 0-D, 1-D and 2-D fillers as well as their hybrids. Firstly, the composites of polyethylene with increasing fraction of psyllium husk were analyzed for the impact of biofiller on the composite properties. This was followed by a comparison in the performance of psyllium husk (0-D), sepiolite (1-D) and thermally reduced graphene (2-D) containing composites. Finally, the physically mixed hybrid fillers in the combinations of graphene/sepiolite, husk/graphene, husk/sepiolite as well as husk/graphene/sepiolite were also incorporated in polyethylene and composite performance was characterized. Figure 6.1 depicts the unit cell/structural representation of sepiolite, thermally reduced graphene and psyllium husk. The complex structure of psyllium husk was described by Fischer *et al.*, Edwards *et al.*, and Marlett & Fischer, among other authors [13,19-22]. The main part is a complex heteroxylan having monosaccharides, arabinose and xylose, collectively called arabinoxylan. It is highly branched non-starch polysaccharide composed of b-(1,4)-linked xylopyranose residues, arabinofuranose and xylopyranose residues, rhamnose and galacturonic acid [21].

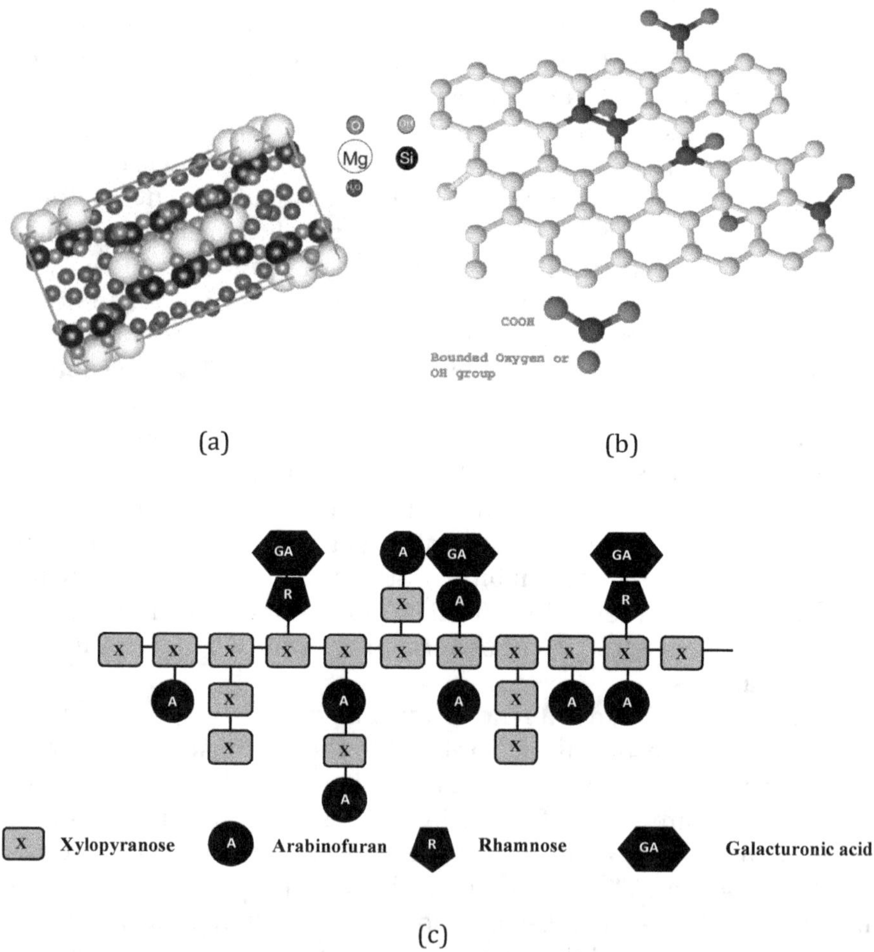

Figure 6.1 Structural representation of (a) sepiolite, (b) graphene and (c) psyllium seed husk.

6.2 Experimental

6.2.1 Materials

HDPE BB2581 was supplied by Abu Dhabi Polymers Company Limited (Borouge), UAE. It was in the form of white pellets with specific gravity of 0.958. The polymer was used as obtained. Thermally reduced graphene was prepared through thermal exfoliation of precursor graphite oxide [23] using modified Hummer's method [24], as reported earlier [6]. A short description for the preparation of graphene is as follows: 5 g of graphite powder was added with 125 ml concentrated H_2SO_4, to which 2.5 g of $NaNO_3$ was subsequently added. After stirring the mixture for 30 min while keeping in an ice-bath (5 °C), 15 g of $KMnO_4$ was added and the temperature was raised gradually to 35 °C. After stirring for 2 h, deionized water was added and the temperature was raised to 100 °C. Subsequently, the mixture was quenched and diluted by pouring it into 1.5 l deionized water. This was followed by addition of 30% H_2O_2 until the evolution of bubbles of hydrogen stopped. The solution was then filtered and the residues were dispersed in 2 l deionized water and were added with dilute HCl (6%) (2 l) to remove the SO_4^{2-} ions. The dispersion was filtered and the filtrate was analyzed for SO_4^{2-} and Cl^- ions using $BaSO_4$ and $AgNO_3$, respectively. The washed graphite oxide was dried under vacuum at 60 °C for 24 h. Graphene was generated via thermal exfoliation of graphite oxide by placing 1 g in a long quartz tube with 25 mm internal diameter and sealed at one end. The sample was flushed with nitrogen, followed by insertion of the tube in a tube furnace preheated to 1050 °C. The tube was held in the furnace for 30 s. The confirmation of graphene formation was achieved through Raman spectroscopy, microscopy and X-ray diffraction. Sepiolite was procured from Sigma Aldrich, whereas psyllium seed husk was procured locally.

6.2.2 Generation of Composites

To form composites, the fillers and polymer pellets were physically mixed before adding to the compounder. One step melt mixing of polymer and fillers was carried out at 190 °C using a mini twin screw extruder (MiniLab HAAKE Rheomex CTW5, Germany). The screw length and screw diameter were 109.5 mm and 5/14 mm conical respectively. Batch size of 5 g was used in all cases and shear mixing was performed for seven minutes at 80 rpm. Pure HDPE was also processed by subjecting it to similar shear and thermal conditions. Table 6.1 also describes the compositions of different composites. Disc- and rectangular bar-shaped test specimens were injection molded using a mini injection molding machine (HAAKE MiniJet, Germany) at a processing temperature of 190 °C. The injection pressure was 700 bar for 6 s, whereas holding pressure was 400 bar for 3 s. The temperature of the mold was kept at 50°C.

Table 6.1 Composition of the composites (in wt%)

Notation	PE content (%)	Husk content (%)	Graphene content (%)	Sepiolite content (%)
PE	100	-	-	-
PE+2.5H	97.5	2.5	-	-
PE+5H	95	5	-	-
PE+10H	90	10	-	-
PE+5G	95	-	5	-
PE+5S	95	-	-	5
PE+2.5G+2.5S	95	-	2.5	2.5
PE+2.5H+5G	92.5	2.5	5	-
PE+2.5H+5S	92.5	2.5	-	5
PE+2.5H+2.5G+2.5S	92.5	2.5	2.5	2.5

6.2.3 Material Characterization

Calorimetric properties of composites were recorded on a Perkin-Elmer Pyris-1 differential scanning calorimeter under nitrogen atmosphere. The scans were obtained from 30 to 190 °C at a heating rate of 10 °C/min (second heating runs). The heat enthalpies were measured with an error of ±0.1% and were confirmed by repeating the runs. Thermal degradation properties of the composites were recorded using Netzsch thermogravimetric analyzer (TGA). Nitrogen was used as a carrier gas and the scans were obtained from 50 to 700 °C at a heating rate of 20 °C/min.

Rheological properties such as storage modulus (G'), loss modulus (G"), viscosity and elasticity of the composites were measured using AR 2000 Rheometer from TA Instruments. The measuring temperature and gap opening were 185 °C and 1.6 mm respectively. Disc-shaped samples of diameter 25 mm and thickness 2 mm were used.

Strain sweeps were recorded at ω 1 rad/s from 0.1 to 100% strain. The shear stability of the samples was observed up to 10% strain. Hence, as a safe approach, frequency sweeps (dynamic testing) were recorded at 4% strain from ω 0.1-100 rad/s. DMA (dynamic mechanical analyzer) RSA3 from TA Instruments was used to study the change in modulus with temperature (25-100 °C). Dynamic temperature ramp test was used at a frequency of 5 rad/s, ramp rate 3 °C/min, and 1% strain for all samples.

Ultrathin sections (30-90 nm) as well as block faces of the composite samples were obtained using a Leica Ultracut E microtome (Leica, Austria) equipped with a diamond knife (Diatome, Switzerland) at -120 °C. The so-obtained sections were collected on 400 mesh formvar electron microscopy grids (coated with holey carbon film), and were examined in a Philips CM 20 (Philips/FEI, Germany) electron microscope at 200 kV at room temperature without additional staining. TEM image processing was performed using DigitalMicrograph software (Gatan, USA). All energy-dispersive X-ray spectroscopy (EDXS) experiments on the composites were performed in STEM mode with a probe current of 20 nA and a beam diameter of 100 nm. EDXS spectra were collected using a HPGe detector (solid angle 0.13 sr). For the calculation of elemental distribution images, jump ratio method was used and EDXS spectra have been visualized using DigitalMicrograph software (Gatan, USA). The block faces of specimens after cryo-ultramicrotomy were also investigated at ambient conditions using ZEISS Axioplan light microscope (Zeiss, Germany) equipped with Zeiss Axio Cam ICc 1 CCD camera in reflected polarized light.

6.3 Results and Discussion

The current study aims to achieve multiple objectives. The first objective is to demonstrate the use of psyllium husk as a useful filler for polyethylene to enhance its properties as well as to reduce cost. Secondly, the performance of the different 0-D, 1-D and 2-D fillers has been compared in achieving enhancement of mechanical, thermal and rheological performance of polyethylene. Finally, the binary and ternary hybrids of 0-D, 1-D and 2-D fillers have been incorporated in polyethylene in order to gain insights into the effect of these combinations on the composite behavior. Along with the conventionally used filler systems, the use of hybrid fillers (both physically mixed and chemically interacting hybrid fillers) has gained recent research interest. Specifically, the use of fillers of different dimensions can lead to synergistic effects on composite properties. In addition, the hybrid fillers are advantageous as one of the components of the hybrid filler mixture may have the main role of reducing the material's cost, while still retaining (or enhancing) the composite properties achieved through the other filler.

Figure 6.2 demonstrates the TGA thermograms of PE composites as a function of psyllium husk content. The husk exhibited degradation around 325 °C and the magnitude of the degradation peak increased with increasing husk content in the composites. It indicated that the filler resulted in lowering of thermal stability of PE, however, the degradation of the filler temperatures >300 °C indicated that the composites were still useful for high temperature applications as well as had stability towards high temperature processing. It was also confirmed by carrying out the dynamic TGA analysis, where the sample weight was recorded as a function of time at the compounding temperature. No significant weight loss was observed for the test duration equivalent to the melt compounding time, thus, confirming the thermal stability of the composites for processing. The thermal stability was also retained even after subjecting the composites to few processing cycles.

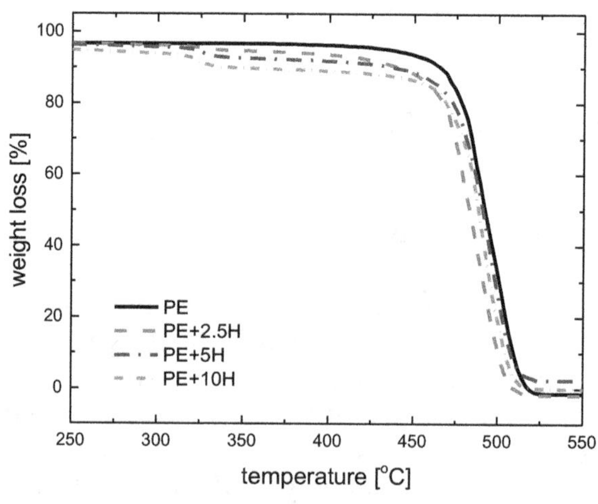

Figure 6.2 TGA thermograms of the PE composites as a function of psyllium husk content.

Table 6.2 and Figure 6.3 show the calorimetric properties of polyethylene composites as a function of husk content. Pure polyethylene

Table 6.2 Calorimetric properties of polyethylene composites with husk

Polymer/Composite	Peak melting temperature, °C	Melt enthalpy, J/g	Peak crystallization temperature, °C
PE	135.0	154	117.0
PE+2.5H	135.4	162	115.2
PE+5H	137.4	161	115.6
PE+10H	137.6	160	115.1

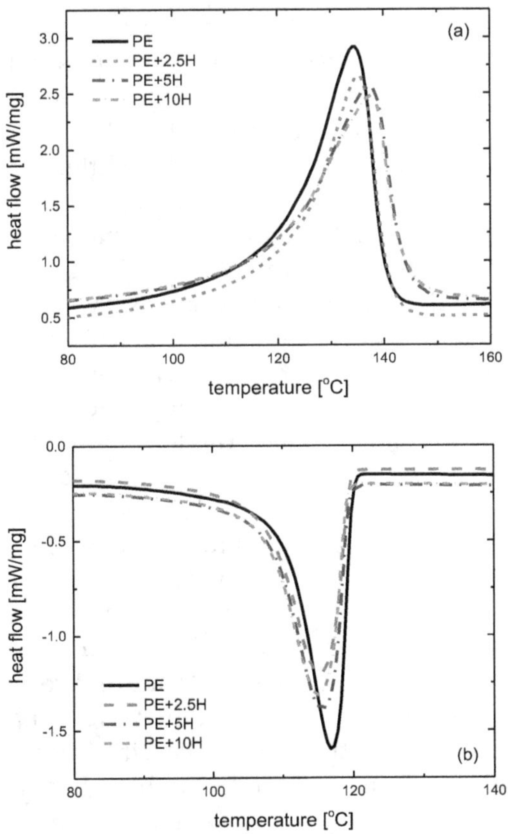

Figure 6.3 DSC thermograms of the composites as a function of psyllium husk content.

exhibited the peak melting point of 135 °C, which was increased slightly on the addition of filler. For instance, the composite with 10% husk had the peak melting temperature of 137.6 °C. It indicated that the crystallization of the polymer was affected by the incorporation of the filler. The melt enthalpy of the polymer was also observed to increase in the range of 4-5% in the composites, thus, confirming the effect of filler on the polymer crystallization and degree of crystallinity. However, the filler did not exhibit any nucleation effect, as the peak crystallization temperature (T_c) was slightly reduced on the addition of filler. Also, the filler content did not have any impact on T_c.

Figure 6.4 presents the rheological analysis of the HDPE composites as a function of psyllium husk content. As evident in Figure 6.4a, the addition of filler exhibited sharp increase in the storage modulus at lower angular frequency. The composite with 10% filler content had the highest storage modulus values. For instance, the storage modulus of the composite with 10% husk was 62940 Pa, in comparison with 39775 Pa for the composites with 2.5% husk. The rate of increment of modulus was observed to slow down at higher angular frequency due to shear thinning effect. The difference between the different composites was retained even at higher frequency, indicating frequency dependent behavior of the composites. The loss modulus of the composites indicated only a slight change till a frequency of 1 rad/s. Subsequently, the modulus was observed to increase exponentially. Similar to the storage modulus, the curves for the different composites did not merge with each other even at high angular frequency, thus, indicating the frequency dependent behavior. For the complete angular frequency range, the loss modulus of the composites was lower in magnitude than the storage modulus, thus, indicating the dominance of the elastic behavior in the composites. Figures 6.4c and d also show the viscosity and elasticity of the samples as a function of angular frequency. Contrary to the modulus, these quantities were observed to decrease on increasing angular frequency. Due to the increase in frequency, polymer structure shows temporary network of entanglements which leads to more flexibility (lower viscosity). In the composites, this allowed more deformation energy to be stored resulting in elastic dominance which leads to increased modulus. Thus, it simultaneously reduced the contribution of lost deformation energy and hence viscous behavior. The viscosity of the composites was almost same at lower frequency whereas higher frequency exhibited some differences depending on the filler content. Also, the slope of decrease in viscosity decreased after a frequency of ~10 rad/s. On the contrary, the elasticity of the composites was a function of filler content at lower frequency, but did not show any difference on increasing the frequency. Similar to modulus, the elasticity of the composites was higher in magnitude than the viscosity for the whole frequency range. It indicated that the psyllium husk was not a passive filler added with an aim to reduce the cost, and contributed significantly to the enhancement of polyethylene properties.

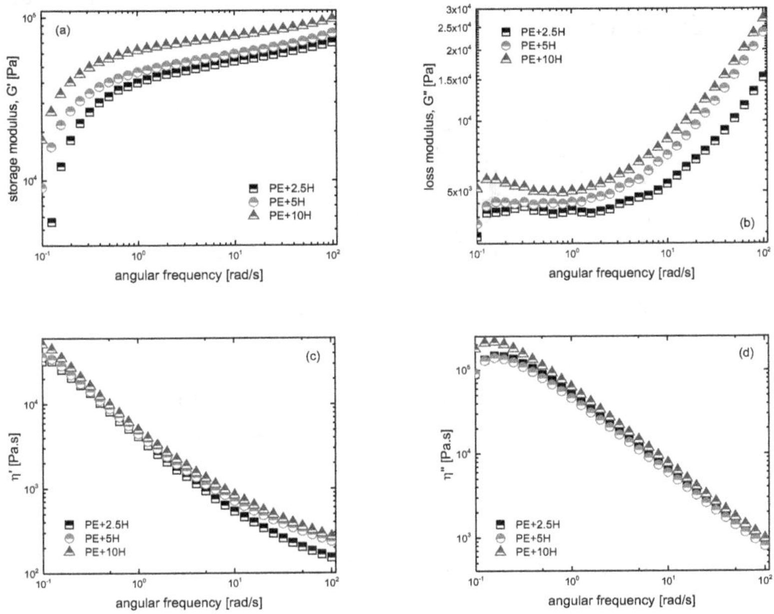

Figure 6.4 Rheological analysis of the composites as a function of psyllium husk content: (a) storage modulus, (b) loss modulus, (c) viscosity, η' and (d) elasticity, η''. The angular frequency range from 0.1 to 100 rad/s was used.

Figure 6.5 also presents the dynamic mechanical properties of the composites as a function of filler content as well as temperature. Similar to the modulus in shear rheology, the storage modulus was observed to increase with filler fraction. There was a gradual drop in modulus with temperature, but the composites exhibited better response than the pure polymer. For instance, the modulus of the 10% husk composite at 50 °C and 70 °C was 5.67E8 Pa and 3.17E8 Pa, in comparison with 3.28E8 Pa and 1.89E8 Pa for pure polymer, respectively. This indicated that the addition of filler enhanced the high temperature performance of polyethylene. Though the incorporation of psyllium husk powder was observed to reduce the thermal degradation temperature of polyethylene, but useful performance at service temperatures was, nonetheless, enhanced. It also confirmed the role of psyllium husk as a reinforcing filler. The loss modulus of the composites was also observed to increase with filler fraction, but was lower than the storage modulus for the complete temperature range. In addition, the drop of loss modulus with temperature also exhibited two steps, first a lower rate of decrease till ~45 °C, followed by faster rate of modulus reduction.

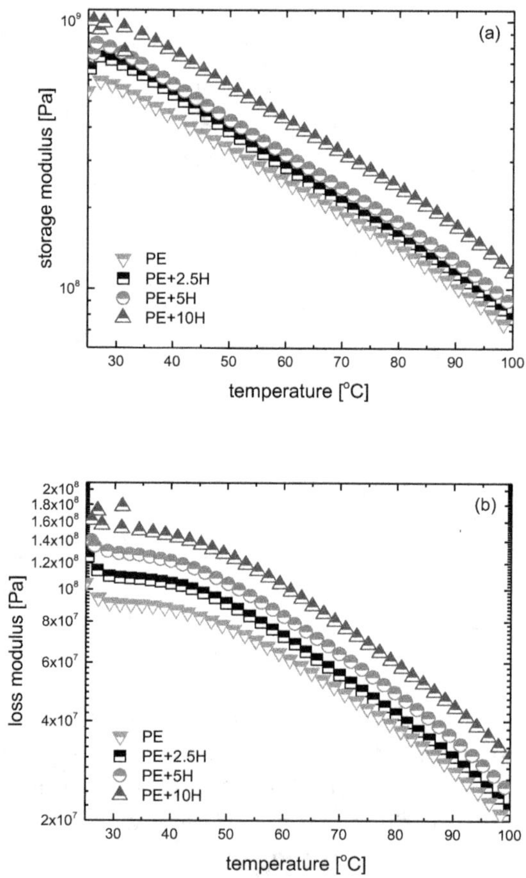

Figure 6.5 Dynamic mechanical analysis of the composites as a function of psyllium husk content and temperature: (a) storage modulus and (b) loss modulus.

The mechanical and rheological performance of the PE composites as a function of husk content confirmed significant reinforcement in the performance on the addition of filler. Further, the impact of husk on the properties of polyethylene was compared with other high performance fillers like 2-D graphene platelets and 1-D sepiolite nanofibers. The extent of filler in the composites was fixed to 5 wt%. In addition, the effect of 1:1 graphene/sepiolite hybrid filler on the composite properties was also studied. All the composites were processed using the same melt mixing parameters so as to relate any property differences to the effect of filler. Figure 6.6 shows the TGA thermograms of the composites. As mentioned earlier, the addition of 5% husk resulted in the thermal degradation peak around 325 °C. Sepiolite incorporation did not impact the onset degradation temperature of the polymer, but the peak degradation temperature was slightly decreased. Addition of graphene resulted in an increase in both onset and peak degradation temperature of polyethylene, indicating better thermal resistance provided by 2-D graphene platelets due to effective heat transfer from the polymer chains to the platelets. Higher surface area and high aspect ratio of graphene platelets would have resulted in enhanced interfacial contact between the polymer chains and the filler particles, thus, leading to enhanced thermal performance. The composite containing the hybrid of graphene and sepiolite exhibited similar degradation pattern as polyethylene, which was intermediate between the behavior of graphene and sepiolite, when added in the composite separately. Similar to the findings from the dynamic TGA analysis mentioned earlier, the weight loss with time at the compounding temperature of 190 °C was not significant, thus, confirming the stability of the composites for high temperature processing as well as applications.

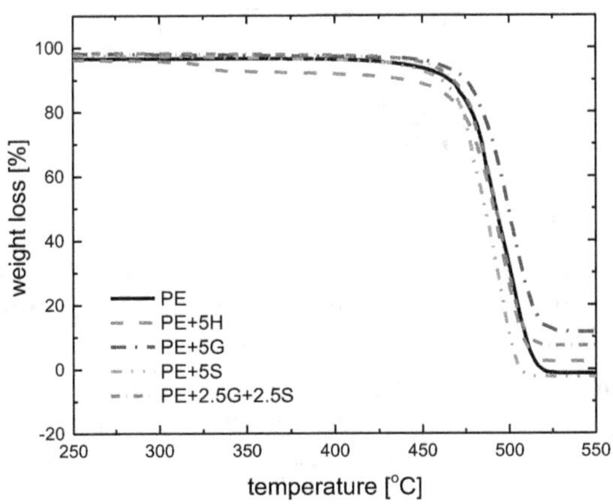

Figure 6.6 TGA thermograms of the composites with 5% filler content.

Table 6.3 and Figure 6.7 demonstrate the calorimetric behavior of the composites with 5% content of husk, graphene, sepiolite and graphene/sepiolite hybrid. As reported earlier, the incorporation of husk resulted in enhanced peak melting temperature as well as melt enthalpy as compared to pure PE, though the filler did not exhibit any nucleation effect. Addition of sepiolite to PE also enhanced the peak melting temperature, however, resulted in slight decrease in the enthalpy indicating the hindrance to crystallization in the presence of filler. Graphene also exhibited decrease in the melt enthalpy, along with peak melting temperature, further confirming the constraint to the polymer chain movement and crystallization due to the 2-D platelets. The hybrid filler exhibited even more pronounced decrease in the degree of crystallization as it was observed to decrease to 139 J/g (normalized to polymer amount in the composite), in comparison with 154 J/g for pure PE. It indicated that the hybrid filler caused more hindrance to polymer chains probably due to specific distribution of the filler particles in the matrix. Presence of both 1-D and 2-D filler particles can also be expected to result in higher degree of anisotropy in the composites. The peak melting point was also deceased the maximum in the composite with hybrid filler.

Table 6.3 Calorimetric properties of the composites with 5% filler content

Polymer/ Composite	Peak melting temperature, °C	Melt enthalpy, J/g	Peak crystallization temperature, °C
PE	135.0	154	117.0
PE+5H	137.4	161	115.6
PE+5G	135.3	149	115.5
PE+5S	137.3	150	115.2
PE+2.5G+2.5S	133.7	139	115.1

The rheological performance of the composites is also compared in Figure 6.8. As observed earlier, the storage modulus enhanced exponentially at the lower frequency, followed by leveling off at the higher frequency values. The storage modulus was observed to be maximum in the case of graphene composite, indicating better stress transfer from the polymer chains to the 2-D platelets, also supporting the earlier observation of better heat transfer due to the dispersed graphene platelets. The storage modulus in the sepiolite containing composite was marginally better than husk as filler, which indicated

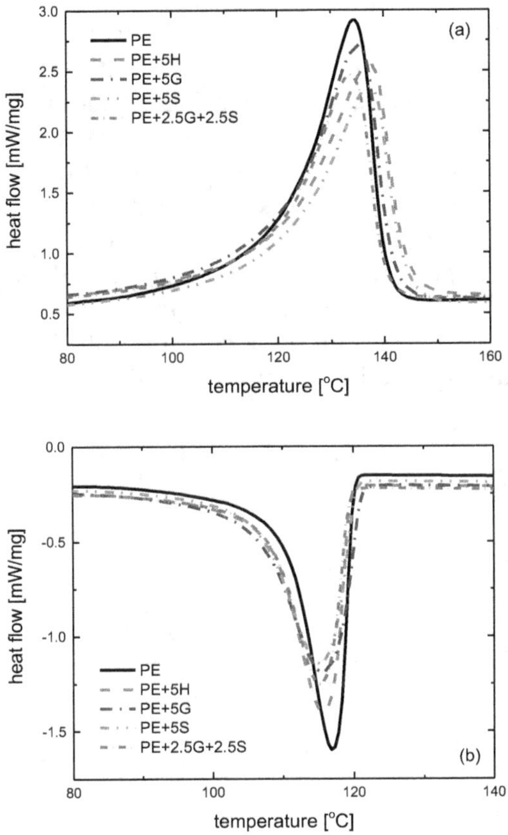

Figure 6.7 DSC thermograms of the composites with 5% filler content.

lower degree of reinforcement of 1-D sepiolite needles as compared to 2-D graphene platelets. Only at higher angular frequency, the resistance of the sepiolite composite was observed to be better than husk based composite. The hybrid filler containing composites had similar modulus as the graphene composite, except at very low and high angular frequency. It demonstrated that higher degree of performance could be achieved in this case as compared to sepiolite alone. The loss modulus of the hybrid filler composite was also observed in between the moduli of graphene and sepiolite containing composites. In addition, the loss modulus of the composites was lower in magnitude than the storage modulus due to dominant elastic character. As earlier, the loss modulus also exhibited slower increase at lower frequencies, followed by higher rate of increment at higher angular frequency values. The viscosity of the composites (as shown in Figure

6.8c) also exhibited similar behavior as loss modulus, with graphene composites demonstrating maximum enhancement in viscosity followed by the composites with hybrid filler and sepiolite. However, it should be noted that even though significant increment in the viscosity was observed in the composites, these were still processable the same way as the pure polymer. The elasticity of the composites was higher in magnitude than the viscosity, similar to the observation made earlier for storage and loss modulus. These findings, thus, indicate that the incorporation of graphene and graphene/sepiolite hybrid filler resulted in highest extent of property increase, however, the organic filler also resulted in considerable improvement and should be further explored as a component of the hybrid filler system.

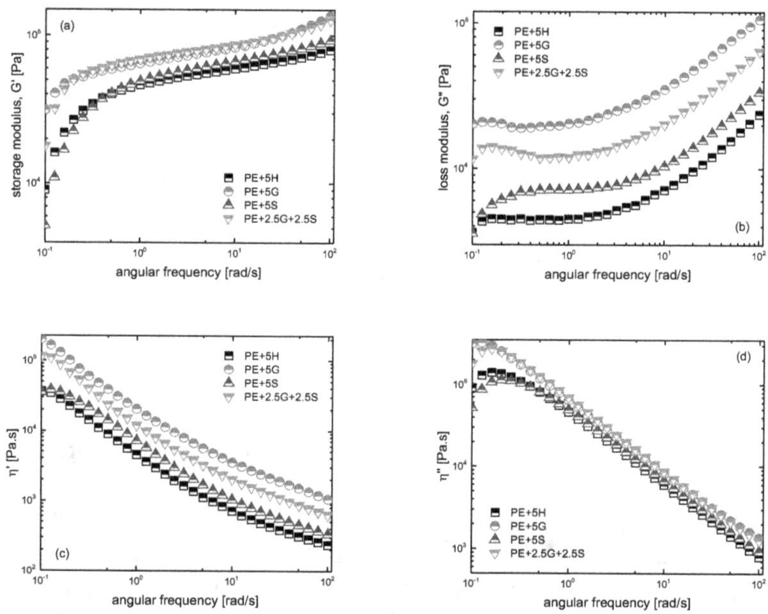

Figure 6.8 Rheological analysis of the 5% filler composites as a function of angular frequency: (a) storage modulus, (b) loss modulus, (c) viscosity, η' and (d) elasticity, η". The angular frequency range was 0.1 to 100 rad/s.

Figure 6.9 also demonstrates the dynamic mechanical analysis of the composite with 5% husk, graphene, sepiolite and graphene/sepiolite hybrid. Similar trends were observed for storage modulus, as observed earlier in rheological analysis. The graphene composite exhibited higher modulus over the whole temperature range due to effective stress transfer. The other composites though exhibited lower modulus values, however, the difference was not significant. Overall, though the modulus decreased by an order of magnitude at 100 °C, however, the composites still exhibited higher modulus, thus, exhibiting improved property profiles. The loss modulus of the composites exhibited similar trend, especially at higher temperatures. The elastic nature of the composites was also confirmed as the loss modulus of these composites was almost an order of magnitude lower than the storage modulus.

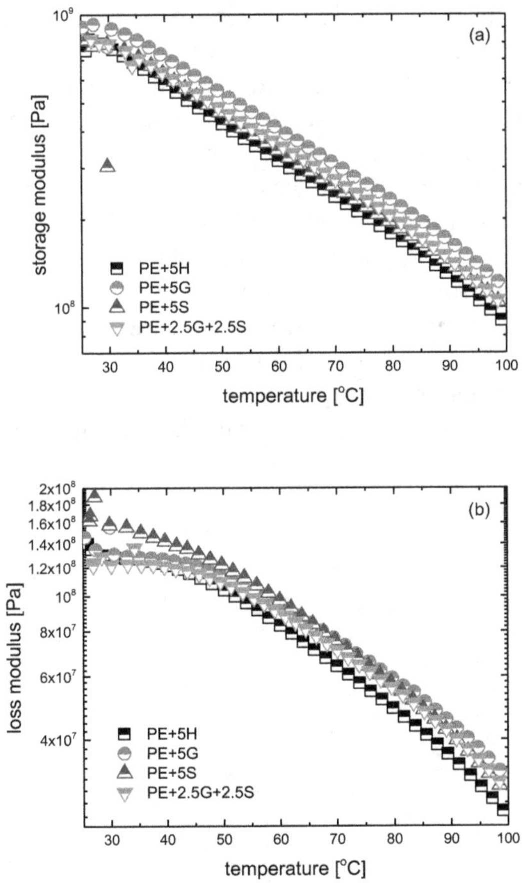

Figure 6.9 Dynamic mechanical analysis of the 5% filler (husk, graphene, sepiolite, graphene/sepiolite hybrid) composites as a function of temperature.

Finally, the effect of hybrid fillers was explored further by incorporating hybrids of husk/graphene, husk/sepiolite as well as husk/graphene/sepiolite. For comparison, the extent of filler was fixed at 7.5 wt% and same processing protocol was used to generate the composites so as to relate the observed morphology and properties solely to the filler phase in the polymer matrix. Figure 6.10 presents the TGA thermograms of the composites in comparison with pure polymer. As observed earlier, the composite with husk/sepiolite hybrid exhibited a degradation peak around 325 °C which was attributed to the degradation of husk. In addition, the composite had reduction in onset and peak degradation temperature of PE phase, which was attributed to the degradation of sepiolite. Due to better thermal resistance, the hybrids containing graphene did not show any negative impact on the onset and peak degradation temperature of PE. In addition, the degradation owing to husk phase was also largely improved in the case of husk/graphene hybrid composite. These findings re-confirmed the earlier observations of slight reduction in the thermal stability of the polymer with husk and sepiolite fillers, whereas enhancement in the thermal performance in the presence of graphene. However, the composites were still suitable for high temperature applications and processing cycles. The thermal stability was maintained even after multiple processing cycles at compounding/processing temperature. As mentioned earlier, the thermal stability of the composites was also confirmed by carrying out the dynamic TGA analysis.

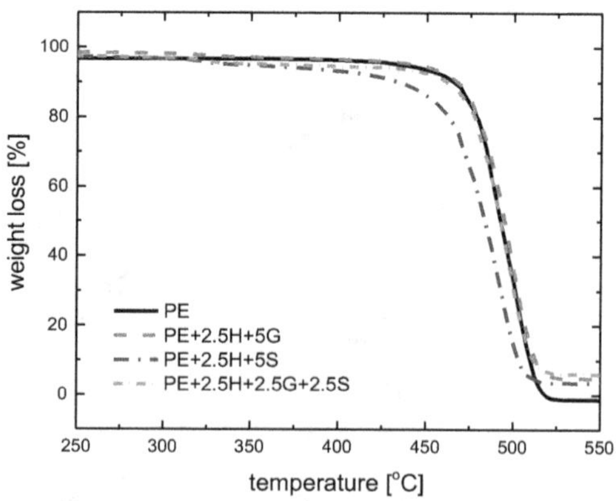

Figure 6.10 TGA thermograms of the composites with 7.5 wt% of hybrid fillers.

The calorimetric response of the composites with hybrid fillers is demonstrated in Table 6.4 and Figure 6.11. The composite with husk/graphene hybrid exhibited a significant reduction in the melt enthalpy to 138 J/g, as compared to 154 J/g for pure PE. Similar reduction was also observed for the composite with hybrid of graphene/sepiolite, indicating hindrance to the polymer chain movement in the presence of graphene. Relatively lower reduction in melt enthalpy was observed for the composite with husk/sepiolite composite, as also observed earlier for sepiolite composite in comparison with graphene. It should also be noted that the enthalpy of the graphene and sepiolite composites was lower when incorporated with husk, as compared to the composites with graphene and sepiolite alone. Though incorporation of husk alone was also observed earlier to enhance the melt enthalpy, but in the hybrid form with graphene and sepiolite, this effect was observed to be opposite, thus indicating a specific morphology of these hybrid fillers in the polymer matrix, as suggested earlier. The composite with hybrid of husk/graphene/sepiolite also exhibited significantly reduced melt enthalpy predominantly due to the presence of graphene. Thus, it indicated that the effect of the hybrid fillers on the crystallization of PE was not directly correlated with the effect of the individual fillers. Specific changes to the composite morphology in the presence of hybrid fillers would have resulted in such effect, even though the fillers were physically mixed with each other and were not expected to have any chemical interaction with each other. Presence of particles of different dimensions are generally envisaged to generate a more stabilized dispersion and intermixing with the polymer. The similar enthalpy was observed in the samples processed for different cycles indicating that the composite morphology was stable.

Similar to the earlier case, the rheological characterization of the composites with hybrid filler systems (husk/graphene, husk/sepiolite and husk/graphene/sepiolite) was also carried out, as shown in Figure 6.12. The composite with husk/sepiolite hybrid exhibited the

Table 6.4 Calorimetric properties of the composites with 7.5 wt% of hybrid fillers

Polymer/Composite	Peak melting temperature, °C	Melt enthalpy, J/g	Peak crystallization temperature, °C
PE	135.0	154	117.0
PE+2.5H+5G	135.5	138	115.2
PE+2.5H+5S	136.9	146	115.0
PE+2.5H+2.5G+2.5S	135.8	137	115.4

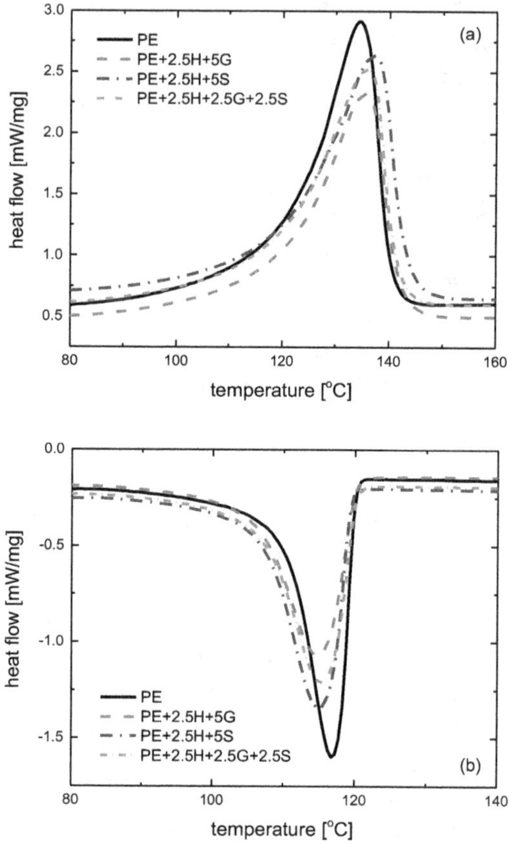

Figure 6.11 DSC thermograms of the composites with 7.5 wt% of hybrid fillers.

lowest modulus as compared to the composites with husk/graphene and husk/graphene/sepiolite hybrid fillers. Specifically, the composite with husk/graphene/sepiolite hybrid exhibited a synergistic enhancement in the modulus as it had higher modulus than both husk/graphene and husk/sepiolite containing composites. Moreover, the incorporation of husk in addition to graphene in the composites led to the increment in the modulus. For instance, the storage modulus of the PE+5G composite at 1 rad/s was 64580 Pa, which was increased to 83630 Pa for the PE+2.5H+5G composite. It indicated the addition of husk did not deteriorate mechanical performance of graphene, and, in fact, enhanced it. The modulus of the sepiolite composite was observed to remain unchanged with the addition of husk, thus, reconfirming that husk did not negatively impact the properties

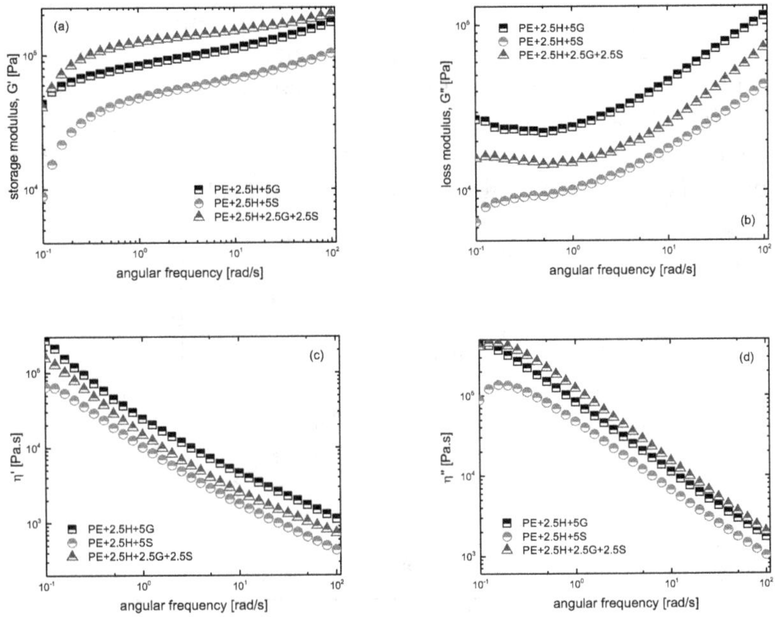

Figure 6.12 Rheological analysis of the 7.5 wt% hybrid filler composites as a function of angular frequency: (a) storage modulus, (b) loss modulus, (c) viscosity, η' and (d) elasticity, η". The angular frequency range was 0.1 to 100 rad/s.

and can be useful in this case to reduce the material cost. Similarly, the modulus of the husk/graphene/sepiolite ternary hybrid filler composite was much higher than the graphene/sepiolite binary hybrid filler composite. For all composites, the modulus had much faster rate of enhancement at the lower frequency, followed by slower rate of modulus increase, as also observed earlier. The loss modulus of the ternary hybrid filler composites was observed to in between the husk/graphene and husk/sepiolite containing composites. As also observed earlier, the loss modulus exhibited faster rate of increment only after a frequency of 1 rad/s. In addition, the loss modulus of the composites was smaller in magnitude than the storage modulus over the whole angular frequency range due to the elastic nature of the composites. The viscosity of the composites also exhibited similar behavior as the loss modulus. Similar to the other composites, the elasticity of the composites was higher in magnitude than the viscosity, thus, confirming the elastic nature of the composites. Similar to storage modulus, the elasticity of the ternary filler composite was the highest among the composites with hybrid fillers as well as the composites with 5% graphene or sepiolite. These findings confirmed the positive impact of husk on the composites properties as no reduction in the rheological performance was observed when husk was added along with graphene and sepiolite.

The dynamic mechanical properties of the hybrid filler containing composites is demonstrated in Figure 6.13. The addition of husk with

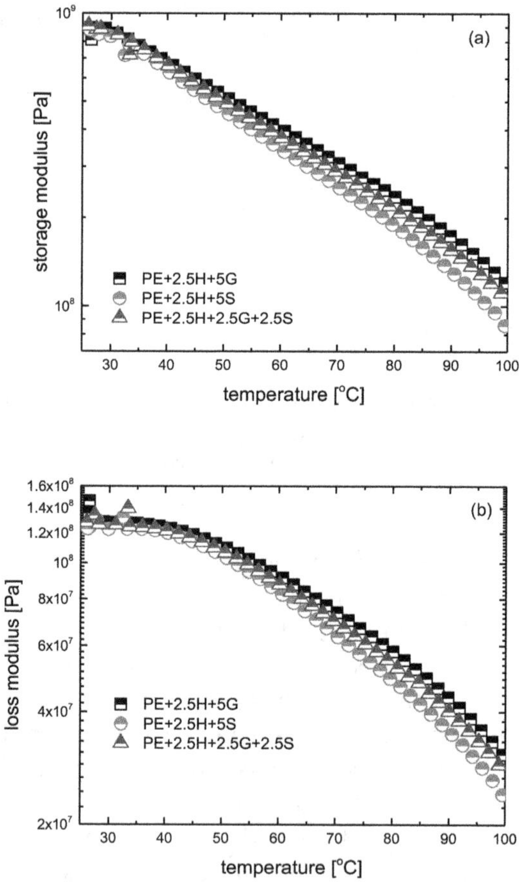

Figure 6.13 Dynamic mechanical analysis of the 7.5 wt% hybrid filler composites as a function of temperature.

other fillers was observed to retain the modulus to the similar values as observed in Figure 6.9 and no negative impact on the mechanical performance was observed over the whole temperature range. Thus, even though the modulus did not enhance the way as observed in the

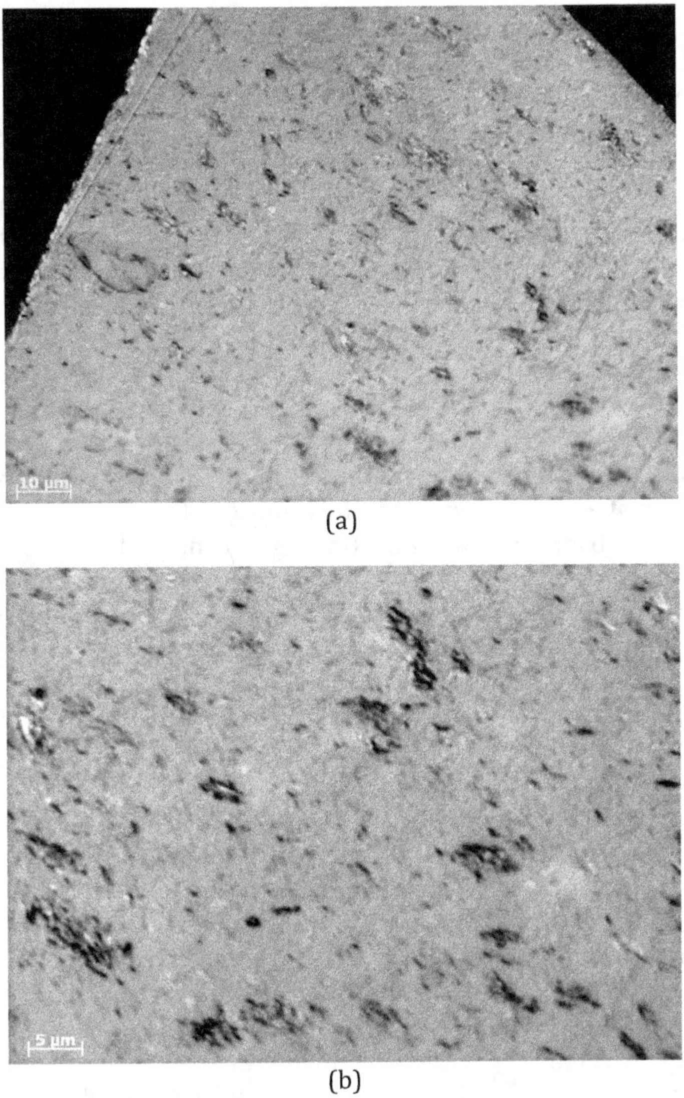

Figure 6.14 Optical micrographs of PE+2.5H+2.5G+2.5S composite.

rheological analysis, however, the overall observed behavior of the husk containing composites confirmed that the properties of the composites with graphene or sepiolite could be retained on the addition of husk, which would, thus, provide cost efficiency due to much lower price of husk. Overall, the composite with husk/graphene exhibited the highest modulus followed by the ternary hybrid filler and husk/sepiolite composites. The modulus of composites decreased gradually with temperature and had an order of magnitude reduction in the storage modulus at 100 °C, but the composites still had appreciable mechanical resistance. Similar to the storage modulus, the loss modulus of the hybrid composites also exhibited similar values as observed earlier in Figure 6.9. As is evident, the husk/graphene reinforced composite exhibited the highest magnitude. The loss modulus of the composites was also significantly lower in magnitude than the storage modulus due to the dominant elastic nature of the composites, which has been observed for all the composites generated in the study.

Further insights in the morphology of the ternary hybrid filler composite were also gained through microscopy analysis. Figure 6.14 shows the optical micrographs of the composite at two different magnifications. Block face preparation for the microscopy analysis was carried out at -120 °C due to the flexibility of the polymer. No filler pullout was observed during sectioning and block face preparation indicating good degree of filler adhesion with the polymer. The filler distribution was observed to be uniform with no presence of large filler aggregates. It has to be noted that the composites were generated without the addition of commonly used compatibilizers, thus, the good degree of polymer-filler intermixing indicated that the processing parameters used to generate the composites were appropriate.

Figure 6.15 demonstrates the TEM analysis of the PE+2.5H+2.5G+2.5S composite. As evident in Figure 6.15a and c, uniform filler distribution was confirmed. As the fillers were physically mixed with each other during compounding and no thermodynamic interaction among these was expected, therefore, the filler particles were observed to be present as separate phases in the composite morphology. Large aspect ratio platelets of graphene were observed in Figure 6.15c, whereas Figure 6.15b demonstrates the fibrillar morphology of sepiolite. The Si ratio image in Figure 6.15d also confirmed the presence of sepiolite (bright signal). The husk particles could not be observed in the composites probably due to the difficulty in generating phase contrast with the organic PE phase.

EDXS analysis was also carried out on the PE+2.5H+2.5G+2.5S composite, as shown in Figure 6.16. Different locations were chosen on the TEM micrograph and were subsequently analyzed for the elemental analysis. The area marked as 1 in the TEM micrograph indicated the presence of Si and Mg signals, thus, further confirming the presence of sepiolite in the composite. Similarly, the area marked as

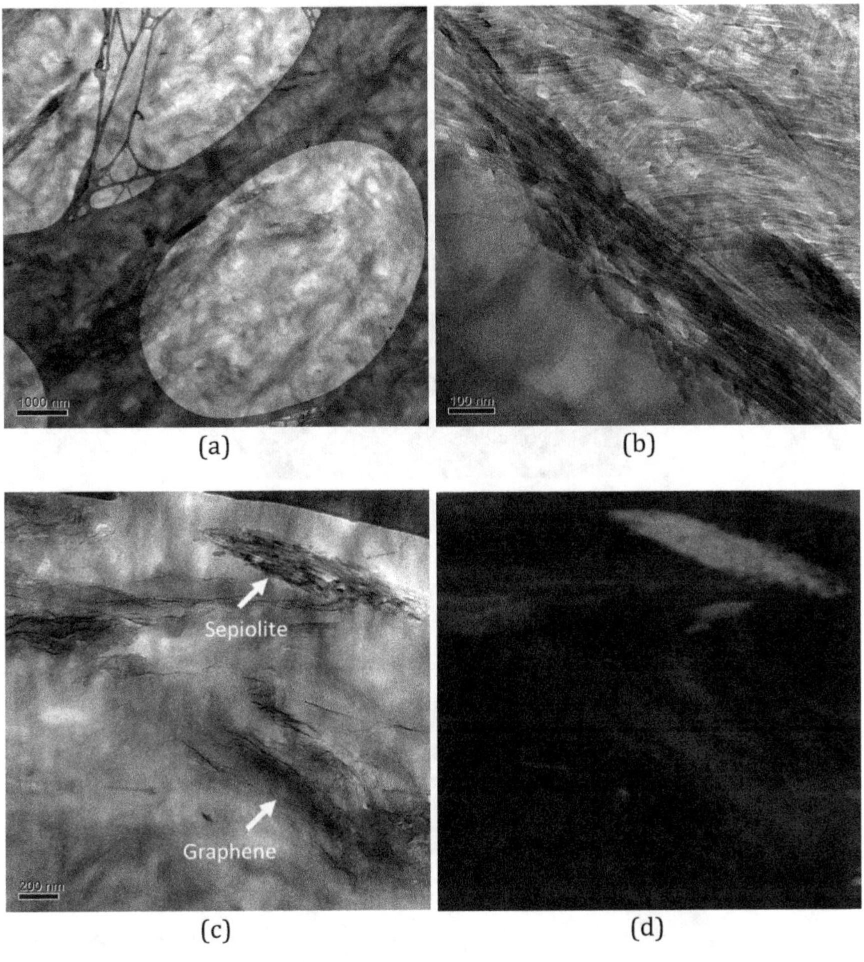

Figure 6.15 Electron microscopy analysis of PE+2.5H+2.5G+2.5S composite: (a) low magnification image, (b) sepiolite distribution, (c) high magnification image depicting the distribution of graphene and sepiolite and (d) corresponding Si ratio image to confirm the presence of sepiolite fibers.

3 on the TEM micrograph exhibited the C signal, due to the presence of graphene. The area marked as 2 on the micrograph exhibited diffuse signals of sepiolite along with C, probably due to the joint presence of sepiolite and graphene due to occasional intermixing. Thus, the hybrid composite exhibited uniform morphology where the three fillers co-existed with each other without leading to any antagonistic effect on the composite properties. These findings also confirmed the

potential of husk as a functional filler which can either enhance or retain the composite properties, along with providing cost efficiency.

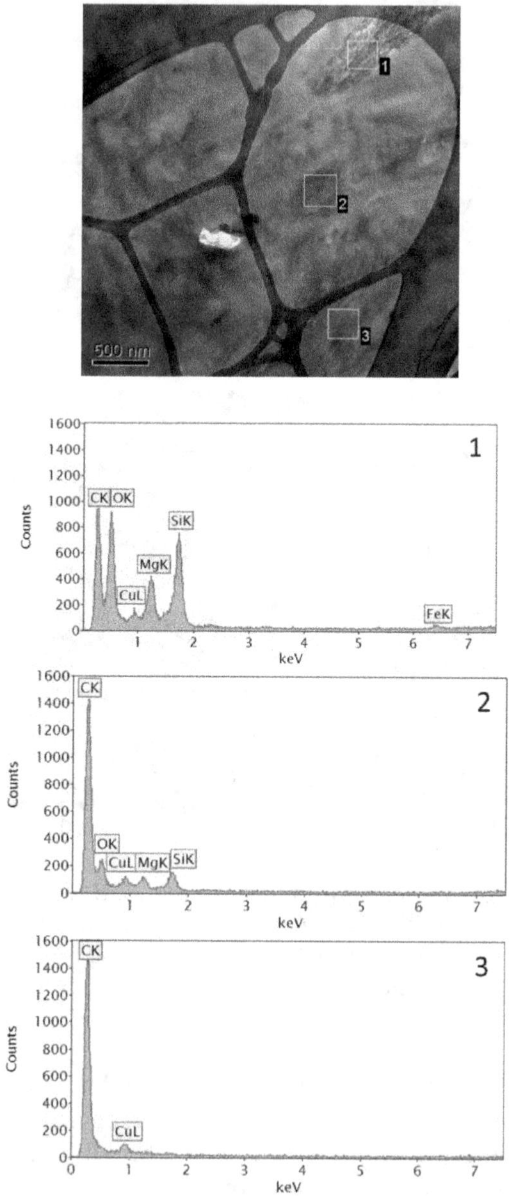

Figure 6.16 (a) TEM image of the PE+2.5H+2.5G+2.5S composite; (c-d) EDXS analysis at the three locations in the composite depicting sepiolite, hybrid filler and graphene (marked at 1, 2 and 3).

6.4 Conclusions

In the current study, the composites of polyethylene were prepared with a variety of 0-D, 1-D and 2-D fillers, along with their hybrid systems in order to analyze their impact on thermal, mechanical and rheological properties. Incorporation of varying amounts of 0-D psyllium husk particles to HDPE led to slight reduction in the thermal stability due to the degradation of husk at 325 °C, however, the composites were still suitable for high temperature applications. The husk particles also enhanced the peak melting temperature and melt enthalpy of polyethylene with increasing filler content, however, the crystallization temperature did not change in the composites. The storage modulus from shear rheology was observed to increase as a function of husk content and was also higher than loss modulus due to dominant elastic nature of the composites. Similarly, the elasticity of the composites was higher than the viscosity due to the elastic nature. The dynamic mechanical analysis also exhibited increase in modulus with increasing filler content, thus, confirming the positive impact of husk on the composite properties. Though the modulus of the composites was observed to decrease by an order of magnitude by 100 °C, however, it was still significant, thus, confirming the enhancement of high temperature performance. On comparing the composites with husk, sepiolite, graphene, and graphene/sepiolite hybrid as fillers, the graphene composite was observed to have the best thermal performance due to efficient heat transfer from the polymer chains to the high aspect ratio 2-D graphene platelets. However, the other filler systems also exhibited composite stability against high temperature processing. The use of graphene also resulted in decreased melt enthalpy and peak melting point of PE due to hindrance to chain movement and crystallization. The storage modulus of both graphene and graphene/sepiolite based composites was observed to be higher than other composites due to effective stress transfer from the polymer chains to the graphene platelets owing to higher degree of interfacial contacts between the polymer and filler phases. The addition of husk particles to graphene, sepiolite and graphene/sepiolite led to further reduction in the melt enthalpy, thus, indicating combined effect of fillers, though the addition of husk alone was observed to increase the melt enthalpy of the polymer. The addition of husk also resulted in significant enhancement of the storage modulus (from shear rheology) in both graphene and graphene/sepiolite composites, whereas the modulus was retained in the case of husk/sepiolite composite.

The dynamic mechanical analysis also exhibited that the incorporation of the husk particles to graphene, sepiolite and graphene/sepiolite retained the storage modulus. Overall, the addition of husk either improved or retained the composite properties, which resulted in the cost-efficiency of the composites due to the lower cost of husk particles. The morphology of the composites with hybrid filler systems was observed to be uniform with filler phases distributed effectively in the polymer matrix, even without the use of externally added compatibilizer. This also indicated that the processing parameters used for the generation of the composites were appropriate. Thus, these findings confirmed the usefulness of employing filler hybrids to achieve synergistic improvement in the composite properties. The obtained composites are effective alternatives to many polymeric materials in terms of properties as well as cost.

References

1. Mittal, V. (2010) *Advances in Polyolefin Nanocomposites*, CRC Press, USA.
2. Jancar, J. (1999) *Mineral Fillers in Thermoplastics I: Raw Materials and Processing*, Springer-Verlag, Germany.
3. Nwabunma, D., and Kyu, T. (2008) *Polyolefin Composites*, Wiley, USA.
4. Mittal, V. (2010) *Polymer Nanotube Nanocomposites: Synthesis, Properties, and Applications*, John Wiley & Sons, USA.
5. Kim, S., Do, I., and Drzal, L. T. (2009) Multifunctional xGnP/LLDPE nanocomposites prepared by solution compounding using various screw rotating systems. *Macromolecular Materials and Engineering*, **294**(3), 196-205.
6. Chaudhry, A. U., and Mittal, V. (2013) High-density polyethylene nanocomposites using masterbatches of chlorinated polyethylene/graphene oxide. *Polymer Engineering and Science*, **53**(1), 78-88.
7. Mukhopadhyay, P., and Gupta, R. K. (2011) Trends and frontiers in graphene-based polymer nanocomposites. *Plastics Engineering*, **67**, 32-42 (2011).
8. Tartaglione, G., Tabuani, D., Camino, G., and Moisio, M. (2008) PP and PBT composites filled with sepiolite: Morphology and thermal behaviour. *Composites Science and Technology*, **68**(2), 451-460.
9. Chen, H., Zheng, M., Sun, H., and Jia, Q. (2007) Characterization and properties of sepiolite/polyurethane nanocomposites. *Materials Science and Engineering: A*, **445-446**(0), 725-730.

10. Ghazanfari, A., Emami, S., Panigrahi, S., and Tabil, L. G. (2008) Themal and mechanical properties of blends and composites from HDPE and date pits particles. *Journal of Composite Materials*, **42**, 77-89.
11. Shnawa, H. A., Abdulah, N. A., and Mohamad, F. J. (2011) Thermal properties of low density polyethylene with oyster shell composite: DSC study. *World Applied Sciences Journal*, **14**(11), 1730-1733.
12. Tong, J. Y., Royan, N. R. R., Ng, Y. C., Ghani, M. H. A., and Ahmad, S. (2014) Study of the mechanical and morphology properties of recycled HDPE composite using rice husk filler. *Advances in Materials Science and Engineering*, Article ID 938961.
13. Singh, B., and Kumar, S. (2008) Synthesis and characterization of psyllium-NVP based drug delivery system through radiation crosslinking polymerization. *Nuclear Instruments and Methods in Physics Research Section B: Beam Interactions with Materials and Atoms*, **266**(15), 3417-3430.
14. Bhatia, M., and Ahuja, M. (2014) Synthesis and evaluation of psyllium-g-poly(N-vinyl-2-pyrrolidone) as mucoadhesive gel. *Der Pharmacia Lettre*, **6**(2), 127-131.
15. Lee, G.-W., Park, M., Kim, J., Lee, J. I., and Yoon, H. G. (2006) Enhanced thermal conductivity of polymer composites filled with hybrid filler. *Composites Part A: Applied Science and Manufacturing*, **37**(5), 727-734.
16. Yu, A., Ramesh, P., Sun, X., Bekyarova, E., Itkis, M. E., and Haddon, R. C. (2008) Enhanced thermal conductivity in a hybrid graphite nanoplatelet - carbon nanotube filler for epoxy composites. *Advanced Materials*, **20**(24), 4740-4744.
17. Lorenz, H., Fritzsche, J., Das, A., Stockelhuber, K. W., Jurk, R., Heinrich, G., and Kluppel, M. (2009) Advanced elastomer nanocomposites based on CNT-hybrid filler systems. *Composites Science and Technology*, **69**(13), 2135-2143.
18. Gwon, J. G., Lee, S. Y., Chun, S. J., Doh, G. H., and Kim, J. H. (2010) Effects of chemical treatments of hybrid fillers on the physical and thermal properties of wood plastic composites. *Composites Part A: Applied Science and Manufacturing*, **41**(10), 1491-1497.
19. Marlett, J. A., and Fischer, M. H. (2002) A poorly fermented gel from psyllium seed husk increases excreta moisture and bile acid excretion in rats. *The Journal of Nutrition*, **132**(9), 2638-2643.
20. Fischer, M. H., Yu, N., Gray, G. R., Ralph, J., Anderson, L., and Marlett, J. A. (2004) The gel forming polysaccharide of psyllium husk (Plantago ovata Forsk). *Carbohydrate Research*, **339**(11), 2009-2017.
21. van Craeyveld, V., Delcour, J. A., and Courtin, C. M. (2009) Extractability and chemical and enzymic degradation of psyllium (Plantago ovata Forsk) seed husk arabinoxylans. *Food Chemistry*, **112**(4), 812-819.

22. Edwards, S., Chaplin, M. F., Blackwood, A. D., and Dettmar, P. W. (2003) Primary structure of arabinoxylans of ispaghula husk and wheat bran. *Proceedings of the Nutrition Society*, **62**(1), 217-222.
23. McAllister, M. J., Li, J. L., Adamson, D. H., Schniepp, H. C., Abdala, A. A., Liu, J., Herrera-Alonso, M., Milius, D. L., Car, R., Prud'homme, R. K., and Aksay, I. A. (2007) Single sheet functionalized graphene by oxidation and thermal expansion of graphite. *Chemistry of Materials*, **19**, 4396-4404.
24. Hummers, W. S., and Offeman, R. E. (1958) Preparation of graphitic oxide. *Journal of American Chemical Society*, **80**, 1339.

7

Polyethylene Nanocomposites with Surface Treated Sepiolite

7.1 Introduction

Polymer nanocomposites have been extensively studied over the last two decades owing to the significant enhancement of a wide spectrum of material properties at low filler volume fractions [1,2]. Use of polar compatibilizers is a most commonly employed technique to achieve polarity in the nonpolar polymer matrices which helps to achieve improvements in filler exfoliation, however, the use of compatibilizers also leads to occasional lowering of mechanical performance [3]. Ideally, if the filler surface is optimally designed and modified, it would avoid the need to alter the polymer polarity and, thus, would help to retain the properties of the homopolymer. Various techniques have been developed in recent years to achieve optimum surface modification of the filler, including grafting of polymer chains 'from' or 'to' the surface [4-10], chemical reactions of the long chain alcohols or acids with the carboxyl or hydroxyl groups ionically bound to the filler surface [11-16], physical adsorption of the organic molecules or polymer chains on the filler surface [17-20], etc. The techniques result in modified fillers which have much better chances of complete exfoliation in the polymer matrices. These advanced fillers do not require the use of any additional low molecular weight compatibilizer for nanocomposite synthesis [21].

Mittal [9] reported filler surface modifications carrying methacrylic functional groups and free radical initiator, which were used to polymerize lauryl methacrylate in order to further improve the organic coverage of the filler platelets by grafting polymer chains either 'to' or 'from' the clay surface. Significance of the amount of monomer or initiator attached on the surface, time, as well as temperature used

M. R. Vengatesan[a], Sevim Isci[b] and Vikas Mittal[a,]*
[a]The Petroleum Institute (part of Khalifa University of Science and Technology), Abu Dhabi, UAE; [b]Istanbul Technical University, Istanbul, Turkey
**Current address: Bletchington, Wellington County, Australia*

for polymerization was successfully quantified. For instance, Figure 7.1 presents the diffraction analysis of organically modified filler with initiator present in an amount corresponding to 70% surface coverage. The filler had the 001 basal plane spacing of 1.9 nm, which remained unchanged when reacted with lauryl methacrylate at 60 °C for 72 h under non-living conditions, as is evident from a minor increase in organic weight loss. However, when reacted at 120 °C under the living conditions, the *d*-spacing was increased to 3.8 nm. It indicated that the effective control over the termination reactions by using the living agent was beneficial in enhancing the basal plane spacing of the filler platelets. Thus, it confirmed the successful generation of the polymethacrylate brushes grafted from the filler surface.

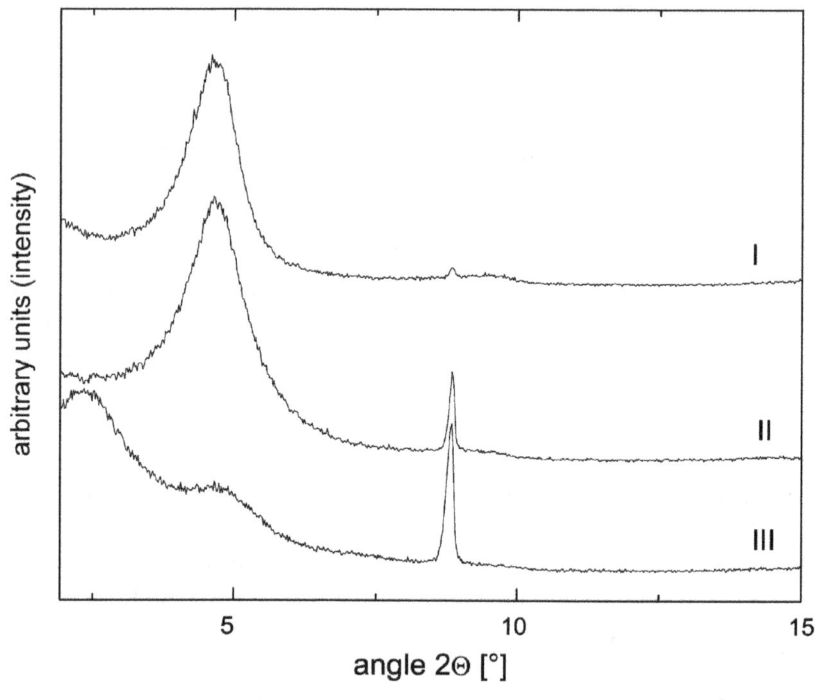

Figure 7.1 X-ray diffractograms of the (I) filler modified by using bicationic initiator in an amount corresponding to 70% surface coverage, (II) filler of I when reacted under non-living conditions at 60 °C, and (III) filler of I when polymerized under living conditions at 120 °C. The peak at 8.84° 2θ corresponds to the 001 basal plane spacing of mica platelets used as internal standard. Reproduced from Reference 9 with permission from Elsevier.

In another study, Mittal [16] further explored the surface reactions approach for the achievement of higher filler basal spacing in order to decrease the electrostatic interactions and, thus, to ease the filler exfoliation in the polymer matrix. Filler platelets were modified with modifiers of different chemical architectures in order to study the effect on the extent of surface esterification reactions with long chain fatty acids, such as dotriacontanoic or lacceroic acid. Figure 7.2 shows the comparison of the modified filler before and after the surface reaction. As observed, the higher extent of organic matter on the clay surface led to a significant improvement in the basal plane spacing, which leads to higher susceptibility of these filler stacks to exfoliate in polymer when compounded together with the polymer under shear in the compounder.

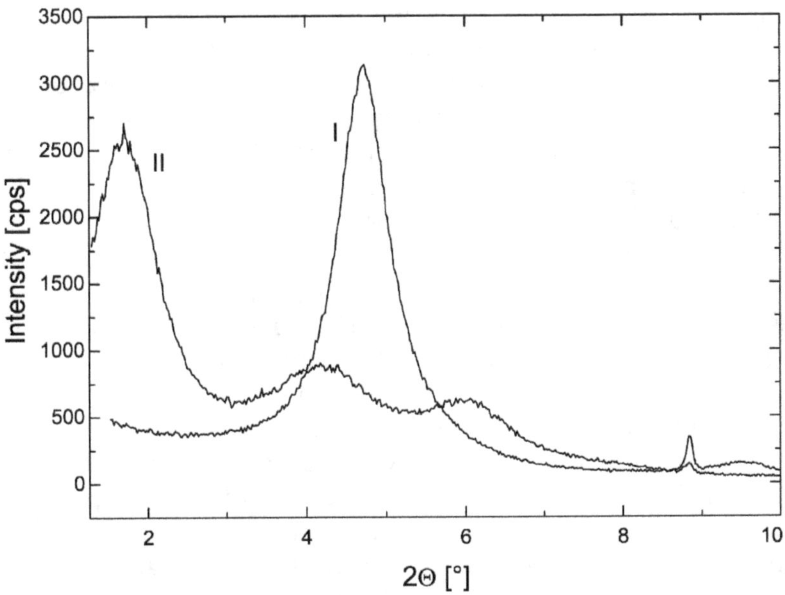

Figure 7.2 X-ray diffractograms of I: modified filler and II: filler of I after surface esterification reaction with dotriacontanoic acid. The peak at 8.84 2θ corresponds to the 001 basal plane spacing of mica platelets used as internal standard. Reproduced from Reference 16 with permission from Elsevier.

Mittal and Herle [20] also reported the potential of physical adsorption of long chain solvent-soluble molecules and polymers on the surface of the filler. These molecules (e.g. long alkyl chain alcohols) could adsorb on the surface by forming H bonds with the OH groups present either in the inside structure of the filler crystals or on the edges of the filler platelets. Figure 7.3 shows the thermal degradation behavior of the pre-modified filler subsequently physically adsorbed with varying amount of poly(vinylpyrrolidone) (PVP). The physical adsorption led to thermal behavior enhancement as the peak degradation temperature of the PVP-modified clay increased to 332 °C (curve II in Figure 7.3a). This increased thermal resistance is particularly beneficial when compounding the filler at high compounding temperatures required by polymer matrices like polyethylene. The mechanism of adsorption was suggested as follows: the more electronegative oxygen atoms in the PVP chains can form hydrogen bonds with OH groups present either on edges or in the internal structure of filler.

High density polyethylene (HDPE) is a well-known thermoplastic material possessing good mechanical properties and has been extensively used in different industrial applications [22]. In order to improve the properties of HDPE polymer, incorporation of various nanomaterials has been reported. This leads to the reinforcement effect on the HDPE performance, thus, resulting in better thermal, mechanical, rheological and barrier properties [23-26]. Specifically, nanomaterials such as silica [27], alumina [28], mica [29], clays [30-32], natural fibers [33,34], carbon black [35], CNT [36,37], graphene [38,39], etc. have been used as filler materials for the polyethylene matrix. Among these, clay materials are more economical and commercially acceptable class of nano-fillers for the reinforcement of polymer matrices so as to expand their spectrum of applications. Clay based polymer nanocomposites have been reported to exhibit a good balance of thermal, mechanical as well as the rheological properties [40]. Different types of clay materials such as montmorillonite, bentonite, hectorite, sepiolite, etc., differing in structural composition and morphology, have been used as reinforcing fillers for thermoplastic materials [41-44].

Sepiolite is one of the most useful mineral in the category of clay materials with 1-D morphology. It is a non-swelling, natural, abundant and low-cost clay having theoretical unit cell formula $(Si_{12}Mg_8O_{30})(OH)_4(OH_2)_4{\cdot}8H_2O$ (crystal structure depicted in Figure 7.4). It belongs to the class of layered silicates comprising of hydrated magnesium silicate having micro-fibrous needle like morphology. As compared to phyllosilicates (BET surface area 82±1 m^2/g), it has high surface area (BET 374±7 m^2/g) and contains large amount of silanol groups present at the surface [45,46]. The silanol groups have been proven effective as sepiolite acts as reinforcing and compatibilizing agent for different polymer matrices owing to coupling reactions. In a related study, Chen *et al.* [47] prepared well dispersed polyurethane/sepiolite nanocomposite using *in-situ* polymerization technique. Sepiolite improved the tensile strength and modulus of polyurethane due to the strong interfacial interaction between the organic

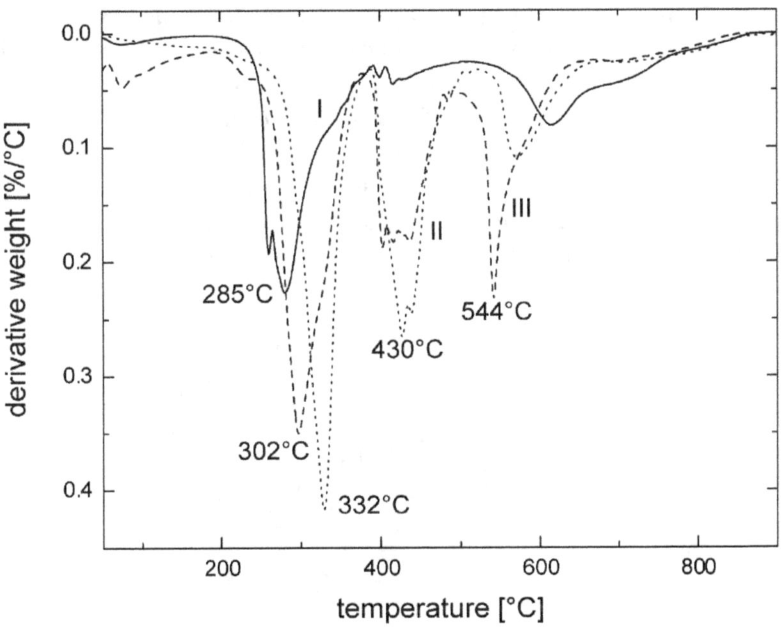

Figure 7.3 (a) Derivative weight thermograms of modified filler before adsorption (I), after adsorption with 1.5 g of initially added PVP (for 2 g of filler) (II) and after adsorption with 1 g of initially added PVP (for 2 g of filler) (III). Reproduced from Reference 20 with permission from Elsevier.

and inorganic phases [47]. Similarly, it has also been utilized as a nano-filler for other polymer matrices [48-51].

The incorporation of the unmodified sepiolite clay into the polymer matrices may lead to an aggregated morphology due to the mismatch of surface polarity between the polymer and inorganic phases, thus, resulting in deterioration of the composite properties. Therefore, the dispersion/defibrillation of the 1-D sepiolite crystals in the polymer matrix and the interfacial interactions between the polymer and sepiolite phases are the major factors required to be optimized to enhance thermal stability and mechanical properties of the nanocomposites [52]. In the case of HDPE based composites, no natural interfacial interaction between the organic and inorganic phases exists, thus, surface modification of the sepiolite particles is solely responsible for the minimization of the contact between sepiolite nanofibers and correspondingly enhanced interfacial interaction/contacts between inorganic and organic phases [53]. The surface of the sepiolite nano-fibrils is generally modified with different types of organic materials such as organosilanes, cationic, anionic and non-ionic surfactants and polymers through covalent, H-bonding and physical adsorption (van der Waals interaction) methods [54-58]. As mentioned earlier, the surface modification of the filler particles through the physical adsorption method is an acceptable strategy for use in polymer nanocomposites, even though the binding of the modified molecules to the filler surface is not covalent in nature. Similarly, physical adsorption method for the surface modification of sepiolite

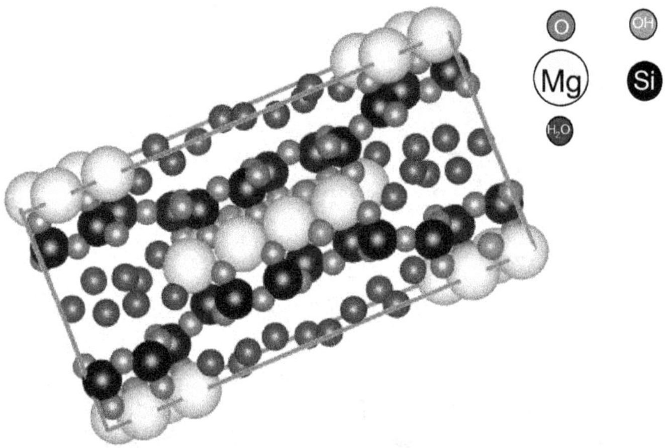

Figure 7.4 Schematic representation of sepiolite crystal structure.

crystals using different electrolytes and uncharged polymers can also be useful in its development as functional filler for nanocomposites [58,59]. The adsorption of the uncharged polymer on the sepiolite surface leads to an 'entropy-driven' process. Therefore, the adsorbed polymer is generally observed to be randomly coated on the clay surface. However, the higher molecular weight of the adsorbing polymer leads to overall significant extent of surface adsorption due to van der Waals interactions, thereby, resulting in organo-modified sepiolite.

By the use of physically adsorbed organic modification molecules of amphiphilic nature, the interfacial interaction of the filler particles with the polymer matrix can be achieved. Thus, the nature of the polymers adsorbed on the sepiolite surface significantly influences the dispersion/defibrillation process and interfacial adhesion with the polymer matrix which, thus, results in the variation in the mechanical, thermal as well the rheological properties of the resulting nanocomposites [60].

In the present work, the influence of a variety of polymeric surface modifications physically adsorbed on the surface of sepiolite was studied on the thermal, mechanical and rheological properties of the HDPE nanocomposites. The various modifications included poly(vinyl pyrrolidone) (PVP), poly(vinyl butyral) (PVB), polyethylene glycol (PEG), polyurethane (PU) and ethylene vinyl acetate-graft-maleic anhydride (EVA*g*MA). The amount of filler was fixed to 5 wt% and the composites were generated using industrially accepted melt mixing method without the use of any external compatibilizers. The compounding conditions were carefully chosen to ensure that good polymer-filler intermixing was obtained in the compounder under shear, however, the thermal degradation of the polymer and surface modification was avoided.

7.2 Experimental

7.2.1 Materials

High-density polyethylene (BB2581) was supplied by Abu Dhabi Polymers Company Limited (Borouge), UAE. It was in the form of white pellets with a specific gravity of 0.958. Sepiolite clay was received from Sigma Aldrich. It was in the form of fine white powder and was used as received for surface modification trials. Polyurethane liquid, polyvinylpyrolidone, polyvinylbutaral and polyethylene glycol with an average M_w of 8000 (PEG8000) were procured from Sigma Aldrich and were used without further purification. Ethylene vinyl acetate-graft-maleic anhydride copolymer (density 0.950 g/cm^3, melt flow index (190 °C/2.16 kg) 16 g/10 min and peak melting temperature 71 °C) with a trade name of Fusabond C190 was obtained from DuPont. The standard solvents such as toluene and acetone for surface modification experiments were also procured from Sigma Aldrich and were used as received. Deionized water with low conductivity was used for the surface modification trials performed in water.

7.2.2 Surface Modification of Sepiolite

The surface of pristine sepiolite (S-P) was modified with EVA*g*MA, PEG, PU, PVB and PVP polymers. For modification with PU, PU was dispersed in deionized water at a concentration of 1 mg/ml at 70 °C. Sepiolite (2% w/w) was added to the polymer dispersion and shaken for 2 h and ultra-sonicated for 5 min. The content was further stirred overnight. The sample was washed with deionized water followed by acetone. The sample was finally dried at 50 °C. The final product was termed as S-PU clay. The same protocol was used to prepare PEG and PVP modified sepiolite in the presence of deionized water and termed as S-PEG and S-PVP respectively. For the modification of sepiolite with EVA-g-MA and PVB, toluene and acetone were used as solvents respectively, while keeping the other parameters constant. The EVA*g*MA and PVB modified materials were denoted as S-EVA*g*MA and S-PVB respectively.

7.2.3 Preparation of HDPE/Sepiolite Nanocomposites

Pristine and surface modified sepiolite were melt mixed with HDPE using mini twin screw conical extruder (HAAKE MiniLab, Thermo Scientific) at a temperature of 180 °C using a rotation speed of 80 rpm for 10 minutes. Before compounding, the fillers and polymer were dried in a vacuum oven at 60 °C overnight so as to remove any traces of surface bound moisture, which can lead to thermal degradation during high temperature compounding. In addition, the materials were physically mixed prior to melt mixing in the compounder. The test specimens for the mechanical and rheological analysis were prepared using mini injection molding machine (HAAKE MiniJet PRO, Thermo Scientific) at 180 °C at 400 bar for 10 s. The mold temperature was maintained as 50 °C. Care was taken to cool all samples by following the same procedure so as not to affect the polymer crystallinity.

7.2.4 Materials Characterization

IR spectra of pristine and modified sepiolite were collected on Nicolet iS10 spectrometer equipped with SmartiTR diamond ATR accessory (angle of incidence of 45°), DTGS KBr detector and KBr beam splitter. It had a diamond ATR crystal (index of refraction 2.4 at 1000 cm^{-1}) and a depth of penetration of 2 μm at 1000 cm^{-1} for a sample with refractive index of 1.5. IR Spectra were recorded by the OMNIC software in the 4000-525 cm^{-1} region with a resolution of 4 cm^{-1} from 16 scans.

Differential scanning calorimetric (DSC) analysis of the nanocomposites was carried out using a Discovery DSC from TA instruments under nitrogen atmosphere. 3-7 mg of sample was heated to 200 °C using a heating rate of 10 °C/min. Subsequently, the sample was cooled to -60 °C using a cooling rate of 5 °C/min and kept under isothermal conditions at -60 °C for 1 min. The sample was re-heated to 200 °C using a heating rate of 10 °C/min and finally cooled to 25 °C using the same rate. The thermal degradation behavior of the samples was studied using a Discovery thermogravimetric analyzed (TGA) from TA instruments under nitrogen atmosphere. Samples weighing 3-7 mg were analyzed from 25 °C to 700 °C by using the heating rate of 10 °C/min.

Tensile properties such as tensile strength, modulus and extension at break of composite samples were evaluated at ambient temperature on universal testing machine (Instron 3345) according to ASTM D638 at a crosshead speed of 10 mm/min. The dumbbell shaped test specimen were used and the reported values represented the average of five test specimens. The melt rheological characteristics of HDPE and its nanocomposites were analyzed in shear mode on AR 2000ex Rheometer from TA Instruments according to ASTM D4440. A parallel plate configuration was used with plate diameter of 25 mm and a geometric gap of 1.4 mm. Injection molded disc shaped samples of 25 mm diameter and 1.5 mm thickness were used for the analysis. The samples were scanned in the dynamic oscillatory mode from 0.1 to 100 rad/s at 180 °C with a constant strain of 4%.

Analytical powder diffractometer (X'Pert PRO) (45 kV, 40 mA) with Cu-Kα radiation (wavelength = 1.5406 A°) was utilized for X-ray diffraction analysis of the polymer and nanocomposites samples. The diffraction of the samples was measured at a rate of 0.0170 deg/s and 2θ scale ranging from 5 to 60°.

For the bright field transmission electron microscopy (TEM) analysis of the nanocomposite samples, Philips CM 20 (Philips/FEI, Eindhoven) electron microscope at 120 kV and 200 kV accelerating voltages was used. Thin sections of 70-90 nm thickness were microtomed from the sample blocks using a Leica Ultracut E microtome (Leica, Austria) equipped with a diamond knife (Diatome, Switzerland) at a temperature of -120 °C. The so-obtained sections were supported on 100 mesh grids sputter coated with a 3 nm thick carbon layer. The modified fillers were analyzed in Hitachi field emission in-lens S-900 high resolution scanning electron microscope (SEM) at accelerating voltages of 10-20 kV. In addition, the block faces of the nanocomposite specimens were also investigated at ambient conditions using ZEISS Axioplan light microscope (Zeiss, Germany) equipped with Zeiss Axio Cam ICc 1 CCD camera in reflected polarized light. Zeta potential measurements of the modified sepiolite materials were performed using ZetaPALS device.

7.3 Results and Discussion

In the current study, 1-D nano-fibrils of sepiolite mineral were surface modified with a variety of polymeric modifiers using physical adsorption approach. The choice of the modifications was based on tuning the polarity of the sepiolite surface so as to achieve good degree of delamination in the polyethylene matrix. In addition to other polymers, copolymers like EVA*g*MA were also used for this purpose as the ethylene component can be expected to have good intermixing with the HDPE matrix, whereas the polar groups would interact with filler surface, thereby leading to phase compatibilization and filler delamination. As mentioned earlier, such an approach of filler modification eliminates the need of the use of conventional low molecular weight compatibilizers, which occasionally negatively affect the crystallinity and mechanical properties of the nanocomposites. Even though the surface of the filler in the case of physical adsorption approach would be partially covered by the polymeric modifier owing to the steric hindrance to the high molecular weight chains, however, it can still be expected to significantly change the surface behavior of the filler particles. In addition, the amount of the modification needed for the purpose is much lower than the amount of the compatibilizer required to attain good filler dispersion.

7.3.1 Properties of Modified Fillers

Table 7.1 demonstrates the zeta potential values of the pristine and modified sepiolite. Pristine sepiolite in water was observed to have the zeta potential value of -19.5 mV, which indicated that the sepiolite particles tend to flocculate in water. In the case of the surface modification with water soluble polymers like PEG and PVP, decrease in the zeta potential values for the modified sepiolite was observed. For instance, the zeta potential was reduced to -13.1 for S-PEG, whereas it was observed to be even lower (-10.2 mV) for S-PVP. It indicated that the polymer chains partially covered the surfaces of the sepiolite particles. The sepiolite particles modified with other polymers such as PU, PVB and EVA*g*MA in water dispersions were observed to flocculate and the zeta potential values of these materials increased.

Table 7.1 Zeta potential values of the pristine and modified sepiolite

Sample	Zeta Potential (mV)
S-P	-19.5±1.0
S-PEG	-13.1±0.5
S-PVB	-28.8±3.7
S-PU	-29.2±0.9
S-PVP	-10.2±1.5
S-EVA*g*MA	-20.6±3.4

The SEM analysis of the modified sepiolite also provided more insights into their morphology. Figure 7.5 shows the SEM images of the pristine sepiolite and sepiolite modified with PEG and PVP. The particles were observed to be uniformly dispersed as primary particles without forming any large sized aggregates. On the other hand, Figure 7.6 presents the SEM images of the sepiolite modified with PVB, PU and EVA*g*MA. Filler particles were observed to form large aggregates in these modified sepiolite materials, thus, indicating that the different surface modifications resulted in varying extent of bridging of primary filler particles. This behavior is also confirmed in the high magnification SEM images of the modified sepiolite, as shown in Figure 7.7. At high magnification too, the PEG and PVP modified sepiolite particles exhibited dispersed morphology, whereas the morphology was more aggregated in the case of PVB, PU and EVA*g*MA modified sepiolite. However, it should be noted the observed aggregates may be only loosely held aggregated structures, which may be easily broken when compounded with polymers under shear. Thus, all modified fillers exhibited potential of uniform dispersion in the filler matrix due to the surface modification making these more susceptible to exfoliation.

The interaction between the adsorbed polymer and clay surface was determined with FTIR. As mentioned earlier, the non-ionic polymers create mostly hydrogen bonding with the clay surface. Any changes in the characteristic filler peaks after modification may indicate the formation of hydrogen bonds between the polymer and sepiolite surface. The band at 3684 cm^{-1} in the IR spectra corresponded

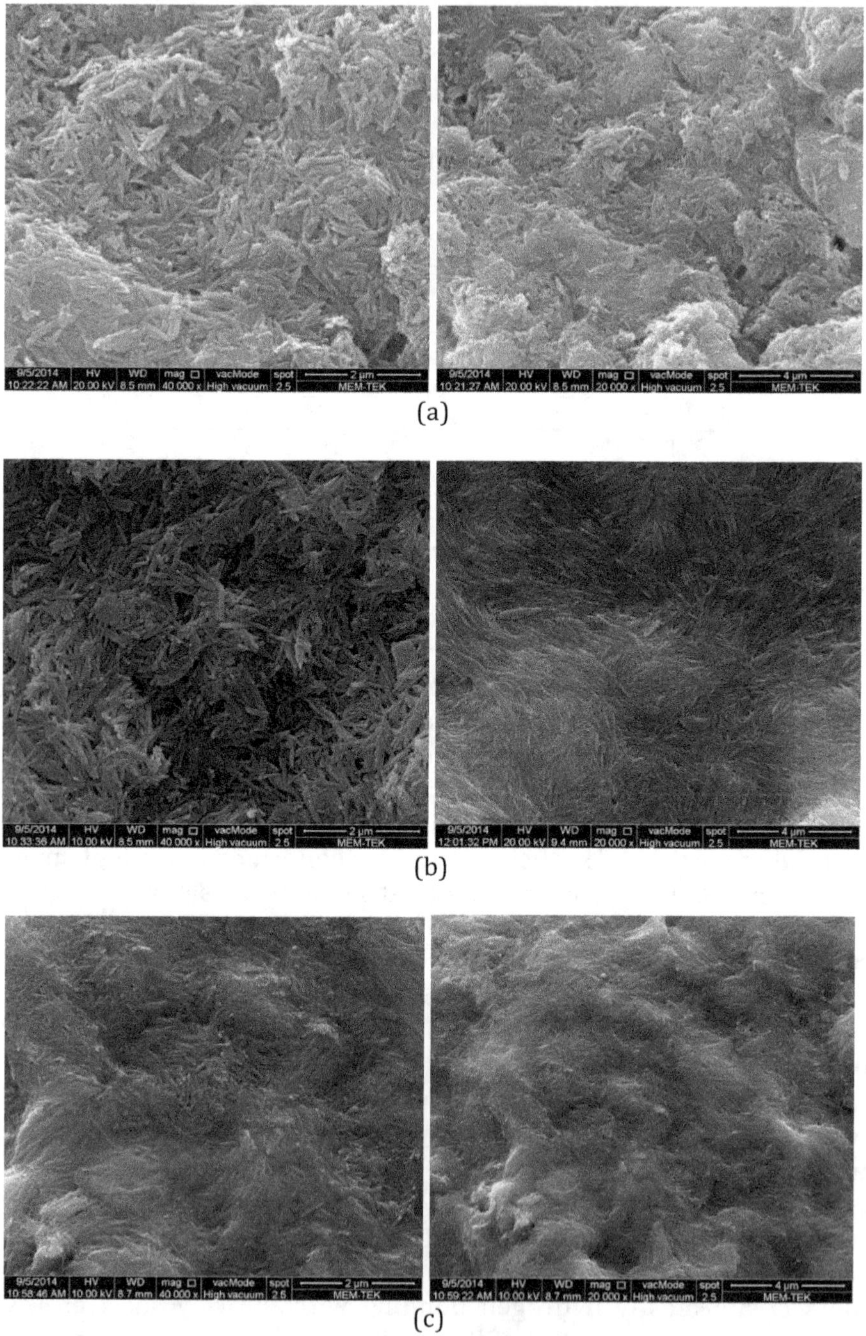

Figure 7.5 SEM images of (a) S-P, (b) S-PEG and (c) S-PVP at two different magnificaitons.

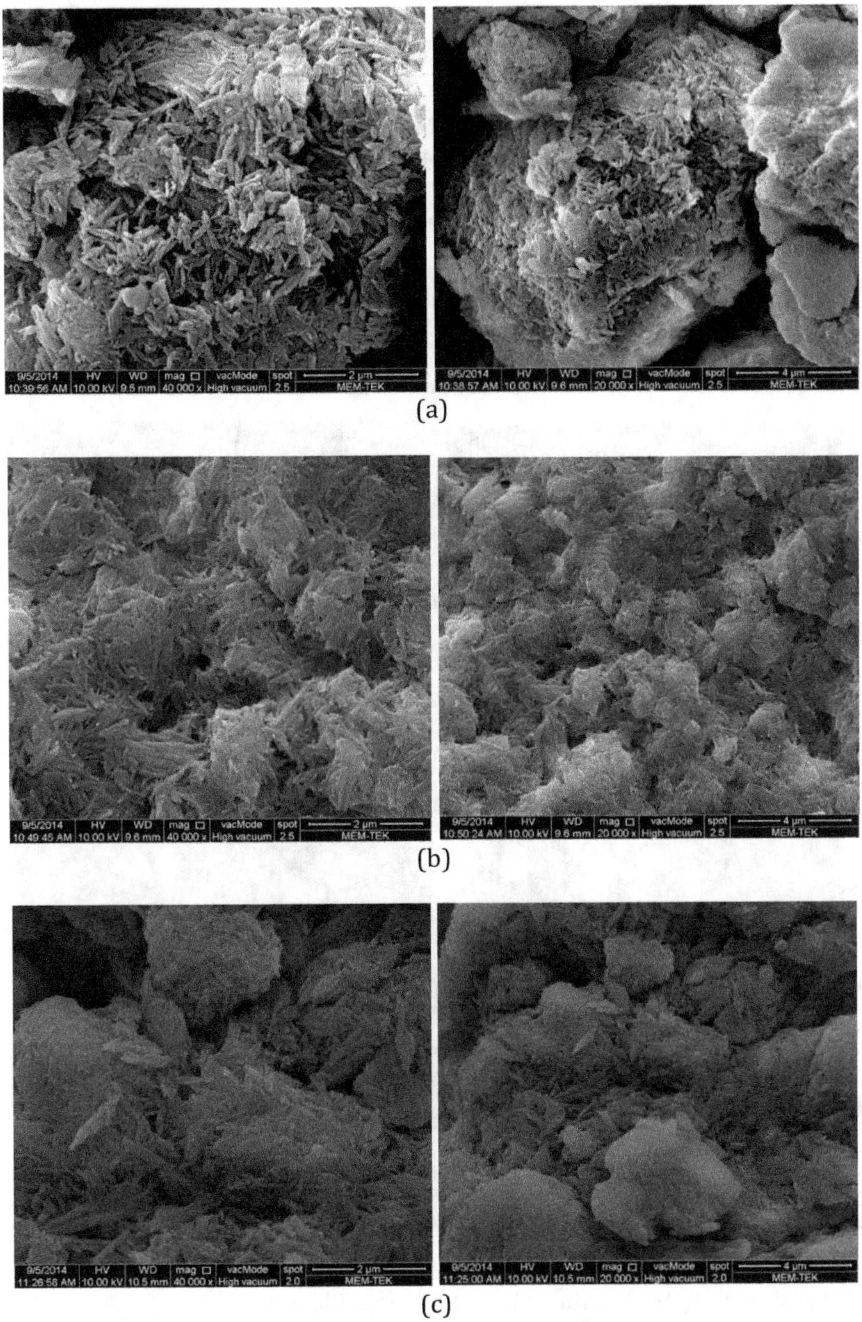

Figure 7.6 SEM images of (a) S-PVB, (b) S-PU and (c) S-EVA*g*MA at two different magnificaitons.

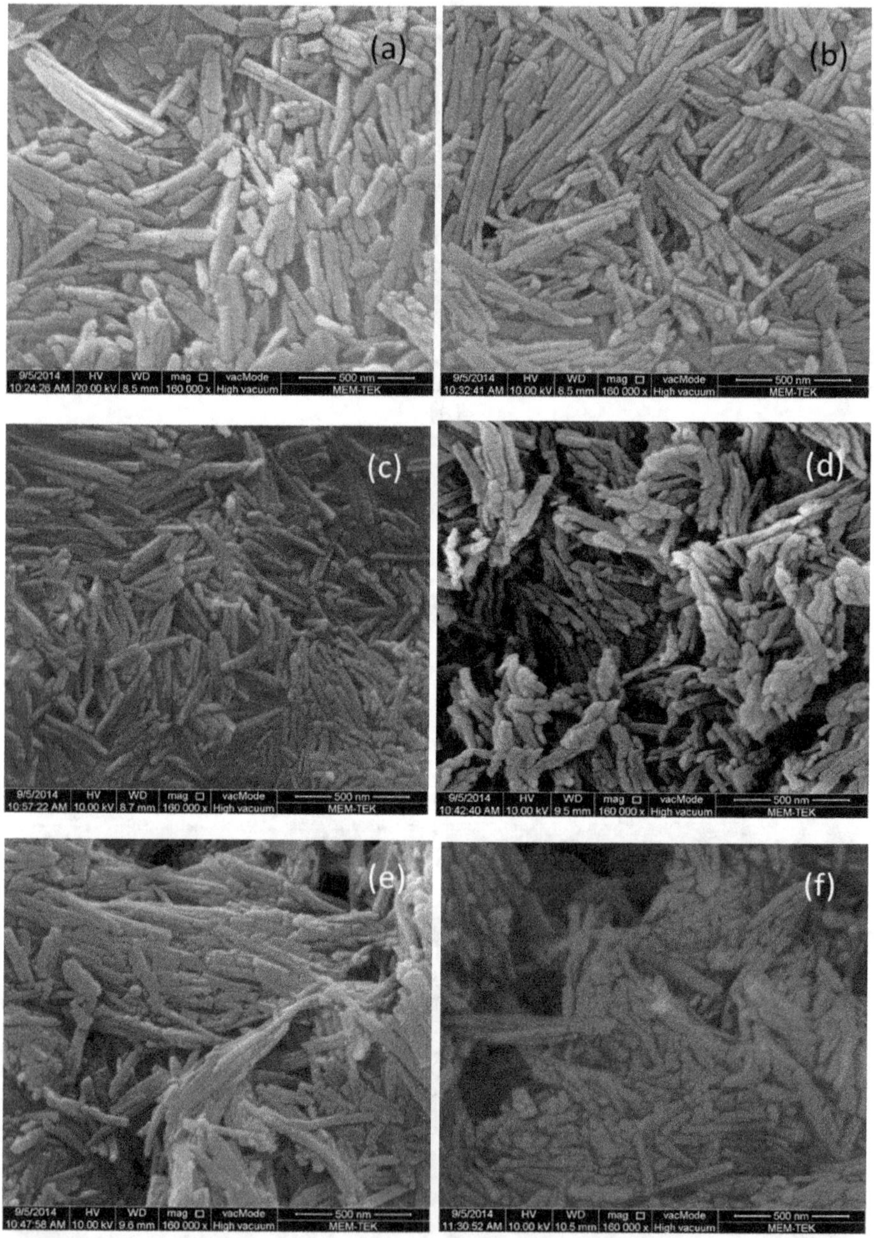

Figure 7.7 SEM micrographs of (a) S-P, (b) S-PEG, (c) S-PVP, (d) S-PVB, (e) S-PU and (f) S-EVA*g*MA minerals at high magnification.

to the stretching (vOH) vibrations of hydroxyl groups (belonging to Mg_3OH) attached to Mg ions located in the interior blocks of natural sepiolite. No change in its position was observed after modification of the surface with polymeric chains. However, the other OH deformation peaks at 3550 cm^{-1} changed significantly with polymer adsorption, as shown in Table 7.2. Also, the H-O-H deformation IR peaks were observed to change with polymer addition on the mineral surface. These observations indicated the formation of H-bonds between the polymer chains and sepiolite surface, which can benefit the composite properties effectively when the modified fillers are incorporated in the HDPE matrix.

Table 7.2 Variation in the IR bands after surface modification of sepiolite with different modifiers

Sample	OH deformation	H-O-H deformation		Si-O stretching
S-P	3550	1693	1658	1022
S-PEG	3562	1742	1657	1015
S-PVB	3565	--	1669	1019
S-PU	3565	1696	--	1018
S-PVP	3564	--	1657	1022
S-EVA*g*MA	3570	1671	--	1019

7.3.2 Nanocomposite Properties

Figure 7.8a presents the X-ray diffraction pattern of pristine sepiolite. Pristine sepiolite exhibited an intense reflection at 2θ = 7.48° (110) with a *d*-spacing of 1.08 nm. It is well known that the *d*-spacing of the clay platelets is influenced by the organic modifier [61]. In the present work, as sepiolite was modified with different polymers through van der Waals interactions, thus, the basal plane spacing is expected to increase for the modified sepiolite samples (data not provided). However, as observed in the SEM micrographs, the surface coating of the filler particles by the polymer chains was not extensive, thus, a significant increment in the basal plane spacing cannot be expected. In addition, the polymer coating on the sepiolite surface might also slightly reduce its diffraction intensity. However, it was of larger interest to study the changes in the basal plane spacing of the modified sepiolite in the composites, the diffraction patterns of which are compared with pure polyethylene in Figure 7.8b. The diffraction pattern of polyethylene exhibited characteristic diffraction peaks at 2θ = 22.0° (110) and 24.3° (200). Both of these characteristic diffraction

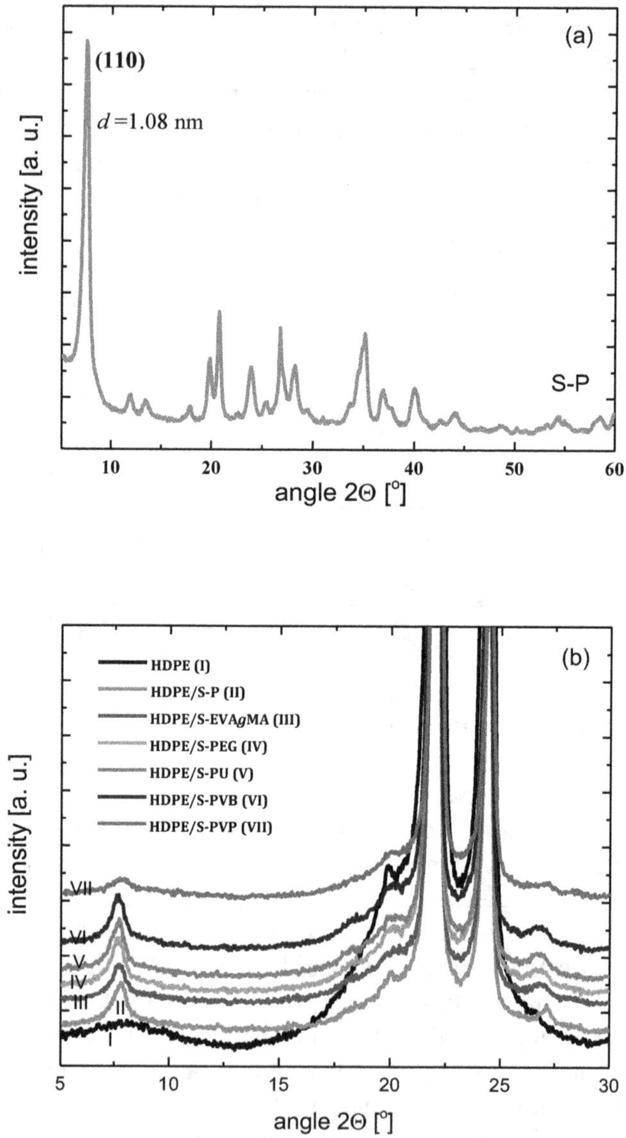

Figure 7.8 X-ray diffractograms of (a) S-P and (b) HDPE/sepiolite composites.

peaks of HDPE were also observed in the nanocomposites, which indicated that the incorporation of sepiolite did not affect the crystal structure of HDPE. The characteristic peak of sepiolite at 2θ = 7.48° was observed to shift by a small extent in the composite with HDPE, thus, indicating some polymer intercalation inside the filler interlayers, even in the absence of any filler surface modification. The peak represented a *d*-spacing of 1.13 nm in the composite, as compared to 1.08 nm in the pristine filler. The characteristic peak of sepiolite was observed to shift further, though only marginally, in the nanocomposites generated with modified sepiolite filler. For example, the sepiolite characteristic peak (110) in the case of S-PEG reinforced HDPE nanocomposite exhibited the *d*-spacing value of 1.16 nm. Similarly, other composites also exhibited slight increase in the basal plane spacing of the filler. In addition, the intensity of the diffraction peak in the S-PVP composite was also significantly reduced, thus, also qualitatively signifying the enhanced degree of filler dispersion. It indicated firstly that the surface modification of sepiolite led to enhanced degree of polymer intercalation. Secondly, the dispersion level of sepiolite was observed to be influenced by the surface coated polymer, though the differences in the resulting basal plane spacing in the composites with different modified fillers were small. As X-ray does not quantify the extent of filler intercalation and exfoliation, thus, it is also likely that varying degrees of filler exfoliation were also achieved as a result of surface modification, which is required to be analyzed through microscopy as well as by correlating with the property enhancements.

Figure 7.9 shows the TGA thermograms of pure HDPE, and nanocomposites with pristine and modified sepiolite. The thermal degradation temperatures for 10% and 40% weight loss and char yield at 700 °C are also presented in Table 7.3. It was observed that the addition of both unmodified and modified sepiolite in HDPE resulted in a significant improvement in the thermal stability. It indicated that effective heat transfer took place from the polymer chains to the sepiolite nano-fibrils, therefore, delaying the thermal degradation of the organic matter. Other studies have also reported the formation of a physical network by sepiolite with HDPE interphase, thus, resulting in enhanced thermal stability [61-63]. Among the nanocomposites with modified sepiolite, S-PVB containing composite exhibited inferior thermal stability, though it was still significantly improved as compared to pure polymer. Composites containing S-PEG and S-PVP exhibited highest temperature for 10% and 40% weight loss. The observed temperatures for these composites were 25 °C and >20 °C higher than the corresponding temperature for 10% and 40% weight

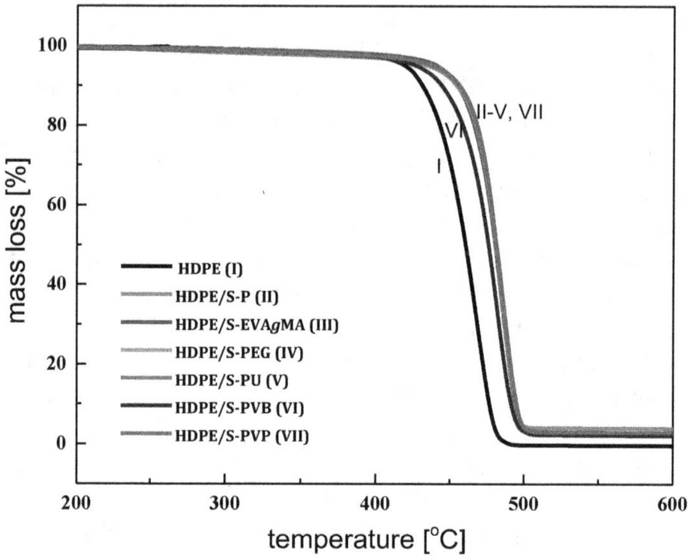

Figure 7.9 TGA thermograms of HDPE and HDPE/sepiolite nanocomposites.

loss of pure HDPE. It indicated that more uniform surface modification observed in SEM for S-PEG and S-PVP (no bridging of the primary particles) would have resulted in uniform filler delamination in the polymer matrix, thus, resulting in enhancement of the thermal performance of the polymer. However, higher thermal stability of the adsorbed polymer may have also resulted in the observed superior performance of the S-PEG and S-PVP containing composites. The char yield of the nanocomposites was also higher than that of pristine HDPE. Thus, sepiolite acted as the thermal degradation barrier at the interface with polymer, which restricted the thermal degradation of polymer at higher temperature, the effect which was further enhanced on the surface modification of sepiolite [64]. The dynamic TGA was also carried out at the compounding temperature of 180 °C as a function of time so as to ascertain the suitability of the modified sepiolite minerals for high temperature processing and applications. No significant weight loss was observed over the test duration (equivalent to the compounding time), thus, indicating that the adsorption of filler surface with the polymer did not result in early initiation of degradation.

Table 7.3 Thermal stability and char yield of HDPE/sepiolite nanocomposites

Sample	T^d_{10} (°C)	T^d_{40} (°C)	Char yield at 700 °C (%)
Pure HDPE	431	456	0.00
HDPE/S-P	454	476	3.9
HDPE/S-EVA*g*MA	454	476	2.9
HDPE/S-PEG	456	478	2.9
HDPE/S-PU	454	477	2.4
HDPE/S-PVB	444	471	2.1
HDPE/S-PVP	456	477	2.9

T^d_{10} (°C): 10% weight loss temperature
T^d_{40} (°C): 40% weight loss temperature

Figure 7.10 represents the melting and crystallization thermograms of pure HDPE, as well as composites with unmodified and modified sepiolite. The percentage of crystallinity of the polymer in the HDPE/sepiolite nanocomposites was calculated by using the following equation and the calorimetric values are also depicted in Table 7.4.

$$\text{Xc (\%)} = \frac{\Delta H_m}{(1-f)\Delta H_m^o} \times 100$$

where X_c is the percentage crystallinity, ΔH_m is the melting enthalpy of the polymer (second heating cycle) and ΔH_m^o is the melt enthalpy of 100% crystalline HDPE, taken as 293 J/g [65]. The addition of 5 wt% of unmodified sepiolite increased the extent of crystallinity of the HDPE (67.8% in the composite as comparison to 61% for pure polymer) which indicated that the sepiolite nano-fibrils acted as a nucleating agent for the crystallization of the HDPE matrix. It was also indicated by the increase in peak crystallization temperature in the composite to 120.6 °C from 119.3 °C for pure polymer. Variations were observed in the calorimetric behavior of nanocomposites with modified sepiolite. Composites with S-PEG, S-PVP and S-PU sepiolite exhibited similar increment in the melt enthalpy and degree of crystallinity as the composite with S-P. On the other hand, composites with S-EVA*g*MA and S-PVB exhibited lower melt enthalpy values (176 J/g) and correspondingly lower degree of crystallinity (63.2%). Same

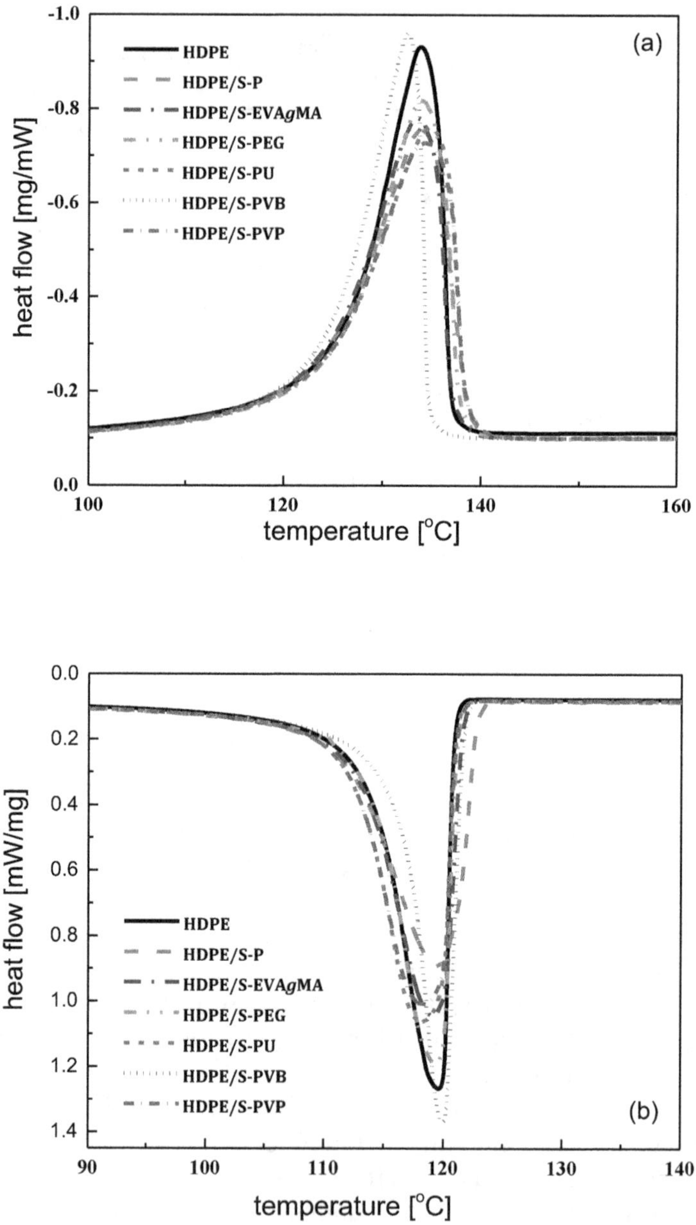

Figure 7.10 DSC (a) melting and (b) crystallization thermograms of HDPE and nanocomposites with pristine and modified sepiolite.

effect has also been observed in PA6 based polymer/organo-modified sepiolite nanocomposites [53]. However, it should be noted that the degree of crystallinity in the composites with S-EVA*g*MA and S-PVB was still higher than the pure polymer. Thus, the presence of EVA*g*MA and PVB chains adsorbed on the sepiolite surface had lower extent of nucleating efficiency of the filler during the polymer crystallization. The observed changes in the crystallinity are also expected to affect the mechanical performance, along with the effect of filler dispersion.

Table 7.4 Calorimetric behavior of HDPE, HDPE/sepiolite HDPE/modified sepiolite composites

Sample	T_m (°C)	ΔH_m (J/g)	T_c (°C)	ΔH_c (J/g)	X_c (%)
Pure HDPE	167.1	181.8	119.3	175.7	61.0
HDPE/S-P	168.2	188.9	120.6	175.2	67.8
HDPE/S-EVA*g*MA	167.1	176.0	119.2	170.0	63.2
HDPE/S-PEG	167.6	188.8	119.7	176.4	67.8
HDPE/S-PU	167.4	188.0	119.0	177.0	67.5
HDPE/S-PVB	166.4	176.0	120.1	164.0	63.2
HDPE/S-PVP	166.2	185.2	119.0	174.5	66.5

The tensile properties of the HDPE/sepiolite nanocomposites are demonstrated in Table 7.5. Pure HDPE had a tensile modulus and strength of 757 and 18.3 MPa. The addition of 5 wt% S-P significantly improved the modulus and strength of HDPE to 930 and 25.2 MPa respectively. Thus, the incorporation of even pristine sepiolite exhibited significant stress transfer and reduced the crack formation in HDPE at higher load which resulted in increased modulus. In addition, the high aspect ratio of the sepiolite nano-fibrils must have underwent some degree of defibrillation due to shear in compounder, thus, resulting in good dispersion in HDPE. The nanocomposites with modified sepiolite (except S-EVAgMA) exhibited even further enhancement in the modulus, thus, indicating that the surface modification of sepiolite surface led to enhanced polymer-filler interactions, thereby, resulting in effective filler dispersion. The modified filler may have also allowed effective delamination of the filler stacks under shear in the compounder. Thus, as also mentioned earlier, even though the diffraction peak of the filler did not change significantly in the polymer composites with modified sepiolite, however, a good degree of filler dispersion could still be achieved owing to combination of polymer-filler interactions as well as shearing in the compounder, thus, resulting in overall superior mechanical performance. Comparing the properties of composites with sepiolite modified with different polymers, it was observed that the mechanical properties of the nanocomposite were affected by the nature of modifier molecules. Maximum enhancement in the modulus was observed for the S-PEG containing composite which exhibited the modulus of 1117 MPa. It was followed by the composite with S-PVP which also exhibited significantly higher modulus of 1039 MPa. S-PVB and S-PU containing composites also enhanced the tensile modulus, though the values

Table 7.5 Mechanical properties of HDPE/sepiolite nanocomposites

Sample	Tensile modulus (MPa)	Tensile stress at yield (MPa)	Elongation at break (%)
Pure HDPE	757 (±10)	18.3 (±3)	10.3 (±2)
HDPE/S-P	930 (±10)	25.2 (±3)	8.4 (±2)
HDPE/S-EVA*g*MA	864 (±10)	25.4 (±2)	8.7 (±2)
HDPE/S-PEG	1117 (±10)	25.3 (±2)	9.4 (±2)
HDPE/S-PU	966 (±10)	24.7 (±2)	10.0 (±2)
HDPE/S-PVB	977 (±10)	24.1 (±2)	9.5 (±2)
HDPE/S-PVP	1039 (±10)	25.0 (±2)	10.3 (±2)

were only slightly higher than the corresponding composite with S-P. Incorporation of S-EVA*g*MA, on the other hand, exhibited the least increase in the modulus of PE. Thus, induced differences in crystallinity as well as polymer-filler interactions resulted in overall good degree of modulus enhancement in most of the nanocomposites. In addition, the observed enhancements were also attained without the addition of any external compatibilizer for filler compatibilization, which generally leads to matrix plasticization, thus, leading to deterioration of the properties. It should also be noted that the polymers were only physically adsorbed on the filler surface and were uncharged and non-reactive in nature. Thus, such non-reactive polymer chains on the filler surface are expected to improve the interfacial adhesion with the HDPE matrix through the van der Waals forces. In addition, the overall structural architecture and polarity of these adsorbed chains would drive such interactions with the non-polar polymer matrix. In cases like S-EVA*g*MA, probably there was a significant polarity mismatch with the polymer matrix, which resulted in lower extent of property enhancement. On the other hand, PEG chains on the filler surface may have resulted in good degree of intermixing with the polymer at the interface. Also, the PEG chains on the sepiolite surface have been reported to increase the free volume of the nanocomposite and superficially endorse the stress transfer at the interface, thus, resulting in a higher modulus value [66]. Overall, it should be noted that the interactions suggested in this study are weak in nature due to non-polar nature of the matrix and the surface modification of filler also contributed to effective kinetic mixing of the phases during shearing. The tensile strength of composites with surface modified sepiolite was observed to be similar to the composite with S-P. The elongation at break in the composites with S-PEG, S-PVP, S-PU and S-PVB minerals was almost similar to the pure polymer, which indicated that the incorporation of the filler did not lead to reduction in the toughness of the polymer. It also signified a good degree of filler dispersion in the matrix which did not result in strain hardening of the polymer chains. On the other hand, the composites with S-P and S-EVA*g*MA exhibited lower elongation due to less optimum filler dispersion and presence of large filler aggregates.

Melt rheological properties of the nanocomposites were studied using shear rheometer at 180 °C with 4% strain rate. Figure 7.11 shows the frequency sweep curves of complex viscosity, storage modulus and loss modulus the HDPE/sepiolite composites, in comparison with pure polymer. The complex viscosity of the HDPE and composites decreased with increasing ω, indicating the existence of shear thinning behavior which arises due to the pseudo-plastic nature [67]. This suggested that the nanofiller was finely dispersed in the HDPE matrix. The HDPE nanocomposites with modified sepiolite exhibited lower η^* values as compared to the composite with unmodified sepiolite. This is attributed to the low surface friction between the HDPE matrix and surface modified sepiolite, which resulted in interlayer slipperiness. Among the composites with surface modified fillers, S-EVA*g*MA containing composite exhibited the lowest viscosity. This might be due to the orientation of the copolymer chains on the sepiolite surface with the flow direction of HDPE matrix and probably disentangling from each other, thus, resulting in the reduced viscosity [68]. As the modification is only physically bound to the filler surface, a fraction of these chains can also be expected to dissociate from the filler surface at melt shearing conditions, thus, disturbing the polymer-matrix interface, which can also result in lower viscosity (and modulus) due to reduced stress transfer to the filler particles. In addition, the materials were observed to exhibit lower values of η^* at

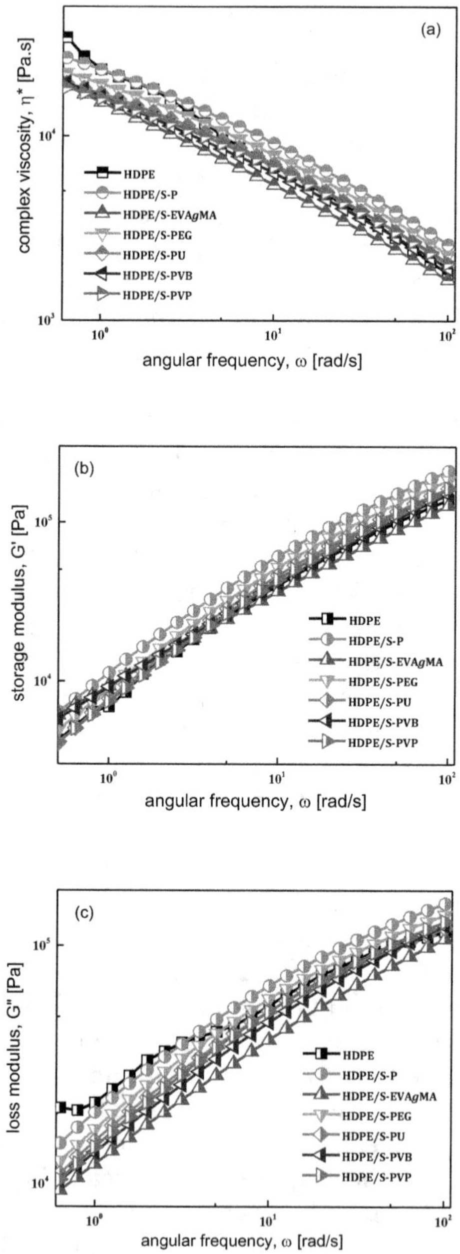

Figure 7.11 (a) Complex viscosity, (b) storage modulus and (c) loss modulus of the HDPE/sepiolite composites as a function of angular frequency.

higher angular frequencies. This is due to the fact that the molecules become completely oriented to the flow direction, thus, resulting in the generation of stable entanglements in shorter time period, which leads to reduction in the resistance to flow [69]. As observed in Figures 7.11 b and c, the G′ and G″ values of the materials increased with increase in angular frequency which indicated that the system behavior was from liquid like to solid like. The composite with unmodified sepiolite exhibited higher values of G' and G" than the composites with modified sepiolite due to the formation of physical network by the unmodified sepiolite inside the HDPE matrix phase. Similar results have also been reported for the composites with pristine and modified carbon nanotubes [70]. On the other hand, in the case of nanocomposites with surface modified sepiolite, the adsorbed polymer chains on the surface of sepiolite may result in lower G' and G" after entanglement with the polymer phase. Such an effect is not observed during the tensile analysis due to the solid state of the sample, where the entanglement of the modifier chains with the polymer matrix results in enhanced stress transfer to the filler particles. Mir *et al.* [71] observed similar G" values of non-crosslinked components compared to crosslinked components at higher frequency region. As mentioned earlier, the melt shearing conditions may also lead to a fraction of the surface adsorbed polymer chains to dissociate from the filler surface, thus, resulting in loss of interfacial contact between the polymer and filler phases. Comparing also the composites with different modified sepiolite minerals, S-PEG and S-PVP containing composites exhibited the highest storage modulus, thus, also corroborating the earlier findings from tensile analysis. It also confirmed that the nature of polymer-filler interface generated due to the type of filler surface modification contributes significantly towards the composite properties. The loss modulus of the composites was also observed to be higher than the storage modulus for the complete angular frequency range, thus, indicating the dominance of viscous nature at the melt shearing conditions.

The morphology of the S-PEG and S-PVP composites was also studied further through microscopy. As observed in the optical micrographs of the composites in Figure 7.12, the filler particles (observed as black domains) were uniformly distributed in the polymer matrix and no large filler aggregates were visible. It, thus, confirmed the earlier findings of enhanced composite properties resulting due to the effective polymer-filler interactions leading to enhanced degree of filler dispersion. Figure 7.13 demonstrates the TEM images of

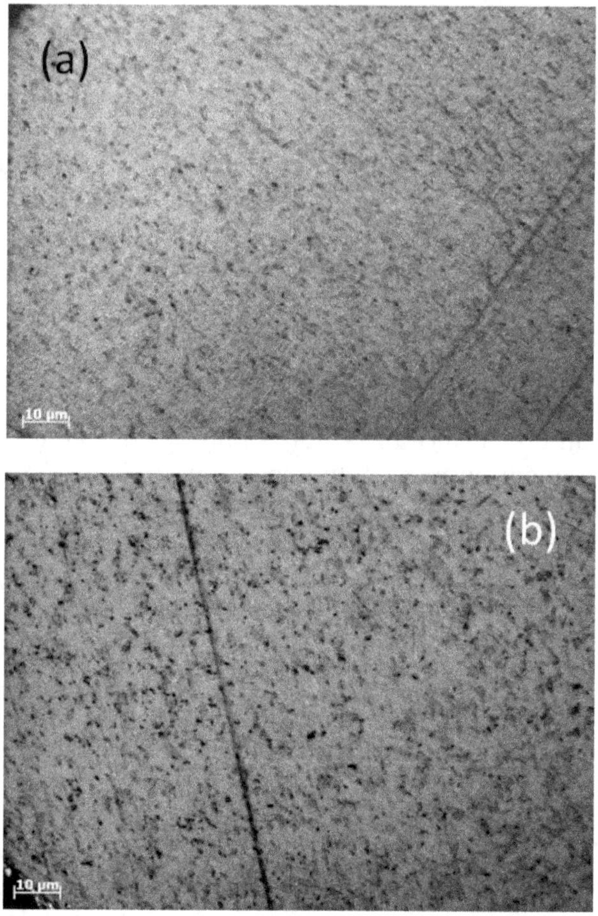

Figure 7.12 Optical micrographs of the (a) HDPE/S-PEG and (b) HDEP/S-PVP nanocomposites.

the composites with S-PEG and S-PVP at two different magnifications, in comparison with the composite with S-P. Though some degree of filler delamination was observed in the case of S-P containing composite, however, it also exhibited the presence of occasional large filler aggregates, as shown in Figure 7.13 a. On the other hand, the composites with both modified sepiolite fillers exhibited large extent of nanoscale filler delamination in the matrix, therefore, confirming the findings from optical microscopy. The enhanced interfacial contacts achieved owing to the observed filler delamination, thus, resulted in superior heat and stress transfer from the polymer chains to the filler

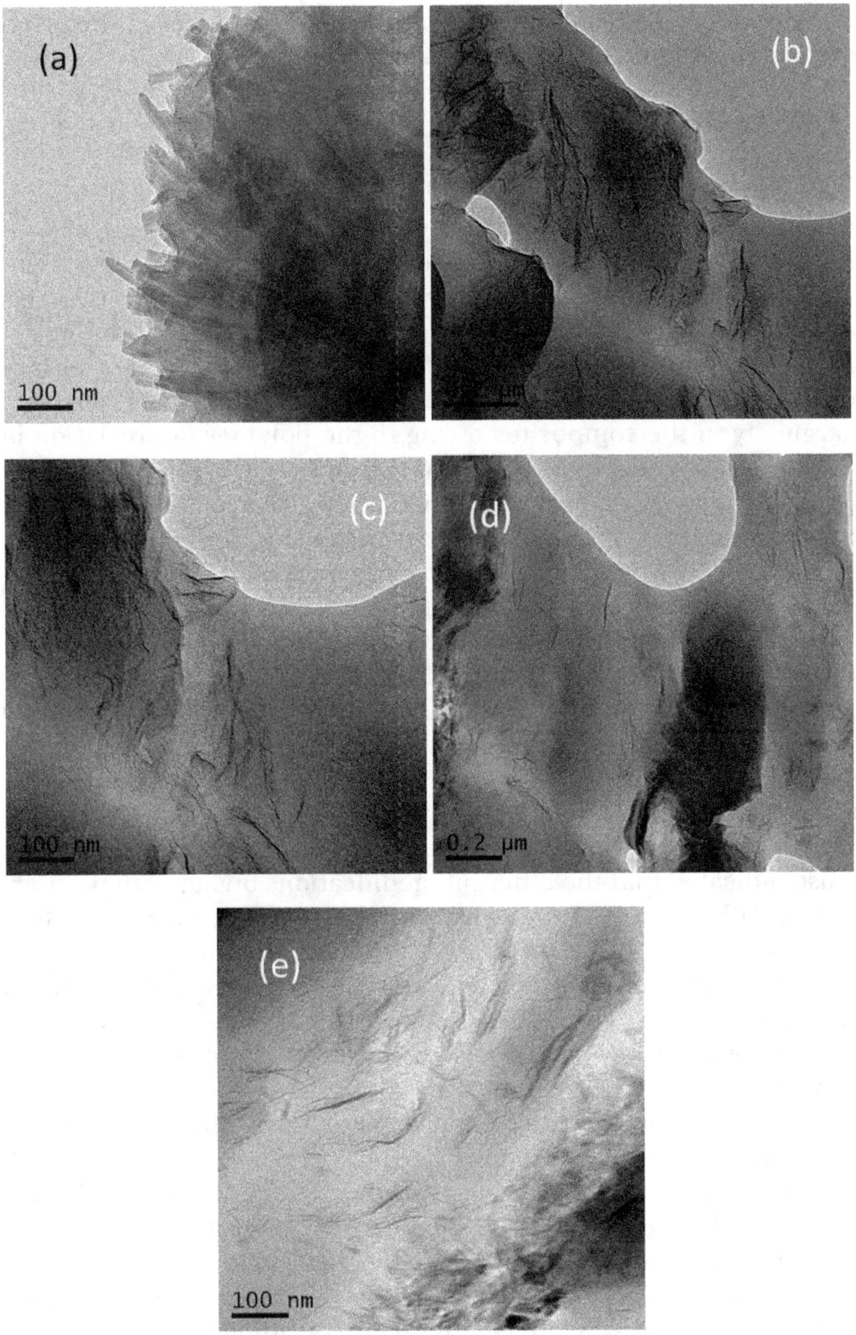

Figure 7.13 TEM images of (a) HDPE/S-P, (b,c) HDPE/S-PEG and (d,e) HDPE/S-PVP.

particles, thereby, resulting in significant improvement of the properties in the S-PEG and S-PVP containing composites.

7.4 Conclusions

In the current study, surface modification of 1-D sepiolite nano-fibrils with a variety of polymers was demonstrated using physical adsorption method. The modification of sepiolite through van der Waals interactions of the polymer chains with the inorganic surface was also confirmed through shifts in IR bands. In addition, the morphology of the sepiolite powder was also confirmed to be affected after surface modification. The basal plane spacing of the modified filler increased marginally in the composites owing to the polymer intercalation in the filler interlayers. In addition, increased extent of filler exfoliation due to the incorporation of modified filler also resulted in efficient heat transfer from the polymer chains to the inorganic filler particles, as demonstrated by the increment in the thermal stability of the polymer in the composites. Composites with S-PEG and S-PVP were observed to have largest increased in the thermal stability. The composites with S-P, S-PEG, S-PVP and S-PU exhibited a significant increase in the degree of crystallinity of the polymer, though the peak melting point of the polymer was observed to be almost constant in the different nanocomposites. S-PEG and S-PVP modified sepiolite also exhibited largest increment in the tensile modulus, thus, indicating good degree of phase intermixing of the modified filler and polymer. It also indicated that the different modifications owing to their interactions with the polymer chains had different effect on overall composite properties. Also, in the absence of any thermodynamic interaction between the polymer and modifier molecules, the effect of shear forces in the compounder as well as changes in crystallinity on the composite properties can also be expected to be significant. The storage modulus of the composites with modified sepiolite was observed to be lower than the composite with unmodified sepiolite owing to chain slipping at the interface and corresponding reduction in the stress transfer at the melt shearing conditions. The morphological analysis of the S-PEG and S-PVP containing composites also revealed uniform dispersion of sepiolite in the polymer matrix, thus, corroborating the earlier findings of enhanced mechanical and thermal properties owing to polymer-filler interactions leading to enhanced degree of filler dispersion. Overall, it could be confirmed that the surface tuning of the sepiolite surface was able to eliminate the need of an external compatibilizer. In specific, the S-PEG and S-PVP containing nanocomposites represent advanced materials with optimized polymer-filler interactions.

References

1. Mittal, V. (2009) Polymer layered silicate nanocomposites: A review. *Materials*, **2**, 992-1057.
2. Mittal, V. (2014) Functional polymer nanocomposites with graphene: A review. *Macromolecular Materials and Engineering*, **299**, 906-931.
3. Reichert, P., Nitz, H., Klinke, S., Brandsch, R., Thomann, R., and Mulhaupt, R. (2000) Poly(propylene)/organoclay nanocomposite formation: Influence of compatibilizer functionality and organoclay modification. *Macromolecular Materials and Engineering*, **275**, 8-17.
4. Uthirakumar, P., Kim, C. J., Nahm, K. S., Hahn, Y. B., and Lee, Y. S. (2004) Preparation and characterization of new difunctional cationic radical initiator-montmorillonite hybrids. *Colloids and Surfaces A: Physicochemical and Engineering Aspects*, **247**, 69-75.
5. Fan, X., Xia, C., Fulghum, T., Park, M. K., Locklin, J., and Advincula, R. C. (2003) Polymer brushes grafted from clay nanoparticles adsorbed on a planar substrate by free radical surface-initiated polymerization. *Langmuir*, **19**, 916-923.
6. Meier, L. P., Shelden, R. A., Caseri, W. R., and Suter, U. W. (1994) Polymerization of styrene with initiator bound to high surface area mica: Grafting via an unexpected mechanism. *Macromolecules*, **27**, 1637-1642.
7. Velten, U., Shelden, R. A., Caseri, W. R., Suter, U. W., and Li, Y. (1999) Polymerization of styrene with peroxide initiator ionically bound to high surface area mica. *Macromolecules*, **32**, 3590-3597.
8. Prucker, O., and Ruhe J. (1998) Mechanism of radical chain polymerizations initiated by azo compounds covalently bound to the surface of spherical particles. *Macromolecules*, **31**, 602-613.
9. Mittal, V. (2007) Polymer chains grafted "to" and "from" layered silicate clay platelets. *Journal of Colloid and Interface Science*, **314**, 141-151.
10. Moet, A. S., and Akelah, A. (1993) Polymer-clay nanocomposites: Polystyrene grafted onto montmorillonite interlayers. *Materials Letters*, **18**, 97-102.
11. Lofas, S., and Johnsson, B. (1990) A novel hydrogel matrix on gold surfaces in surface plasmon resonance sensors for fast and efficient covalent immobilization of ligands. *Journal of the Chemical Society, Chemical Communications*, 1990, 1526.

12. Wilner, I., Riklin, A., Shoham, B., Rivenson, D., and Katz, E. (1993) Development of novel biosensor enzyme electrodes: Glucose oxidase multilayer arrays immobilized onto self-assembled monolayers on electrodes. *Advanced Materials*, **5**, 912.
13. Pan, S., Castner, D. G., and Ratner, B. D. (1998) Multitechnique surface characterization of derivatization efficiencies for hydroxyl-terminated self-assembled monolayers. *Langmuir*, **14**, 3545-3550.
14. Persson, H. H. J., Caseri, W. R., and Suter, U. W. (2001) Versatile method for chemical reactions with self-assembled monolayers of alkanethiols on gold. *Langmuir*, **17**, 3643-3650.
15. Guiomar, A. J., Guthrie, J. T., and Evans, S. D. (1999) Use of mixed self-assembled monolayers in a study of the effect of the microenvironment on immobilized glucose oxidase. *Langmuir*, **15**, 1198-1207.
16. Mittal, V. (2007) Esterification reactions on the surface of layered silicate clay platelets. *Journal of Colloid and Interface Science*, **315**, 135-141.
17. Greenland, D. J., Laby, R. H., and Quirk, J. P. (1965) Adsorption of amino-acids and peptides by montmorillonite and illite. Part 2. Physical adsorption. *Transactions of the Faraday Society*, **61**, 2024-2035.
18. Zhou, Q., Frost, R. L., He, H., Xi, Y., and Liu, H. (2007) Adsorbed para-nitrophenol on HDTMAB organoclay - A TEM and infrared spectroscopic study. *Journal of Colloid and Interface Science*, **307**, 357-363.
19. Zhou, Q., He, H., Frost, R. L., and Xi, Y. (2007) Adsorption of p-nitrophenol on mono-, di-, and rrialkyl surfactant-intercalated organoclays: A comparative study. *Journal of Physical Chemistry C*, **111**, 7487-7493.
20. Mittal, V., and Herle, V. (2008) Physical adsorption of organic molecules on the surface of layered silicate clay platelets: A thermogravimetric study. *Journal of Colloid and Interface Science*, **308**, 295-301.
21. Mittal, V. (2010) Clay exfoliation in polymer nanocomposites: Specific chemical reactions and exchange of specialty modifications on clay surface. *Philosophical Magazine*, **90**(17), 2489- 2506.
22. Chaudhry, A. U., and Mittal, V. (2013) High-density polyethylene nanocomposites using masterbatches of chlorinated polyethylene/graphene oxide. *Polymer Engineering and Science*, **53**, 78-88.
23. Mahboube, M., Hamid, G., and Fatemeh, K. (2014) Structural characterization of high-density polyethylene nanocomposites: effect of compatibilizer type and high aspect ratio organoclay platelets. *Polymer Bulletin*, **71**, 2709-2731.
24. Zebarjad, S. M., and Sajjadi, S. A. (2008) On the strain rate sensitivity of HDPE/$CaCO_3$ nanocomposites. *Materials Science and Engineering A*, **475**, 365-367.
25. Singh, V. P., Vimal, K. K., Kapur, G. S., Sharma, S., and Choudhary, V.

(2016) High-density polyethylene/halloysite nanocomposites: morphology and rheological behaviour under extensional and shear flow. *Journal of Polymer Research*, **23**, 43.

26. Doriga, A., D'Amato, M., and Pegoretti, A. (2012) Thermo-mechanical properties of high density polyethylene – fumed silica nanocomposites: effect of filler surface area and treatment. *Journal of Polymer Research*, **19**, 9889.
27. Chrissafis, K., Paraskevopoulos, K. M., Pavlidoua, E., and Bikiaris, D. (2009) Thermal degradation mechanism of HDPE nanocomposites containing fumed silica nanoparticles. *Thermochimica Acta*, **485**, 65-71.
28. Liao, C. Z., and Tjong, S. C. (2013) Mechanical and thermal performance of high-density polyethylene/alumina nanocomposites. *Journal of Macromolecular Science, Part B: Physics*, **52**, 812-825.
29. Oliveira, P. L., Araujo, R. S., Oliveira, V. R., Morais, J. S., and Marques, M. F. V. (2016) Nanocomposites of polyethylene blends using organomica. *Macromolecular Symposia*, **367**, 143-150.
30. Rezanavaz, R., Razavi Aghjeh, M. K., and Babaluo, A. A. (2010) Rheology, morphology, and thermal behavior of HDPE/clay nanocomposites *Polym Composites*, **31**, 1028-1036.
31. Lee, Y. H., Park, C. B., Sain, M., Kontopoulou, M., and Zheng, W. (2007) Effects of clay dispersion and content on the rheological, mechanical properties, and flame retardance of HDPE/clay nanocomposites. *Journal of Applied Polymer Science*, **105**, 1993-1999.
32. Sanchez-Valdes, S. (2016) Influence of maleated elastomer on filler dispersion, mechanical and antimicrobial properties of hybrid HDPE/clay/silver nanocomposites. *Journal of Adhesion Science and Technology*, **30**, 1006-1016.
33. Mokhena, T. C., and Luyt, A. S. (2014) Investigation of polyethylene/sisal whiskers nanocomposites prepared under different conditions. *Polymer Composites*, **35**, 2221-2233.
34. Lei, Y., Wu, Q., Yao, F., and Xu, Y. (2007) Preparation and properties of recycled HDPE/natural fiber composites. *Composites, Part A*, **38**, 1664-1674.
35. Lee, M.-G., and Nho, Y. C. (2001) Electrical resistivity of carbon black-filled high-density polyethylene (HDPE) composite containing radiation crosslinked HDPE particles. *Radiation Physics and Chemistry*, **61**, 75-79.
36. Jeon, K., Lumata, L., Tokumoto, T., Steven, E., Brooks, J., and Alamo, R. G. (2007) Low electrical conductivity threshold and crystalline morphology of single-walled carbon nanotubes e high density polyethylene nanocomposites characterized by SEM, Raman spectroscopy and AFM. *Polymer*, **48**, 4751-4764.
37. Xiang, D., Jones, E. H., and Linton, D. (2015) Characterization and structure-property relationship of melt-mixed high density

PE/multi-walled carbon nanotube composites under extensional deformation. *RSC Advances*, **5**, 47555-47568.

38. Jiang, X., and Drzal, L. T. (2012) Multifunctional high-density polyethylene nanocomposites produced by incorporation of exfoliated graphene nanoplatelets 2: Crystallization, thermal and electrical properties. *Polymer Composites*, **33**, 636-642.
39. Fim, F. C., Basso, N. R. S., Graebin, A. P., Azambuja, D. S., and Galland, G. B. (2013) Thermal, electrical, and mechanical properties of polyethylene-graphene nanocomposites obtained by in-situ polymerization. *Journal of Applied Polymer Science*, **128**, 2630-2637.
40. Ray, S. S. (2013) *Clay-Containing Polymer Nanocomposites*, Elsevier, USA.
41. Shiraz, N. Z., Enferad, E., Monfared, A., and Mojarrad, M. A. (2013) Preparation of nanocomposite based on exfoliation of montmorillonite in acrylamide thermosensitive polymer. *ISRN Polymer Science*, Article ID 280897.
42. Sapalidis, A. A., Katsaros, F. K., Steriotis, T. A., and Kanellopoulos, N. K. (2012) Properties of poly(vinyl alcohol)-bentonite clay nanocomposite films in relation to polymer-clay interactions. *Journal of Applied Polymer Science*, **123**, 1812-1821.
43. Velasco, J. I., Antunes, M., Ayyad, O., Saiz-Arroyo, C., Rodriguez-Perez, M. A., Hidalgo, F., and de Saja, J. A. (2007) Foams based on low density polyethylene/hectorite nanocomposites: Thermal stability and thermo-mechanical properties. *Journal of Applied Polymer Science*, **105**, 1658-1667.
44. Bilotti, E., Duquesne, E., Deng, H., Zhang, R., Quero, F., Georgiades, S. N., Fischer, H. R., Dubois, P., and Peijs, T. (2014) In situ polymerised polyamide 6/sepiolite nanocomposites: Effect of different interphases. *European Polymer Journal*, **56**, 131-139.
45. Tartaglione, G., Tabuani, D., Camino, G., and Moisio, M. (2008) PP and PBT composites filled with sepiolite: Morphology and thermal behaviour. *Composites Science and Technology*, **68**(2), 451-460.
46. Vengatesan, M. R., Singh, S., Stephen, S., Prasanna, K., Lee, C. W., and Mittal, V. (2017) Facile synthesis of thermally reduced graphene oxide-sepiolite nanohybrid via intercalation and thermal reduction method. *Applied Clay Science*, **135**, 510-515.
47. Chen, H., Zheng, M., Sun, H., and Jia, Q. (2007) Characterization and properties of sepiolite/polyurethane nanocomposites. *Materials Science and Engineering A*, **445-446**, 725-730.
48. Yazdi, Z. T., and Ataei, S. M. (2015) Poly (ether-imide) and related sepiolite nanocomposites: investigation of physical, thermal, and mechanical properties. *Polymers for Advanced Technologies*, **26**, 308-314.
49. Liu, Y., Zhao, J., Deng, C.-L., Chen, L., Wang, D.-Y., and Wang, Y.-Z. (2011) Flame-retardant effect of sepiolite on an intumescent flame-

retardant polypropylene system. *Industrial & Engineering Chemistry Research*, **50**, 2047-2054.

50. Mittal, V., Chaudhry, A. U., Cole, M., and Matsko, N. (2017) Polyethylene nanocomposites with psyllium husk, graphene, sepiolite, and their hybrids. In: *Functional Polymer Blends and Nanocomposites*, Mittal, V. (ed.), Central West Publishing, Australia, pp. 165-194.
51. Nohales, A., Munoz-Espi, R., Felix, P., and Gomez, C. M. (2011) Sepiolite-reinforced epoxy nanocomposites: Thermal, mechanical, and morphological behavior. *Journal of Applied Polymer Science*, **119**, 539-547.
52. Pinnavaia, T. J., and Beall, G. W. (2000) *Polymer-Clay Nanocomposites*, Wiley, USA.
53. Lopez, D. G., Fernandez, J. F., Merino, J. C., Santaren, J., and M.Pastor, J. (2010) Effect of organic modification of sepiolite for PA 6 polymer/organoclay nanocomposites. *Composites Science and Technology*, **70**, 1429-1436.
54. Lemic, J., Tomasevic-Canovic, M., Djuricic, M., and Stanic, T. (2005) Surface modification of sepiolite with quaternary amines. *Journal of Colloid and Interface Science*, **292**, 11-19.
55. Ozdemir, O., Cinar M., Sabah, E., Arslan, F., and Celik, M. S. (2007) Adsorption of anionic surfactants onto sepiolite. *Journal of Hazardous Materials*, **147**, 625-632.
56. Mejia, A., Garcia, N., Guzman, J., and Tiemblo, P. (2014) Surface modification of sepiolite nanofibers with PEG based compounds to prepare polymer electrolytes. *Applied Clay Science*, **95**, 265-274.
57. Alan, N., and Isci, S. (2014) Surface modification of sepiolite particles with polyurethane and polyvinyl alcohol. *Progress in Organic Coatings*, **77**, 444-448.
58. Turhan, Y., Turan, P., Dogan, M., Alkan, M., Namli, H., and Demirbas, O. (2008) Characterization and adsorption properties of chemically modified sepiolite. *Industrial & Engineering Chemistry Research*, **47**, 1883-1895.
59. Vengatesan, M. R., and Mittal, V. (2015) Surface modification of nanomaterials for application in polymer nanocomposites: An overview. In: *Surface Modification of Nanoparticle and Natural Fiber Fillers*, Mittal, V. (ed.), Wiley VCH, Germany, pp. 1-28.
60. Isci, S. (2015) Surface treatment of sepiolite particles with polymers. In: In: *Surface Modification of Nanoparticle and Natural Fiber Fillers*, Mittal, V. (ed.), Wiley VCH, Germany, pp. 151-170.
61. Mittal, V. (2010) Polymer nanocomposites: Synthesis, microstructure, and properties. In: *Optimization of Polymer Nanocomposite Properties*, Mittal, V. (ed.), Wiley VCH, Germany, pp. 1-20.
62. Chivrac, F., Pollet, E., Schmutz, M., and Averous, L. (2010) Starch nano-biocomposites based on needle-like sepiolite clays. *Carbohydrate Polymers*, **80**, 145-153.

63. Roy, N., and Bhowmick, A. K. (2010) Novel in-situ polydimethylsiloxane-sepiolite nanocomposites: Structure-property relationship. *Polymer*, **51**, 5172-5185.
64. Bidsorkhi, H. C., Soheilmoghaddam, M., Pour, R. H., Adelnia, H., and Mohamad, Z. (2014) Mechanical, thermal and flammability properties of ethylenevinyl acetate (EVA)/sepiolite nanocomposites. *Polymer Testing*, **37**, 117-122.
65. Joshi, M., Butola, B. S., Simon, G., and Kukaleva, N. (2006) Rheological and viscoelastic behavior of HDPE/octamethyl-POSS nanocomposites. *Macromolecules*, **39**, 1839-1849.
66. Nanjegowda, L. M., Bommulu, R., Juikar, V., and Hatna, S. (2013) Investigation on the influence of different compatibilizers on polycarbonate and high density polyethylene blends: Mechanical properties, thermal properties, morphology, and chemical resistance. *Industrial & Engineering Chemistry Research*, **52**, 5672-5682.
67. Zhu, L., Fan, H.-N., Yang, Z.-Q., and Xu, X.-H. (2010) Evaluation of phase morphology, rheological, and mechanical properties based on polypropylene toughened with poly(ethylene-co-octene). *Polymer-Plastics Technology and Engineering*, **49**, 208-217.
68. Wu, D., Wang, X., Song, Y., and Jin, R. (2004) Nanocomposites of poly(vinyl chloride) and nanometric calcium carbonate particles: Effects of chlorinated polyethylene on mechanical properties, morphology, and rheology. *Journal of Applied Polymer Science*, **92**, 2714-2723.
69. Kumar, M., Priya, N. S., Kanagaraj, S., and Pugazhenthi, G. (2016) Melt rheological behavior of PMMA nanocomposites reinforced with modified nanoclay. *Nanocomposites*, **2**, 109-116.
70. Ahmad, A. A., Al-Juhani, A. A., Thomas, S., De, S. K., and Atieh, M. A. (2013) Effect of modified and nonmodified carbon nanotubes on the rheological behavior of high density polyethylene nanocomposite. *Journal of Nanomaterials*, Article ID 731860.
71. Mir, S., Yasin, T., Halley, P. J., Siddiqi, H. M., Ozdemir, O., and Nguyen, A. (2013) Thermal and rheological effects of sepiolite in linear low-density polyethylene/starch blend. *Journal of Applied Polymer Science*, **127**, 1330-1337.

8

Polyethylene/Graphene Nanocomposites: A Review of Morphological, Mechanical and Thermal Properties

8.1 Introduction

Polymer nanocomposites, with at least one of the dimensions of the dispersed phase at the nanoscale (<100 nm), are an important area of research because of their potential use in a wide range of applications. There are a variety of uses in industry as well as in daily life that devices made out of neat polymers or metals cannot be put to because they do not meet the requisite thermal, mechanical and electrical properties. The compounding of polymers with nanomaterials, most of which have been found to have excellent mechanical, thermal and electrical properties, has enabled the production of nanocomposites whose properties are significantly better than those of the neat polymers due to the synergetic interaction between the polymers and nanomaterials. Nanocomposites have, due to these improved properties, strong potential of application in various commodity and engineering sectors. Out of various reinforcement materials used to develop nanocomposites, silica, alumino-silicate clays, carbon nanotubes and graphene have been extensively reported to impart significant improvement in a variety of polymer properties. For instance, Figure 8.1 shows the height, amplitude, and phase images of the compatibilized polyethylene (PE) nanocomposites, using silica and alumino-silicate as fillers and ethylene vinyl acetate-graft-maleic anhydride (EVA-*g*-MA) as compatibilizer. In the phase images, the filler particles were white in color signifying harder regions, whereas the polymer phase was dark. The phase of the polymer near the interface with the filler surface as well as in the bulk was similar indicating that the filler did not have any specific effect on the crystallization process. The filler-polymer interface in the micrographs also indicated that the compatibility between the phases was significantly improved in the presence of EVA-*g*-MA as compared to the uncompatibilized

Baron Rubahamya and Vikas Mittal, The Petroleum Institute (part of Khalifa University of Science and Technology), Abu Dhabi, UAE*
**Current address: Bletchington, Wellington County, Australia*

composites, which thus impacted the mechanical and thermal properties of the nanocomposite.

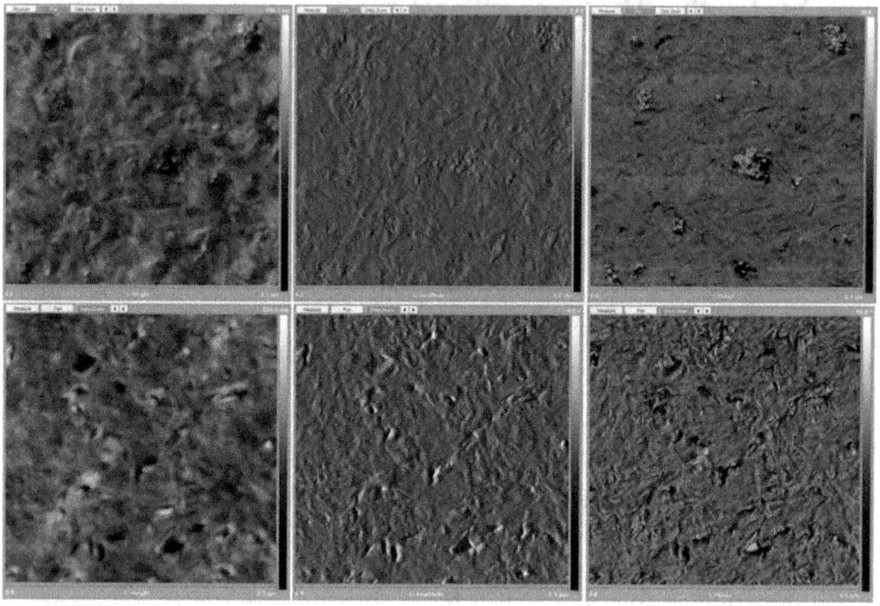

Figure 8.1 AFM height, amplitude and phase images of PE+EVA-*g*-MA+Silica (upper row) and PE+EVA-*g*-MA+Silicate (lower row) composites at lower magnification. Reproduced from Reference 1 with permission from Sage Publications.

Graphene has, since its discovery, gained significant interest from researchers because of its superior properties [2]. Graphene is a two-dimensional nanomaterial made up of sp^2 carbon atoms, has the thickness of one atom and a honeycomb structure. It is the strongest single layer material that has ever been measured having a Young's modulus of 1 TPa and a tensile strength of 130 GPa [3]. Graphene also shows excellent electrical and thermal conductivity values, being 6000 S/cm and 5000 W/(m.K) respectively [4]. These properties coupled with its vast surface area make graphene a nanomaterial of great interest for research concerning its application in making nanocomposites that leverage its individual properties to produce materials of greatly improved mechanical, thermal and electrical properties compared to the neat polymers. This increased interest in graphene can be evidenced by the great volume of research studies on polymer nanocomposites with graphene reported in recent years [2,5].

The synthesis of graphene has been achieved through various routes including bottom-up as well as top-down processes. Some of the top-down processes include the exfoliation of graphite and the derivatives of graphite, most commonly graphite oxide, into sheets of graphene or modified graphene [6]. The exfoliation route of graphene synthesis by directly exfoliating graphite has transitioned from the earlier techniques of micromechanically breaking down graphite [7] to the recent techniques of exfoliating graphite into single layer and multiple layer graphene by dissolving it in superacids [8], electrochemically functionalizing it using ionic liquids [9] as well as the use of sonication [10]. On the other hand, the exfoliation of graphite into sheets of graphite oxide and their subsequent reduction, provides a great potential for the production of graphene on a large scale. The most commonly used methods for the production of graphene from graphite oxide are the Hummer's method [11] and a number of variations of this method, in which graphite is subjected to a process of oxidation in the presence of nitric acid using strong oxidants such potassium permanganate. Bottom-up methods though lead to better control on the structure and properties of graphene, however, these have not been as widely used as top-down methods because of the costly process and the fact that these lead to production of low amounts of graphene. Some of the methods that have been used include chemical vapor deposition [12], surfactant self-assembly [13], reduction of carbon monoxide [14] and arc discharge [15].

To be able to obtain the desired properties that result from compounding polymers with nanomaterials such as graphene, the dispersion of the nanomaterial in the polymer matrix is at nanoscale is of great importance. Thus, a higher degree of graphene dispersion in the polymer matrix has been commonly reported to result in greater improvement of properties [16]. One important advantage of the nanofillers such as graphene is the possibility to achieve improvements in the properties of the polymers with low amounts of filler loading due to the high surface areas and high aspect ratios of these nano-fillers [17]. The interfacial interaction between the polymer matrix and graphene nano-sheets is also influenced by various factors including surface reactive groups, hydrophobicity and molecular weight of the polymer chains, functional groups in the polymer chains, among others. In addition, the dispersion of graphene in the polymer matrices is also influenced by the method of compounding of the polymer with graphene [4] and compounding techniques include, but are not limited to, the following commonly used ones [18]:

In-situ intercalative polymerization: In this technique, graphene or its modified version is simultaneously compounded with the polymer matrix during the process of polymerization. This is performed by mixing graphene with the monomer in the presence of an initiator and subsequently providing the optimum conditions for polymerization.

Intercalation with the polymer or pre-polymer in solution: In this method, a solvent in which the polymer or a pre-polymer (such as an oligomer) can dissolve and in which graphene or its modified form can swell is used. The dissolved polymer and swollen graphene are mixed in solution allowing for the graphene sheets to be intercalated by the polymer chains and a nanocomposite is achieved when the solvent is removed, by means of evaporation. Even though the entropy of the system is reduced due to the restraining of the polymer chains within the graphene sheets, this process is thermodynamically possible due to the increase in entropy achieved by the desorption of solvent molecules from the surface of the graphene sheets.

Melt intercalation: In this method, the polymer forms a melt at a raised temperature and in the molten state is mixed with graphene or modified graphene under shear, such as by using a rotor. The mixture is then cooled to obtain a nanocomposite in which the graphene sheets are exfoliated and intercalated by the polymer chains. The temperature of compounding should be carefully chosen to achieve thorough mixing without degradation of the polymer as well to achieve the optimum viscosity at which the polymer melt can withstand the shear stress in the compounder. This method has been widely used to synthesize nanocomposites of graphene with thermoplastic polymers [5].

As mentioned earlier, the interfacial interaction between the polymer matrix and graphene plays an important role in the extent of dispersion of the graphene sheets. The graphene sheets have the tendency to agglomerate by means of van der Waals forces and through π-π stacking, especially when added in large amounts [19]. This phenomenon is usually irreversible to a large extent and the agglomerates formed lead to significant reduction in the surface area and aspect ratio. To achieve improved interfacial interactions between graphene and polymer matrices, and consequently improved dispersion and nanocomposite properties, graphene sheets' surfaces are usually chemically functionalized. For the dispersion of graphene in polar polymers such as polyurethane, polyamide and poly (vinyl alcohol), the surfaces of the nano-sheets are modified with polar groups including epoxies, hydroxyl groups and carboxylic acid groups [20-22], while for dispersion in non-polar polymer matrices such as polyolefins, the surface is usually rendered organophilic by the incorporation of groups including maleic acid grafted polyolefins, alkylamines, phenyl isocyanate, silanes, and others [23,24].

PE is a widely used thermoplastic polyolefin due to its large number of beneficial properties. The most common forms of PE include low density polyethylene (LDPE), high density polyethylene (HDPE), ultra-high molecular weight polyethylene (UHMWPE) and linear low density polyethylene (LLDPE). LDPE has a melting temperature between 100 and 120 °C, a specific gravity ranging from 0.91 to 0.94 and has partial crystallinity ranging between 50 to 70%. LDPE exhibits a greater level of branching compared to HDPE, with 10 to 20 branches for every 1000 carbon atoms compared to HDPE's average 1 branch for every 200 carbon atoms in the main chain. Existence of branches in the carbon chains is a defect and affects the degree of crystallinity, which along with molecular weight and molecular weight distribution determines the structural properties of polyethylene. Mechanical properties including tensile strength, rigidity and others are observed to be directly proportional to crystallinity, and consequently inversely proportional to the amount of branching within the chain. Toughness of polyethylene, however, decreases with increasing degree of crystallinity [25]. Polyethylene finds application in a variety of areas including electrical appliances, industrial equipment, construction, body part replacements, packaging, transportation and agriculture, among others [25]. The wide range of applications of polyethylene, therefore, makes it a polymer of interest for further developing its nanocomposites, thus, expanding its spectrum of applications further.

This study aims to review recent studies reporting the development of polyethylene/graphene nanocomposites with a focus on the composite properties, especially morphological, mechanical and thermal properties. In addition, these properties have also been correlated with the synthesis methods as well as filler type, content, compatibilization and surface functionalization. Specifically, various mechanical properties such as Young's modulus, yield stress, yield strain, tensile strength, elongation at break, impact strength and dynamic mechanical response have been reviewed. On the other hand, thermal properties such as glass transition temperature, melting temperature, degree of crystallinity, thermal degradation, etc., have been studied. In addition, some insights into the rheological performance of nanocomposites have also been provided.

8.2 Review of Nanocomposite Properties

Along with other applications, UHMWPE has been extensively studied for its use in replacement of human body parts such as the joints of knee, shoulder, hip and spine. This requires for it to have specific mechanical properties such as high tensile strength to ensure increased wear resistance and prevention of early failure as well as high yield stress to ensure that the material does not deform easily. To this end, attempts have been made to improve the inherent mechanical properties of UHMWPE for it to be able to meet the requirements of this application and some of the techniques that have been used include increasing its degree of crosslinking. UHMWPE/graphene nanocomposites present a potential alternative for meeting these goals. A range of synthesis methods have been reported to generate these UHMWPE nanocomposites in the recent years. For instance, Lahiri *et al.* [26] synthesized a nanocomposite of UHMWPE and graphene nano-platelets (GNP) by an electrostatic spraying procedure. In this procedure, GNPs were uniformly dispersed in acetone using ultrasonication and UHMWPE powder was added subsequently with further ultrasonication. It was followed by removal of the solvent by means of evaporation. Electrostatic spraying was then used to deposit films of the nanocomposite on non-stick substrate coated with Teflon. The films were then cured, cooled and peeled off the substrate. The compositions of the synthesized films were neat UHMWPE, UHMWPE with 0.1 wt% as well as 1 wt% of GNPs. Analysis of the mechanical properties of these films exhibited that the elastic modulus of neat UHMWPE increased from 0.53 GPa to 0.69 GPa for its nanocomposite containing 0.1 wt% GNPs and increased further to 1.19 GPa when the GNPs concentration was 1 wt%. The same trend was observed for yield stress which increased from 22 to 35 and then to 42 MPa respectively. Tensile strength and toughness, on the other hand, showed different trends with the tensile strength increasing from 45 MPa for neat UHMWPE to 77 MPa when the nanocomposite contained 0.1 wt% GNPs, but then reduced to 68 MPa when the GNPs content was 1 wt%. Toughness, on the other hand, was initially 98 $J.m^{-3}$, increasing to 151 $J.m^{-3}$ before reducing to 30 $J.m^{-3}$ respectively.

These trends were explained by the fact that the stress transfer from UHMWPE to the platelets resulted in the continuous increase in elastic modulus and yield stress with increase in weight percentage of the platelets. However, due to the agglomeration that resulted from increased amounts of GNPs, the interfacial interactions between UHMWPE and the platelets deteriorated making the nanocomposites brittle with reduced tensile strength and toughness. In another study, Suner *et al.* [27] prepared nanocomposites of UHMWPE with graphene oxide (GO) using a ball milling technique in which a slurry of UHMWPE powder and GO powder in ethanol was mixed using a planetary ball mill. Afterwards, the solvent was evaporated and the nanocomposite sheets were molded from the powders using a hot press. The nanocomposites were prepared with GO content of 0.1, 0.3, 0.5, 0.7, 1 and 2 wt%. The mechanical properties exhibited the same trends as noted by Lahiri *et al.* [26] where improved properties as compared to neat UHMWPE were achieved with optimum loading of 0.5 wt% of GO and higher amounts of loading resulting in decrease in the values. Thus, the toughness, fracture stress and fracture strain of the nanocomposite with 2 wt% GO were observed to be lower than those of neat UHMWPE. The melting temperature, temperature of crystallization and degree of crystallization values were constant with increasing GO content and were observed to be same as neat UHMWPE. The nanocomposites, however, exhibited improved thermal stability where the oxidation temperature, the linear weight loss initiation temperature and the temperature of complete volatilization increased with increasing GO content.

Sturzel *et al.* [28] synthesized UHMWPE/graphene nanocomposite via an *in-situ* polymerization route (Figure 8.2). GO was prepared

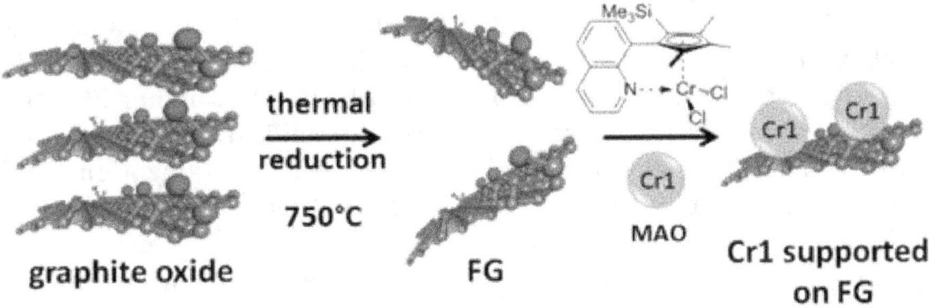

Figure 8.2 Synthesis of UHMWPE/graphene nanocomposite via *in-situ* polymerization. Reproduced from Reference 28 with permission from American Chemical Society.

from graphite by means of a modified Hummer's method before being thermally reduced to obtain functionalized graphene (FG). By dispersing FG/methylaluminoxane (MAO) in n-heptane, the authors immobilized catalyst sites composed of MAO-activated chromium on the FG to enable the dispersion of the nano-sheets in UHMWPE to happen simultaneously with the polymerization process under optimum conditions. Nanocomposite samples were obtained using hot press molding with varying FG compositions of 0.5, 1, 2.5, 5 and 10 wt%. The Young's modulus increased with increasing FG content of the nanocomposite, increasing from 420 MPa for the commercial UHMWPE used as the reference point through 630 MPa for the composite with 0.5 wt% FG content up to 991 MPa for 10 wt% FG content. The elongation at break and tensile strength, on the other hand, increased with increasing FG content until 1 wt% FG content, but then reduced as the content of filler increased, with the 10 wt% FG content nanocomposites exhibiting even lower properties than the commercial UHMWPE. This observation further evidenced that maximization of the mechanical properties of UHMWPE/graphene nanocomposites can be attained at low graphene loadings and deterioration happens with large loading amounts of the nanomaterial. It should also be noted that Sturzel *et al.* [28] achieved greater values of tensile strength for the nanocomposites at the same graphene loading as Lahiri *et al.* [26] and Suner *et al.* [27]. This could be due to the use of FG and *in-situ* polymerization technique that improved interfacial interaction between the polymer matrix and filler phases as well as reduced the agglomeration tendency of the filler nano-sheets. In addition, the authors indicated that the melting temperature of the nanocomposites was almost constant with increasing filler content with a difference of only a few degrees (~3 °C), while the temperature of crystallization increased when FG was added to the polymer with the highest temperature being observed at 0.5 wt% fraction. Lower values of crystallization temperature were observed at higher FG content, but these were still greater than the crystallization temperature of neat UHMWPE. The observed increase in the crystallization temperature was due to the nucleation sites that were provided by the FG filler.

In the study reported by Chen *et al.* [29], powdery GO generated using modified Hummer's method from graphite was uniformly dispersed by means of ultrasonication in alcohol to form a suspension to which UHMWPE powder was added and the solvent evaporated after further ultrasonication. Nanocomposite samples of 0, 0.1, 0.3, 0.5 and

1 wt% GO content were then molded from the solid using hot pressing. The yield strength was observed to increase with increasing GO content while the ultimate tensile strength and elongation at break both reached maximum at 0.5 wt% GO content before showing a tremendous decrease for 1 wt% GO composite. These findings were also similar to the observation reported by Suner *et al.* [27] for the composites of UHMWPE with derivative of graphene (GO) using the similar route of nanocomposite synthesis. The authors also reported that the melting point of the nanocomposites was constant, while the degree of crystallization increased with addition of GO with the highest observed degree of crystallization obtained at 0.1 wt% GO content. It was observed to subsequently reduce as more GO was added. The filler provided nucleation sites, but high filler content led to agglomeration, thus, resulting in poor dispersion.

Kuila *et al.* [30] functionalized graphite oxide (GO) using dodecylamine (DA) through the nucleophilic substitution of the DA moieties onto the epoxy groups of GO and subsequently reduced the product and dried it. The functionalized GO was then compounded with linear low density polyethylene (LLDPE) by means of solution dispersion using xylene as the solvent. Nanocomposite samples with varying amounts of filler were compression molded. Analysis of the dynamic mechanical properties of the nanocomposites indicated that the storage modulus increased with increasing amount of functionalized GO in the polymer matrix, whereby the nanocomposite sample was observed to have a storage modulus 118% higher than that of virgin LLDPE. The observed improvement in the properties was attributed to the filler decreasing the mobility of the LLDPE chains. Loss modulus measurements were not reported by the authors, but the loss factor at 80 °C showed a general decrease with increasing filler content in the nanocomposite. The nanocomposites exhibited constant melting temperature, while the crystallization temperature increased with increasing graphene content. It was observed to reduce at 8 wt% of functionalized reduced graphene oxide content, yet was still greater than that of neat LLDPE. TGA also revealed that the onset of degradation occurred at lower temperatures for the nanocomposites than for neat LLDPE because of the degradation of the dodecylamine. However, the thermal stability of the nanocomposites increased with increasing filler content as was seen based on the increasing temperatures required for 50% and 80% weight loss as the filler content increased. This was attributed to the filler nano-sheets reducing the rate of degradation by preventing the movement of radicals and flow of oxygen through the polymer matrix, thereby, leading to a shielding effect.

In another study, Yang *et al.* [19] studied the influence of different chain length alkylamine functionalizing agents on the properties of the nanocomposites of functionalized graphene oxide and LDPE. Octadecylamine (ODA), dodecylamine (DDA) and octylamine (OTA) were the three different alkylamines utilized to functionalize reduced graphene oxide, which was used to make nanocomposites with LDPE by means of solution dispersion. The nanocomposites with 0.5 wt% filler loading of reduced graphene oxide functionalized with each of the different chain length alkylamines were analyzed for their properties. It was observed that in each of the three cases, the Young's modulus and tensile strength were higher for the nanocomposite than that of neat LDPE, whereas the elongation at break reduced. Good filler dispersion and improved interfacial interaction were the reasons for stress transfer from the polymer matrix to the filler and resulted in improved mechanical properties. However, the reduction in elongation at break was attributed to certain extent of agglomeration. The filler modified with DDA had higher tensile strength than the other two nanocomposites at the same filler loading, with the nanocomposite with OTA functionalization leading to the least tensile strength. Characterization of the functionalized reduced graphene oxide showed that DDA functionalized reduced graphene oxide contained more primitive graphene than ODA which was reported to be the reason for the higher tensile strength. The temperature of maximum weight loss of the nanocomposites was reported to be greater than that of neat LDPE owing to the barrier effect and the slowing down of oxygen permeation. These results were also similar to those reported by Kuila *et al.* [30].

Mittal and Chaudhry [31] studied the effect of amphiphilic compatibilizers on the properties of nanocomposites of high density polyethylene and thermally reduced graphene. Nanocomposites were prepared by melt mixing the polymer and various types of compatibilizers with thermally reduced graphene with a composition of 5 wt% compatibilizer and 1 wt % thermally reduced graphene. Different compatibilizers resulted in different interactions between the polymer matrix and filler phases. Specifically, the molecular weight, extent of grafting, and polarity in the compatibilizers were the factors which determined their interactions with the filler surface. It was observed that the highest improvements in tensile modulus occurred in the nanocomposites generated using 25% chlorinated polyethylene

(CPE25), ethylene-α-olefin copolymer (E-AO) and ethylene acrylic acid copolymer (EAA), while 35% chlorinated polyethylene (CPE35) resulted in increased plasticization because of its amorphous nature. Dynamic mechanical properties also exhibited improvement for the nanocomposites whereby storage modulus was reported to increase to a varying degree. To attain an optimum amount of compatibilizer in the nanocomposites for property enhancement, the amount of best performing compatibilizer EAA was varied in the composites. As observed in Figure 8.3, the extent of filler exfoliation was observed to increase with increasing EAA content, thus, confirming positive interactions between EAA and thermally reduced graphene, though no specific chemical interactions could be detected. The observed

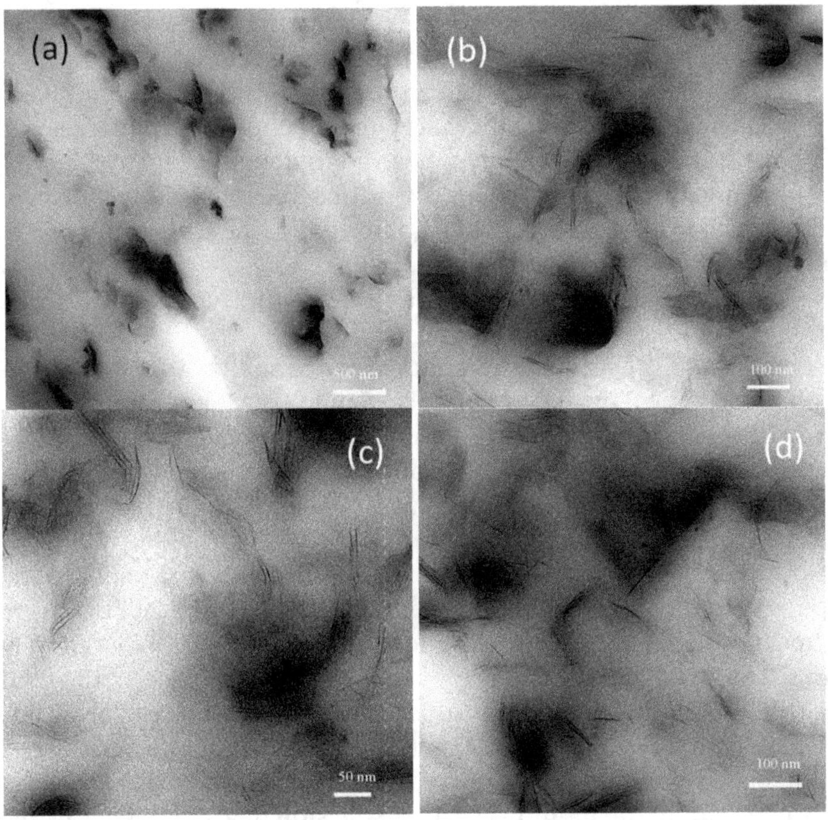

Figure 8.3 TEM micrographs of (a) and (b) PE/G/5%EAA nanocomposite at different magnifications; (c) PE/G/10% EAA nanocomposite and (d) PE/G/15%EAA nanocomposite. Reproduced from Reference 31 with permission from Wiley.

properties from these composites revealed that 7.5 wt% was the optimum amount of EAA in the composites as the mechanical and rheological properties exhibited improvement due to filler exfoliation. Beyond 7.5 wt% content of EAA in the composites, the extent of filler exfoliation though was observed to increase, the performance of the nanocomposites deteriorated due to matrix plasticization. The authors also observed that the compatibilized nanocomposites had higher thermal stability than the non-compatibilized HDPE/G nanocomposite, which was itself more thermally stable than neat HDPE. Nanocomposites synthesized using ethylene acrylic acid copolymer (EAA) as compatibilizer were reported to have the highest thermal stability while those synthesized using 25% chlorinated polyethylene (CPE25) showed increased melt enthalpy because of its semi-crystalline nature. Other nanocomposites, including non-compatibilized HDPE/G nanocomposite, were reported to show lower melt enthalpy than neat HDPE because of reduced crystallinity. Nanocomposites synthesized using ethylene methacrylic acid copolymer with zinc ion, ethylene butyl acrylate copolymer and ethylene methyl acrylate copolymer had dual peak in DSC micrographs due to possible incompatibility, similar to the composites generated using ethylene-*g*-maleic anhydride copolymer, ethylene-α-olefin copolymer and ethylene vinyl acetate-*g*-maleic anhydride due to lack of miscibility on the molecular level.

Park *et al.* [32] studied the effect of functionalization of graphene on the mechanical properties of polyethylene/graphene nanocomposites. Unfunctionalized graphene oxide (GO), thermal reduced graphene oxide (TRG) and 3-isopropenyl-α,α-dimethylisocyanate (TMI)/octadecylamine functionalized graphene oxide (FGO) were dispersed in HDPE, LLDPE and oxidized polyethylene (OPE) by means of solvent blending. The tensile modulus of all polyethylene types was enhanced upon incorporation of FGO. The FGO/HDPE nanocomposite exhibited the highest tensile modulus, whereas TRG and GO showed less increase and a slight decrease was reported for the OPE/graphene nanocomposites. Due to the low molecular weight, viscosity and oxide groups of OPE, formation of unstable heterogeneous nanocomposites was suggested, which led to the observed lack of improvement in the modulus upon dispersion of graphene in the matrix. FGO and TRG were well dispersed in polyethylene whereas GO was poorly dispersed, therefore, inconsistent increase in normalized modulus was reported for polyethylene/GO nanocomposites. In another study, Kim *et al.* [33] also reported the effect of polyethylene functionalization and blending methods on the properties of polyethylene/graphene nanocomposites. TRG was blended with linear low density polyethylene (LLDPE) and its amine, nitrile, and isocyanate functionalized analogs using both melt and solution blending. Graphene platelets were well exfoliated in functionalized LLDPE matrices, whereas phase separation was observed for the unmodified LLDPE. Solvent based blending was also observed to be more effective in dispersing the graphene platelets as compared to melt compounding.

Jiang and Drzal [34] also reported that the thermal stability and thermal conductivity of PE/exfoliated graphene nano-platelets based nanocomposites enhanced with increasing graphene concentration. In another study, Jiang and Drzal [35] coated the graphene nanoplatelets (GNPs) with paraffin wax before dispersing in HDPE by means of melt extrusion and injection-molding to make the test sample coupons. A nanocomposite of graphene without paraffin wax coating was generated by the same method to be used as a reference and samples with varying paraffin wax to GNP weight ratio were prepared. For all the nanocomposites, a 5 vol% filler loading was used. The flexural strength and flexural modulus were observed to initially increase with increased paraffin wax to GNPs ratio, but then reduced when the paraffin wax content of the nanocomposites exceeded 5 wt%. Further tests to affirm the reason for this occurrence attributed the observed effect to the presence of excess low molecular weight paraffin wax in the nanocomposites, thus, leading to the deterioration of the mechanical properties after the 5 wt% threshold.

El Achaby *et al.* [36] melt-compounded reduced graphene oxide nano-sheets (GNs) with HDPE in a twin screw mixer without modifying the surface of the GNs and obtained the test samples by means of hot press molding. The nanocomposites had filler loadings of 0.5, 1 and 3 wt%. Dynamic mechanical analysis showed that both storage and loss moduli of the nanocomposites increased with increasing filler loading and that the storage modulus increased more than the loss modulus. The tensile strength and Young's modulus were also observed to increase with filler loading. The tensile strength was observed to increase by 77% and the Young's modulus by 87% as compared to neat HDPE when the polymer was compounded with 3 wt% GNs. The nanocomposites also exhibited increasing thermal stability with increasing filler content. TGA revealed that the initiation of weight loss occurred at a higher temperature and the temperature required to degrade 5% of polymer weight significantly increased from 431 °C for neat HDPE to 466 °C for the nanocomposite with 3 wt% GNs content.

Mittal and Chaudhry [37] studied the effect of graphene content on the properties of PE/graphene nanocomposites. Nanocomposite with 10% filler content had significantly low elongation indicating the presence of filler aggregates causing stain hardening of polymer chains. The tensile modulus exhibited gradual enhancement with filler fraction and at 10% filler content, an increment of 62% in tensile modulus as compared to pure polymer was observed. The yield stress was observed to remain constant with filler fraction except at 10% content where a small decrease was observed. The yield strain remained unchanged till 3% filler content and exhibited significant reduction on further enhancing the filler content and attained a plateau value at 7% graphene oxide amount. Peak stress was also observed to increase as a function of filler fraction, except for 10% filler composite. The enthalpy of HDPE in the composites (normalized to 100% polymer) was observed to decrease gradually with increasing filler content indicating that the filler stacks hindered the polymer crystallization. As shown in Figure 8.4, the peak melting point was observed to remain largely unchanged. Due to hindrance to crystallization on graphene addition, the peak crystallization temperature was also observed to decrease by 2-3 °C as a function of filler content. The onset and peak degradation temperatures were observed to increase with filler content confirming the efficient heat transfer from polymer chains to graphene.

Chaudhry and Mittal [38] also reported graphene-based nanocomposites generated by solution mixing of the graphene filler with chlorinated polyethylene (CPE) compatibilizers (of different chlorination levels of 25% and 35%, thus, denoted as CPE25 and CPE35) followed by melt mixing of the solution mixed masterbatches with high density polyethylene (Figure 8.5). The CPE35 masterbatch was observed to have better graphene dispersion as compared with CPE25 containing masterbatch. Though complete nanoscale delamination of the graphene platelets was not observed in the composites, however, the composites with CPE35 compatibilizer also had much better filler dispersion as compared with corresponding CPE25 nanocomposites. Higher extent of chlorination in the matrix resulted in higher magnitude of interfacial interactions between the polymer and the filler surface, resulting in increased extent of filler delamination. Peak melting point as well as overall crystallinity was affected by graphene oxide as well as compatibilizers. The mechanical performance of the

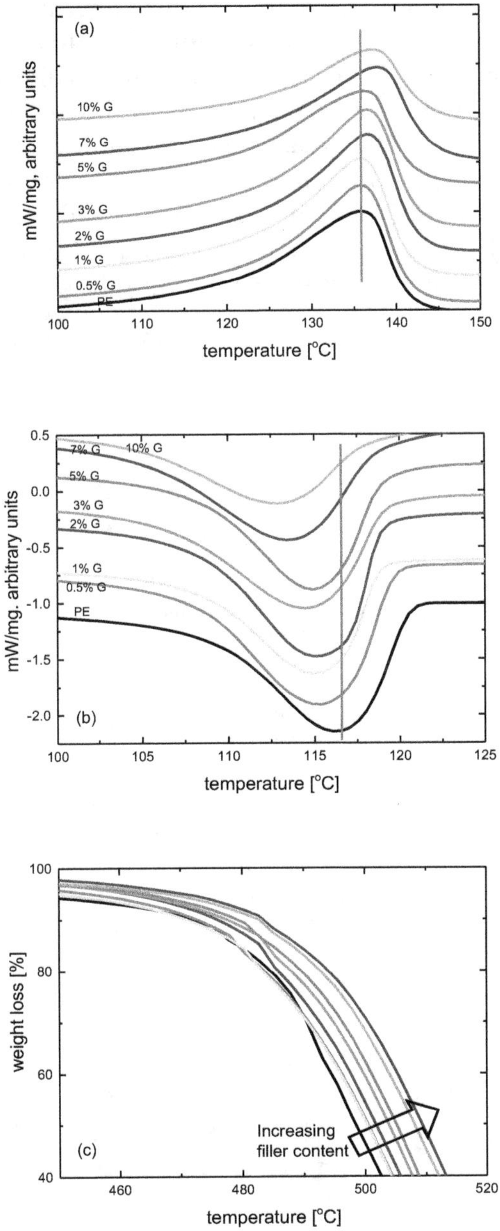

Figure 8.4 DSC (a) melting and (b) crystallization thermograms of HDPE and its nanocomposites with increasing filler content; (c) TGA thermograms of the polymer and nanocomposites. Reproduced from Reference 37 with permission from Wiley.

nanocomposites was affected by the factors like filler delamination, plasticization of the matrix, compatibilizer content as well as their chlorination level and amorphous/semi-crystalline nature. An increase of 16% in tensile modulus and 21% in tensile strength of the composites with 5 wt% of 25% chlorinated compatibilizer at 0.5 wt% graphene content was observed. The nanocomposites were also observed to have reduced low-temperature brittleness on the addition of CPE (especially CPE35) which was also retained even on graphene addition [39]. The height corrugation analysis at different places of the scanned area in the atomic force micrographs (25 mm^2) indicated

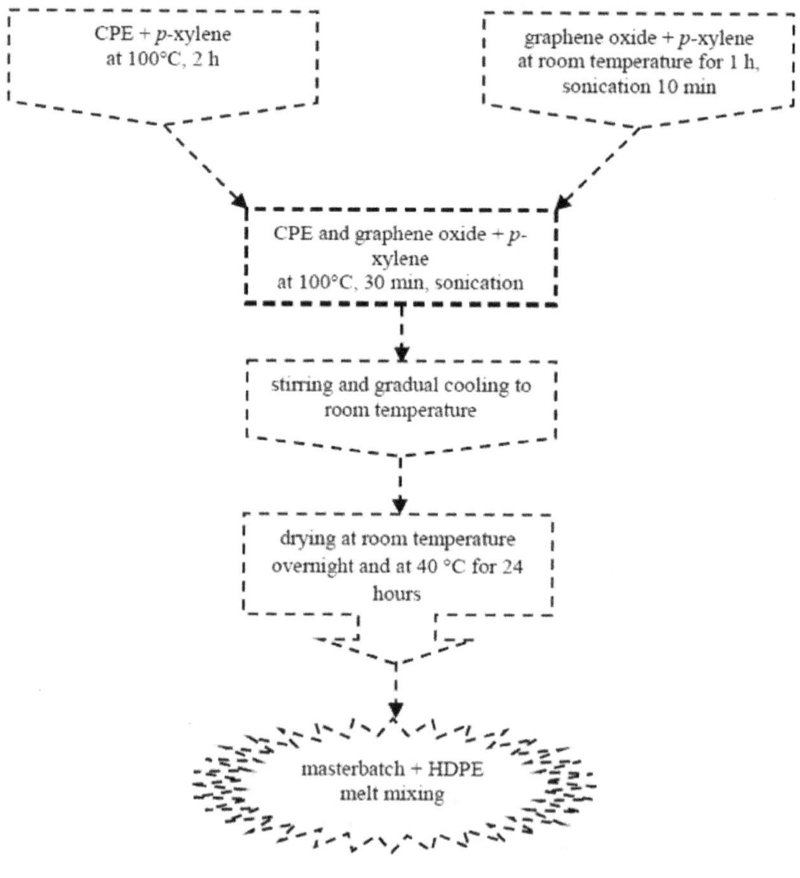

Figure 8.5 Schematic representation of the synthesis of CPE-graphene oxide masterbatches by solution mixing and subsequently HDPE nanocomposites by melt mixing. Reproduced from Reference 38 with permission from Wiley.

that the composites followed similar height variation trend as the PE-CPE blends, however, the height variation was even higher in the composite samples. For example, for PE-G sample, a value of 113 nm for surface variation was calculated which was enhanced markedly to 198 nm for PE-G-CPE25 nanocomposite and 564 nm for PE-G-CPE35 nanocomposite. The samples of atomic force microscopy were prepared at -140 °C, whereas the analysis was performed at room temperature.

In other studies reporting the use of *in-situ* polymerization, Fim *et al.* [40] synthesized HDPE/graphene nanocomposites using graphene nano-sheets (GNSs) with methylaluminoxane as cocatalyst and Cp_2ZrCl_2 as catalyst. Dispersion of the GNSs took place simultaneously along with the polymerization process. Dynamic mechanical analysis exhibited that both storage and loss moduli were higher for the nanocomposites than neat HDPE. The tensile strength and elongation at break values of the nanocomposites were, however, lower than those of neat HDPE. Since the filler loadings used in this study were above 1 wt%, the observation about reduction of tensile strength and elongation at break also mirrored the results of Sturzel *et al.* [28], where *in-situ* polymerization too was used for the composites generation and the tensile strength and elongation at break reduced beyond 1 wt% filler loading due to filler agglomeration. Elastic modulus, on the other hand, was reported to increase with increasing graphene content, as also reported by Sturzel *et al.* [28]. The authors also reported an increase in the thermal stability of the nanocomposites with increase in content of graphene nano-sheets (GNSs). This was indicated by an increase in the temperatures for weight loss onset and maximum weight loss. Comparing different synthesis methods, the observed thermal findings indicated that increased thermal stability was observed in the nanocomposites, irrespective of the different synthesis methods. Synthesis of polyethylene/graphene nanocomposites by *in-situ* polymerization was also reported by Zhang *et al.* [41,42]. In both cases, filler/$MgCl_2$-supported titanium-based Ziegler-Natta catalysts prepared by a facile spray-coagglomeration method were used to synthesize polyethylene and distribute the filler within the polymer matrix. Polyethylene/graphene and polyethylene/graphene oxide nanocomposites prepared by this method [41] exhibited improvement of tensile strength, modulus and elongation at break with increasing filler content. Low amounts of filler (up to 0.39 wt% for graphene and 0.37 wt% for graphene oxide) were used in the study. The results indicated a good stiffness-toughness balance for polyethylene/graphene nanocomposites. Similar improvements in tensile strength and modulus upon addition of graphene were reported by the authors in the other study [42]. Though tensile strength and modulus increased with increasing filler content among the nanocomposites, elongation at break reduced with increasing filler content, as was reported by Fim *et al.* [40]. The authors reported that melting temperature and degree of crystallization reduced with increase in filler content, while crystallization temperature increased [41,42]. Reduction in melting temperature and degree of crystallization was attributed to the hindrance in crystallization due to the presence of the filler particles resulting in faults and imperfections. On the other hand, the increase in crystallization temperature was attributed to the filler acting as a nucleating agent. Thermal stability of the nanocomposites was reported to increase with increase in filler content due to the filler particles acting as barrier to the transport of heat, oxygen and pyrolysis products through the nanocomposite matrix. Comparing polyethylene/graphene oxide (PE/GO) and polyethylene/graphene (PE/G) nanocomposites, it was reported that PE/GO had higher thermal stability than PE/G nanocomposites [41].

In another study, Mittal *et al.* [43] synthesized three different polyethylene samples (molecular weights PE1 < PE2 < PE3) using coordination polymerization, in order to study the effect of the variation in molecular weight on the properties of the nanocomposites, using thermally reduced graphene as filler. Due to efficient heat transfer to the filler platelets owing to their better dispersion in PE1, the highest extent of improvement in the initiation of degradation temperatures was observed in this case. For instance, the PE1 composite with 4 wt% filler fraction was observed to be have initiation delayed by ~40 °C as compared to pure polymer. Due to the higher degree of crystallinity, PE1 had higher tensile modulus of 1334 MPa, which was improved to the maximum extent of 35% in the composite with 4 wt% filler content. Thus, the mechanical properties were enhanced due to the combination of crystallinity and filler effect. The peak stress as well as elongation of the composites was improved in most cases as compared to pure polymers. As shown in Figure 8.6, the PE1 composites also exhibited lowest melt viscosity among the composites probably due to the lower degree of entanglements, which indicated ease of processing similar to the pure polymer. The corresponding PE2 and PE3 composites exhibited an order of magnitude higher viscosity at lower range of angular frequency. For PE1, the degree of crystallinity enhanced on adding 2 and 4 wt% of graphene due to nucleation

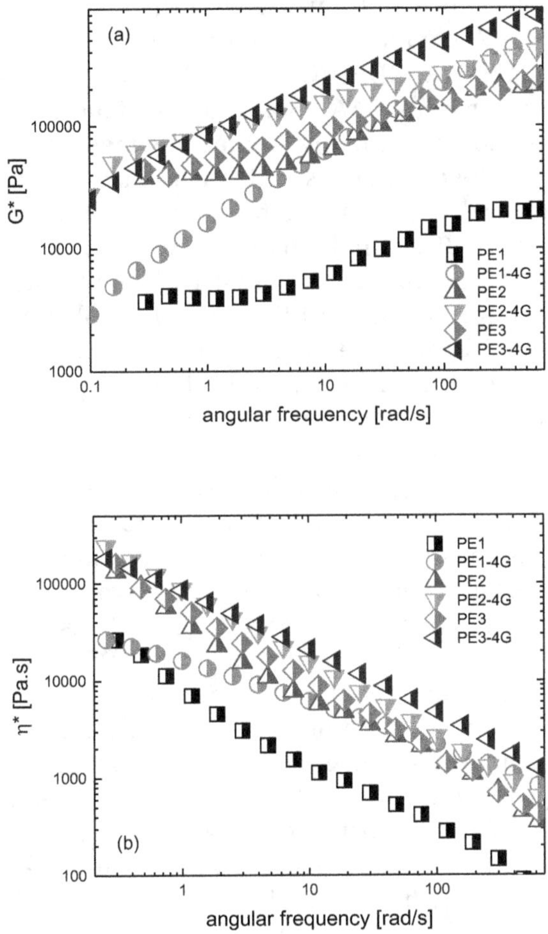

Figure 8.6 (a) Complex modulus and (b) complex viscosity of the pure polymers and nanocomposites (4 % graphene content) as a function of angular frequency. Reproduced from Reference 43 with permission from Springer.

induced by the graphene nano-platelets, however, the crystallinity was observed to decrease in PE2 and PE3.

Mittal and Chaudhry [44] also compared the two main routes of nanocomposite generation, viz., the masterbatch method and direct melt mixing, in relation with the properties of the polyethylene-thermally reduced graphene nanocomposites compatibilized with functional polymers. Filler dispersion was observed to improve in the masterbatch method due to preferential interactions of compatibilizer with the filler, before melt mixing with the matrix polymer. For instance, Figure 8.7 compares the morphology of the HDPE/G, HDPE/G/5 EAA/MM, and HDPE/G/5 EAA/MB composites at different magnifications (where EAA, MM and MB denote ethylene-co-acrylic acid copolymer, melt mixing method and masterbatch method respectively). Both the composites had much improved filler dispersion as compared to HDPE/G composite. Comparing the two, the composite generated with direct melt mixing exhibited less optimal filler dispersion than the composite generated with the masterbatch approach. Filler tactoids of higher thickness were observed for the HDPE/G/5 EAA/MM composite, whereas the tactoid thickness was relatively reduced for the HDPE/G/5 EAA/MB composite. No single exfoliated graphene platelets were observed for the HDPE/G/5 EAA/MM composite, whereas the HDPE/G/5 EAA/MB composite exhibited presence of single graphene layers. Also, the extent of intercalation was observed to be qualitatively higher in the composite generated with the masterbatch approach. This also translated into improved mechanical properties as more than two times increment in tensile modulus was observed in the HDPE/G/5 EAA/MB composite (41%, masterbatch method) as compared to HDPE/G/5 EAA/MM (20%, melt mixing method). In addition, the HDPE/G/5 EAA/MB and other MB based composites did not show any decrease in degree of crystallinity as compared to pure polymer, whereas the corresponding composites generated with melt mixing exhibited slight decrease in the crystallinity. The composites were observed to have enhanced thermal stability as compared to HDPE. Five percent EAA-based composites exhibited even higher thermal stability than HDPE/G composite due to improved filler dispersion.

In the work reported by Wang *et al.* [45], reduced graphene oxide was functionalized with vinyl triethoxysilane (VTES) and compounded with LDPE by means of solution dispersion. The tensile strength and Young's modulus of the nanocomposites increased with increasing filler content, showing an increment of 27% and 92.8% respectively for the nanocomposite with 7 wt% filler content, as compared to neat LDPE. The elongation at break increased until a peak was achieved at filler content of 3 wt% and then started reducing with further increase in graphene content. The nanocomposites of LDPE with un-functionalized graphene had lower elongation at break than neat LDPE even with 1 and 3 wt% filler loading. The chemical bonds formed as a result of reaction of the vinyl groups in VTES with

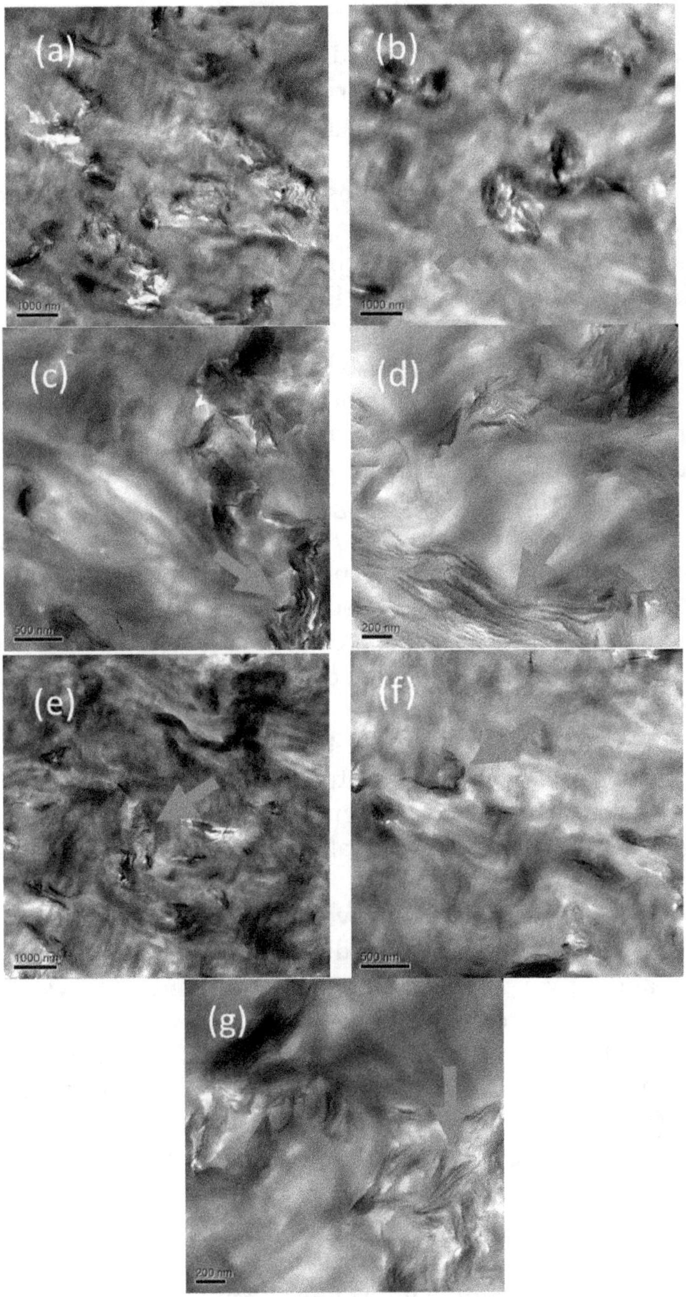

Figure 8.7 TEM micrographs of (a) HDPE/G; (b-d) HDPE/G/5 EAA MM and (e-g) HDPE/G/5 EAA MB composites. Reproduced from Reference 44 with permission from Springer.

the radicals in the LDPE improved the interfacial interaction of graphene with the polymer matrix and, hence, resulted in enhanced mechanical properties. Improved thermal stability of LDPE was also reported by the authors, as the weight loss initiation temperature increased with addition of the graphene filler. In another study, Huang *et al.* [46] prepared nanocomposites with varying fractions of octadecylamine (ODA) functionalized graphene oxide nano-sheets (GONs) and LDPE by means of solution coagulation. Improvement in yield strength upon addition of the functionalized GONs was reported to be limited, while the Young's modulus increased significantly with an increase of up to approximately 59% being observed on the addition of 4 wt% GONs. The observed enhancement was attributed to interfacial transfer of stress that resulted from the fully exfoliated uniformly dispersed GONs and their strong interfacial interaction with the LDPE matrix. However, the increase in rigidity resulted in the loss of ductility as evidenced by the decrease in elongation at break of the nanocomposites with increasing filler content. The results were also similar to the nanocomposites reported by Wang *et al.* [45]. Crystallization peak temperature was reported to increase with enhancing the filler content due to the heterogeneous nucleation effect. The thermal stability was also observed to increase with filler content due to the barrier effect. No change in the melting temperature and melting enthalpy was observed, which indicated that the filler only acted as nucleating agent, thus, accelerating crystallization kinetics but did not have significant effect on the crystalline structure and crystallinity of LDPE nanocomposites.

Patwary and Mittal [47] also developed polyethylene nanocomposites with 5 wt% silica, alumino-silicate and thermally reduced graphene by adding varying amounts of pro-oxidant additive. The graphene particles were observed to be uniformly dispersed in the polymer matrix. In the graphene composite with 2.5 wt% additive content, the tensile modulus was increased by 1.95 times that of pure polymer (Figure 8.8). Increasing the additive content was also observed to enhance the mechanical performance. Graphene nanocomposite with 1% additive content had also 40% and 33% increment in storage modulus at 50 °C and 70 °C respectively as compared to pure PE. Degree of embrittlement of the samples with 2.5% additive was calculated from the carbonyl index measurements as a function of exposure time adjusted to real environment in a constant temperature of 30 °C in sunlight. PE sample containing only 2.5 wt% additive exhibited the fastest rate of degradation reaching 100% embrittlement after ~795 accelerated ageing hours (equivalent to around 8.7 months at 30 °C). Graphene based composite did not reach 100 % embrittlement during the test period, but was still ~48.2% embrittled at the end of the test. Graphene platelets slowed the photo-oxidation due to known property of strong UV absorption by the graphene platelets. However, the observed results confirmed that the addition of pro-oxidant in the composites was necessary to attain their

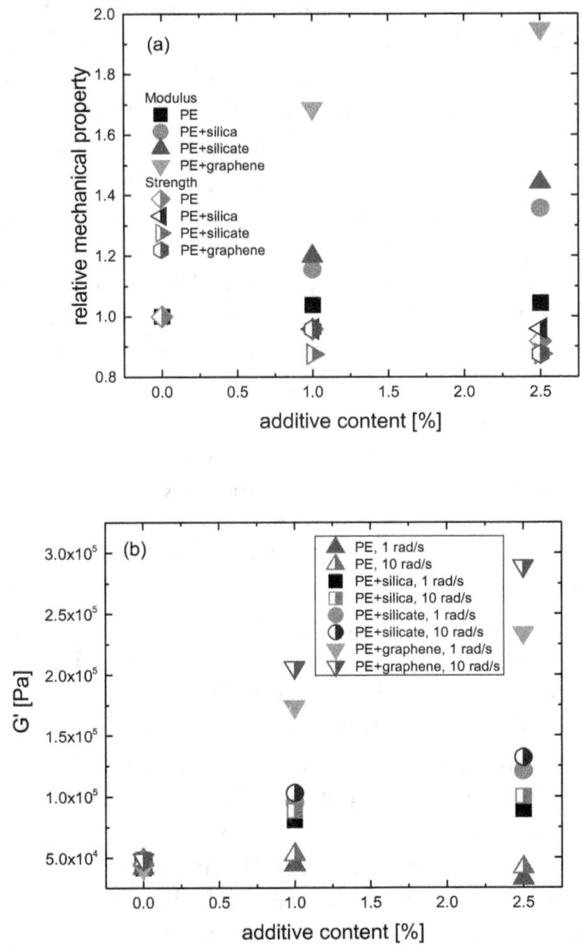

Figure 8.8 (a) Relative tensile modulus and strength of the materials plotted as a function of additive content; (b) comparison of storage modulus in the nanocomposites as a function of additive content at an angular frequency of 1 rad/s (solid dots) and 10 rad/s (half solid dots). Reproduced from Reference 47 with permission from Elsevier.

degradation and a controlled and fairly fast degradation could still be achieved in varying degrees even in thick composite plaques with 5 wt% filler content.

In another study, Mittal and Krauss [48] also reported the mapping of the component distribution in the nanocomposites by using hyperspectral imaging. Ethylene-co-acrylic acid (EAA) and ethylene-comethacrylic acid ionomer (EMAZ) copolymers were used as compatibilizers for polyethylene-graphene nanocomposites generated by melt mixing. Figure 8.9 presents the mapping of the distribution of graphene and EAA at two different locations in the PE-5%EAA-G nanocomposite. Pixels in red represented the presence of graphene peak, whereas pixels in green represented the EAA peak in the nanocomposite. The dark area in the micrographs without any colored signal corresponded to PE. EAA in the PE-EAA-G nanocomposite sample was observed to bond/mix well with the graphene causing spectral shifting and mixing. Several areas were observed to contain characteristic peaks of both EAA and graphene in the nanocomposite. The EAA compatibilizer was, thus, observed to be effective in dispersing the graphene throughout the PE matrix. On the other hand, nanocomposite with EMAZ had no spectral mixing as the components did not mix optimally with each other. The hyperspectral imaging and mapping was, thus, demonstrated to be a useful method for the determination of component distribution in complex nanocomposite systems.

Recently, different chemical routes to functionalize graphene with short-chain polyethylene were reported which influenced the mechanical properties of graphene-based high density polyethylene (HDPE) nanocomposites [49]. The authors observed that the polyethylene grafting on the graphene through thiol-yne click chemistry showed a significant effect on the Young's modulus and elongation at break of the resultant composites. In another study, Vasileiou *et al.* also studied the dispersion and properties of polyethylene-graphene composite using maleated linear low-density polyethylene (LLDPE) derivatives and thermally reduced graphene oxide (TRG) through a non-covalent compatibilization approach [50]. It was observed that the pyridine functionalized polyethylene matrix exhibited a better dispersion with TRG due to the π-π interactions, thereby resulting in significantly improved Young's moduli without reduction of tensile stress and elongation at the break.

In another study, PE based blend nanocomposites with ethylene acrylic acid (EAA) copolymer (40:60) and graphene (1 wt%) could be

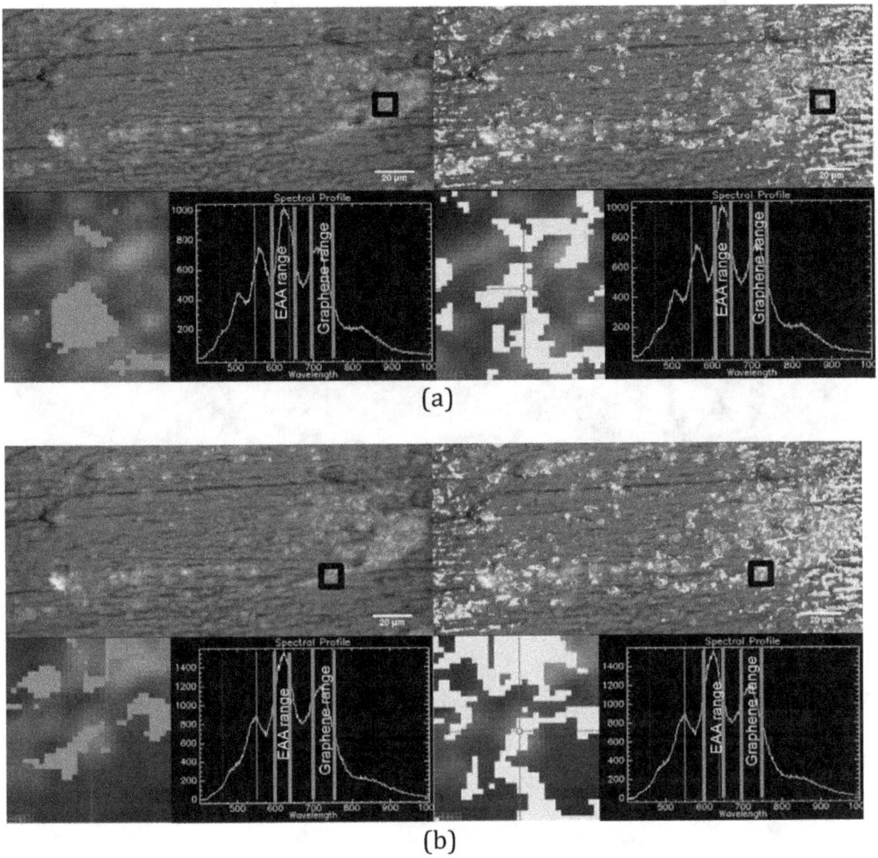

Figure 8.9 Mapping of distribution of graphene and EAA at two different locations in PE-5%EAA-G nanocomposite. Pixels in red represent the presence of graphene peak, whereas pixels in green represent the presence of EAA peak in the nanocomposite. Reproduced from Reference 48 with permission from Springer.

generated by melt mixing. Figure 8.10 presents the morphological, thermal, rheological and mechanical properties of the composite with 1 wt% filler. Phase immiscibility was observed in the PE and EAA phases from the microscopy analysis. Good degree of graphene distribution and delamination was also observed, as shown in Figure 8.10 a and b. In addition, graphene was observed to be dispersed in both PE and EAA phases. The DSC thermogram of the polymer blend exhibited individual melt transitions of PE and EAA, thus, confirming the phase immiscibly (Figure 8.10 c). The peak melting points of both phases were enhanced in the nanocomposite due to the efficient heat

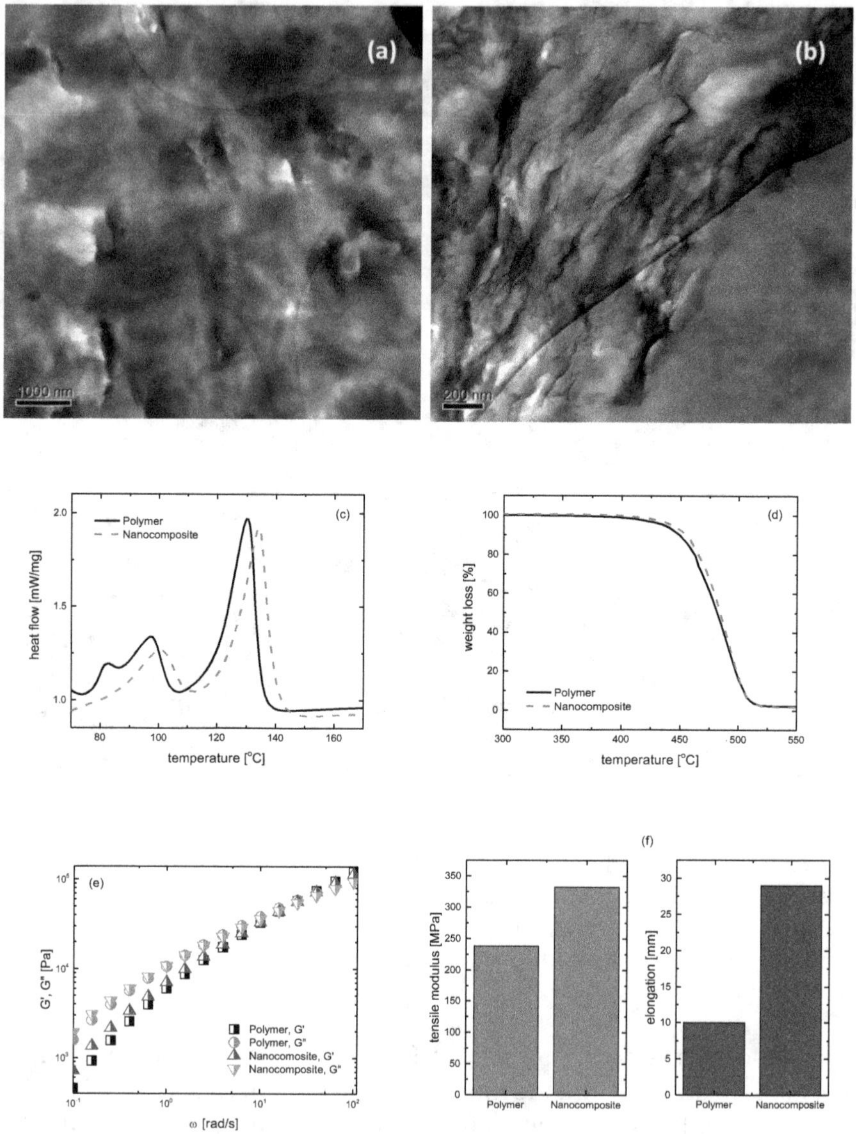

Figure 8.10 Characterization of the morphology as well as thermal, rheological and mechanical properties of the 40%PE+60%EAA+1%G nanocomposite: (a,b) TEM micrographs at different magnifications, (c) DSC thermograms, (d) TGA thermograms, (e) storage and loss modulus and (f) tensile modulus and elongation of the composite in comparison with pure polymer blend.

transfer to the graphene platelets dispersed in both phases. Similarly, the thermal degradation behavior of the nanocomposites also exhibited slight improvement in the onset and peak degradation temperatures. The storage modulus of the composite was observed to be better than the pure polymer, though the loss modulus was higher than the storage modulus at lower angular frequencies. The tensile behavior also exhibited increased tensile modulus of 332 MPa with incorporation of 1 wt% graphene, in comparison to 238 MPa for the polymer blend, thus, indicating effective stress transfer to the uniformly dispersed graphene platelets. As a result, the elongation of the nanocomposite was also enhanced significantly as compared to the PE/EAA blend, thus, indicating synergistic enhancement of toughness and stiffness of the polymer.

Tarani *et al.* [51] also studied the effect of the size of graphene platelets (GNPs) on HDPE melting behavior and crystallization kinetics. It was reported that the crystallization temperature decreased with increase in the diameter of the GNPs. This was reported to be due to the hindrance to lamella formation by the higher diameter GNPs. Higher diameter GNPs were also observed to be rejected from the crystalline phase during the segregation that took place in the process of crystallization, thus, resulting in the difficulty in packing of the HDPE polymer chains. The authors also reported reduction in melting enthalpy and crystallinity attributed to restriction of HDPE chain mobility by the GNPs during the generation of an ordered crystal structure. Melting enthalpy was reported to reduce with increase in GNP diameter.

In another study, Cheng *et al.* [52] investigated the crystallization of PE induced by reduced graphene oxide (RGO) (Figure 8.11) and reported increase of both thermal stability and rate of crystallization in the presence of RGO. The synthesis method involved the dispersion, in two separate vessels, of low amounts of PE and RGO in dichlorobenzene (DCB) (0.05 wt% and 0.01 wt%, respectively) followed by crystallization, mixing the product with a concentrated solution of PE in DCB and then subsequent crystallization to obtain the nanocomposite. Non-isothermal DSC revealed that the melting temperature of the nanocomposites remained the same with increasing RGO content, whereas the crystallization temperature and degree of crystallization increased with increasing the RGO content in the nanocomposites due to the nucleation effect provided by the RGO nano-sheets. Isothermal crystallization analysis exhibited that the time required to complete 50% of the crystallization process reduced with increasing the filler content and the crystallization rate constant as calculated based on the Avrami equation [53] increased with increasing the filler content. Both observations indicated that the rate of crystallization increased with the addition of graphene.

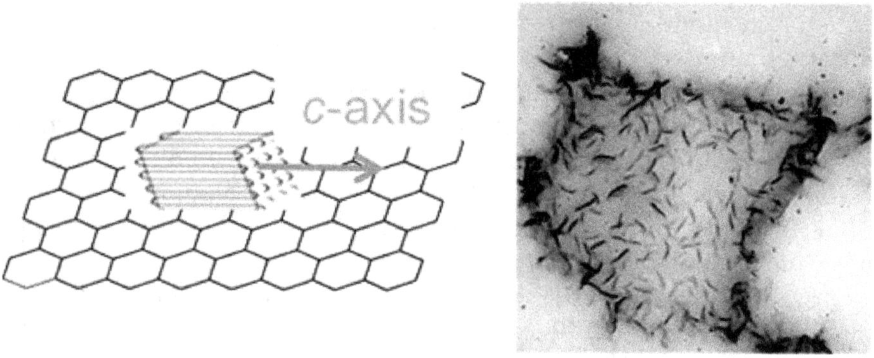

Figure 8.11 PE/RGO nanocomposites (PGNs) fabricated through solution crystallization/precipitation using the PE-decorated RGO hybrid. Reproduced from Reference 52 with permission from American Chemical Society.

Hu *et al.* [54] prepared hyperbranched flame retardant functionalized GO (FGO) using N-aminoethylpiperazine, hydroxyethyl acrylate, methyldichlorophosphate and GO via *in-situ* polymerization technique. FGO was incorporated into cross-linked PE matrix and its flammability and thermal stability were studied. It was observed that the incorporation of FGO in to the cross-linked PE endowed a highly flame retardant and thermally stable PE composite. In a similar work, Han *et al.* [55] studied the flame retardancy behavior of polyethylene/alumina trihydrate (PE/ATH) composites with the effect of incorporation of graphene nano-platelets (GNPs). The composites were prepared through melt blending method. The incorporation of 0.2 wt% of GNPs into the PE/ATH composite decreased the heat release peak by 18%.

Ran *et al.* [56] developed high density polyethylene/brominated polystyrene/graphene nano-platelet (GNPs) composites *via* melt blending and studied the effect of Friedel-Crafts reaction on their thermal stability and flammability. Anhydrous aluminium trichloride ($AlCl_3$) was used as a Lewis catalyst and was added to initiate the Friedel-Crafts reaction for effective dispersion of GNPs into the polymer blends. The results revealed that the GNPs were partly unfolded and the domains became smaller in the presence of $AlCl_3$. The incorporation of $AlCl_3$ into polymer-graphene nanocomposites reduced the flammability and slowed down the heat release rate.

In another study, Carotenuto *et al.* [57] fabricated an electrically conductive polyethylene supported graphene film based on simple mechanical approach. The film exhibited a semiconducting nature at lower temperature. It was observed that the film exhibited good conductivity with increasing graphene content, but it also lowered the optical transparency. Wang *et al.* [58] developed electrically conductive ultrahigh molecular weight polyethylene (UHMWPE)/graphene nanocomposites with segregated structure. The authors found that the UHMWPE powder could generate a static electricity after high speed mechanical friction and adsorb graphene on the surface. The composite exhibited high electrical conductivity with percolation threshold of 0.1 vol%. In another study, Du *et al.* [59] prepared multi-walled carbon nanotube (MWCNT) and graphene nano-sheets (GNS) incorporated HDPE nanocomposites with segregated network structure and compared their electrical properties. HDPE/ MWCNT composites exhibited higher conductivity than that of HDPE/GNS composites at the same filler content. This was due to different conductivity mechanism between the two composites, while the HDPE/MWCNT composites possessed three-dimensional conductive system, the HDPE/GNS composites exhibited a two-dimensional conductive system.

8.3 Conclusions

The excellent mechanical and thermal properties of graphene, along with its large surface area have made it a nanomaterial of wide research interest. The incorporation of graphene or its derivatives in polymer matrices enables the synergetic combination of its properties with those of the polymers to produce nanocomposites that are suitable for applications for which the neat polymers are not useful. One such polymer that is used widely is polyethylene (of different densities) and the mechanical and thermal properties of polyethylene/graphene nanocomposites comprise some of the areas of interest of recent research.

In this review, the properties of polyethylene/graphene nanocomposites reported in recent studies have been summarized. The incorporation of graphene to the polymer was observed to result in the improvement of mechanical and thermal properties as compared to pure PE, irrespective of the synthesis method of the composites. The review summarized the studies reporting the generation of the nanocomposites using different synthesis methods, such as solution dispersion, melt mixing and *in-situ* polymerization. The mechanical properties studied in the review included Young's modulus, yield stress, yield strain, elongation at break, tensile strength and dynamic mechanical properties. The Young's modulus, yield stress and yield strain increased in the reported studies with increasing graphene content in the nanocomposites. This was attributed to the effective stress transfer from the polyethylene chains to the graphene platelets, which resulted in the improved properties. The elongation at break, however, exhibited an increase with increasing graphene content until a certain optimum loading, after which a decrease was observed. This was attributed to the agglomeration of the graphene nano-sheets when added in large contents, hence reducing the effective aspect ratio as well as elongation at break. The dynamic mechanical properties such as storage modulus and loss modulus were also observed to increase with increasing graphene content. In the cases when the graphene functionalization was performed, the properties such as tensile strength were reported to be superior than the composites with un-functionalized graphene due to interfacial interaction and effective filler dispersion. Similar to the mechanical properties, the thermal properties of the nanocomposites were also reported to be better than those of neat polyethylene. The melting temperature of the polymer in nanocomposites was reported to remain constant irrespective of the filler content. The crystallization temperature and degree of crystallinity, on the other hand, were generally reported to increase with increasing graphene content in the nanocomposites because of the nucleation effect of the filler platelets. The rate of crystallization was also observed to increase with addition of graphene. Overall, it can be concluded that significant increment in composite properties have been gained in the polyethylene/graphene nanocomposites by optimizing the synthesis method, filler content, surface functionalization, etc. The obtained nanocomposites represent advanced materials with a high potential of use in a wide variety of applications.

Acknowledgement

The authors are indebted to Dr. Ali Chaudhry for generating and characterizing the PE-EAA-G composites.

References

1. Mittal, V., and Matsko, N. (2014) Characterization of polyethylene-based multiphase systems containing zero-and two-dimensional nanoparticulate reinforcement materials by analytical electron and atomic force microscopy. *Journal of Thermoplastic Composite Materials*, **27**(6), 845-864.
2. Hu, K., Kulkarni, D. D., Choi, I., and Tsukruk, V. V. (2014) Graphene-polymer nanocomposites for structural and functional application. *Progress in Polymer Science*, **39**, 1934-1972.
3. Lee, C., Wei, X., and Kysar, J. W. (2008). Measurement of the elastic properties and intrinsic properties of monolayer graphene. *Science*, **321**, 385-388.
4. Kim, H., Abdala, A. A., and Macosko, C. W. (2010) Graphene/polymer nanocomposites. *Macromolecules*, **43**, 6515-6530.
5. Mittal, V. (2014) Functional polymer nanocomposites with graphene: a review. *Macromolecular Materials and Engineering*, **299**(8), 906-931.
6. Worsley, K. A., Ramesh, P., Mandal, S. K., Niyogi, S., Itkis, M. E., and Haddon, R. C. (2007) Soluble graphene derived from graphite fluoride. *Chemical Physics Letters*, **445**, 51-56.
7. Novoselov, K. S., Geim, A. K., Morozov, S. V., Jiang, D., Zhang, Y., Dubonos, S. V., Grigorieva, I. V., and Firsov, A. A. (2004) Electric field effect in atomically thin carbon films. *Science*, **306**, 666-669.
8. Behabtu, N., Lomeda, J. R., Green, M. J., Higginbotham, A. L., Sinitskii, A., Kosynkin, D. V., Tsentalovich, D., Para-Vasquez, A. N. G., Schmidt, J., Kesselman, E., Cohen, Y., Talmon, Y., Tour, J. M., and Pasquali, M. (2010) Spontaneous high-concentration dispersions and liquid crystals of graphene. *Nature Nanotechnology*, **5**, 406-411.
9. Liu, N., Luo, F., Wu, H., Liu, Y., Zhang, C., and Chen, J. (2008) One-step ionic-liquid-assisted electrochemical synthesis of ionic-liquid functionalized graphene sheets directly from graphite. *Advanced Functional Materials*, **18**, 1518-1525.
10. Hernandez, Y., Nicolosi, V., Lotya, M., Blighe, F. M., Sun, Z., De, S., McGovern, I. T., Holland, B., Byrne, M., Gun'Ko, Y. K., Boland, J. J., Niraj, P., Duesberg, G., Krishnamurthy, S., Goodhue, R., Hutchison, J., Scardaci, V., Ferrari, A. C., and Coleman, J. N. (2008). High-yield production of graphene by liquid-phase exfoliation of graphite. *Nature Nanotechnology*, **3**, 563-568.
11. Hummers, W. S., and Offeman, R. E. (1958) Preparation of graphitic oxide. *Journal of American Chemical Society*, **80**, 1339-1339.
12. Dervishi, E., Li, Z., Watanabe, F., Biswas, A., Xu, Y., Biris, A. R., Saini, V., and Biris, A. S. (2009) Large-scale graphene production by RF-cCVD method. *Chemical Communications*, 4061-4063.
13. Zhang, W., Cui, J., Tao, C., Wu, Y., Li, Z., Ma, L., Wen, Y., and Li, G.

(2009) A strategy for producing pure single-layer graphene sheets based on a confined self-assembly approach. *Angewandte Chemie International Edition*, **48**, 5864-5868.

14. Kim, C., Min, B., and Jung, W. (2009) Preparation of graphene sheets by the reduction of carbon monoxide. *Carbon*, **47**, 1605-1612.
15. Li, N., Wang, Z., Zhao, K., Shi, Z., Gu, Z., and Xu, S. (2010) Large scale synthesis of N-doped multi-layered graphene sheets by simple arc-discharge method. *Carbon*, **48**, 255-259.
16. Schaefer, D. W., and Justice, R. S. (2007). How nano are nanocomposites? *Macromolecules*, **40**, 8501-8517.
17. Potts, J. R., Dreyer, D. R., Bielawski, C. W., and Ruoff, R. S. (2011) Graphene-based polymer nanocomposites. *Polymer*, **52**, 5-25.
18. Kuilla, T., Bhadra, S., Yao, D., Kim, N. H., Bose, S., and Lee, J. H. (2010) Recent advances in graphene based polymer composites. *Progress in Polymer Science*, **35**, 1350-1375.
19. Yang, X., Mei, T., Yang, J., Zhang, C., Lv, M., and Wang, X. (2014) Synthesis and characterization of alkylamine-functionalized graphene for polyolefin-based nanocomposites. *Applied Surface Science*, **305**, 725-731.
20. Lewandowska, K. (2012) Surface studies of microcrystalline chitosan/poly (vinyl alcohol) mixtures. *Applied Surface Science*, **263**, 115-123.
21. Tan, Y., Fang, L., Xiao, J., Song, Y., and Zheng, Q. (2013) Grafting copolymers onto graphenes by miniemulsion polymerization for conductive polymer composites: improved electrical conductivity and compatibility induced by interfacial distribution of graphene. *Polymer Chemistry*, **4**, 2939-2944.
22. Liao, K. H., Park, Y. T., Abdala, A., and Macosko, C. (2013). Aqueous reduced graphene/thermoplastic polyurethane nanocomposites. *Polymer*, **54**, 4555-4559.
23. Stankovich, S., Piner, R. D., Nguyen, S. T., and Ruoff, R. S. (2006) Synthesis and exfoliation of isocyanate-treated graphene oxide nanoplatelets. *Carbon*, **44**, 3342-3347.
24. Yang, X. Y., Wang, X. B., Yang, J., Li, J., and Wan, L. (2013) Functionalization of graphene using trimethoxysilanes and its reinforcement of polypropylene nanocomposites. *Chemical Physics Letters*, **570**, 125-131.
25. Ebewele, R. O. (2000) *Polymer Science and Technology*, CRC press, USA.
26. Lahiri, D., Dua, R., Zhang, C., Socarraz-Novoa, I., Bhat, A., Ramaswamy, S., and Agarwal, A. (2012) Graphene nanoplatelet-induced strengthening of ultra-high molecular weight polyethylene and biocompatibility in vitro. *Applied Materials & Interfaces*, **4**, 2234-2241.
27. Suner, S., Joffe, R., Tipper, J. L., and Emami, N. (2015) Ultra-high mo-

lecular polyethylene/graphene nanocomposites: thermal, mechanical and wettability characterization. *Composites Part B*, **78**, 185-191.

28. Sturzel, M., Kempe, F., Thomann, Y., Mark, S., Enders, M., and Mulhaupt, R. (2012) Novel graphene UHMWPE nanocomposites prepared by polymerization filling using single-site catalysts supported on functionalized graphene nanosheet dispersions. *Macromolecules*, **45**, 6878-6887.
29. Chen, Y., Qi, Y., Tai, Z., Yan, X., Zhu, F., and Xue, Q. (2012) Preparation, mechanical properties and biocompatibility of graphene oxide/ultra-high molecular weight polyethylene composites. *European Polymer Journal*, **48**, 1026-1033.
30. Kuila, T., Bose, S., Mishra, A. K., Khanra, P., Kim, N. H., and Lee, J. H. (2012) Effect of functionalized graphene on the physical properties of linear low density polyethylene nanocomposites. *Polymer Testing*, **31**, 31-38.
31. Mittal, V., and Chaudhry, A. U. (2015) Effect of amphiphilic compatibilizers on the filler dispersion and properties of polyethylene – thermally reduced graphene nanocomposites. *Journal of Applied Polymer Science*, **132**(35), doi: 10.1002/app.42484.
32. Park, S., He, S., Wang, J., Stein, A., and Macosko, C. W. (2016). Graphene - polyethylene nanocomposites: Effect of graphene functionalization. *Polymer*, **104**, 1-9.
33. Kim, H., Kobayashi, S., AbdurRahim, M. A., Zhang, M. J., Khusainova, A., Hillmyer, M. A., Abdala, A. A., and Macosko, C. W. (2011) Graphene/polyethylene nanocomposites: Effect of polyethylene functionalization and blending methods. *Polymer*, **52**, 1837-1846.
34. Jiang, X., and Drzal, L. T. (2012) Multifunctional high-density polyethylene nanocomposites produced by incorporation of exfoliated graphene nanoplatelets 2: Crystallization, thermal and electrical properties. *Polymer Composites*, 33, 636-642.
35. Jiang, X., and Drzal, L. T. (2011) Improving electrical conductivity and mechanical properties of high density polyethylene through incorporation of paraffin wax coated exfoliated graphene nanoplatelets and multi-wall carbon nano-tubes. *Composites: Part A*, **42**, 1840-1849.
36. El Achaby, M., and Qaiss, A. (2013) Processing and properties of polyethylene reinforced by graphene nanosheets and carbon nanotubes. *Materials and Design*, **44**, 81-89.
37. Mittal, V., and Chaudhry, A. U. (2015) Polymer-graphene nanocomposites: Effect of polymer matrix and filler amount on properties. *Macromolecular Materials and Engineering*, **300**, 510-521.
38. Chaudhry, A. U., and Mittal, V. (2013) High-density polyethylene nanocomposites using masterbatches of chlorinated polyethylene/graphene oxide. *Polymer Engineering and Science*, **53**(1), 78-

88.
39. Mittal, V., and Matsko, N. (2013) PE-CPE blends and their graphene oxide nanocomposites with reduced low temperature brittleness, *Colloid and Polymer Science*, **291**(8), 1949-1961.
40. Fim, F. C., Basso, N. R. S., Greabin, A. P., Azambuja, D. S., and Galland, G. B. (2013) Thermal, electrical and mechanical properties of polyethylene-graphene nanocomposites obtained by in-situ polymerization. *Journal of Applied Polymer Science*, **128**, 2630-2637.
41. Zhang, H., Park, J., Ko, E., Moon, Y., Lee, Y., Hu, Y., Zhang, X., and Yoon, K. (2016) Comparison of the properties of graphene- and graphene oxide- based polyethylene nanocomposites prepared by an in-situ polymerization method. *RSC Advances*, **6**, 73013-73019.
42. Zhang, H., Ko, E., Park, J., Moon, Y., Zhang, X., and Yoon, K. (2016) Fabrication of polyethylene/graphene nanocomposites through in-situ polymerization with a spherical graphene/$MgCl_2$-supported Ziegler-Natta catalyst. *Composites Science and Technology*, **136**, 61-66.
43. Mittal, V., Kim, S., Neuhofer, S., Paulik, C. (2016) Polyethylene/graphene nanocomposites: effect of molecular weight on mechanical, thermal, rheological and morphological properties. *Colloid and Polymer Science*, **294**, 691-704.
44. Mittal, V., and Chaudhry, A. U. (2016) Polyethylene-thermally reduced graphene nanocomposites: comparison of masterbatch and direct melt mixing approaches on mechanical, thermal, rheological, and morphological properties. *Colloid and Polymer Science*, **294**, 1659-1670.
45. Wang, J., Xu, C., Hu, H., Wan, L., Chen, R., Zheng, H., Liu, F., Zhang, M., Shang, X., and Wang, X. (2011) Synthesis, mechanical, and barrier properties of LDPE/graphene nanocomposites using vinyl triethyoxysilane as a coupling agent. *Journal of Nanoparticle Research*, **13**, 869-878.
46. Huang, H., Zhou, S., Ren, P., Ji, X., and Li, Z. (2015) Improved mechanical and barrier properties of low-density polyethylene nanocomposite films by incorporating hydrophobic graphene oxide nanosheets. *RSC Advances*, **5**, 80739-80748.
47. Patwary, F., and Mittal, V. (2015) Degradable polyethylene nanocomposites with silica, silicate and thermally reduced graphene using oxo-degradable pro-oxidant. *Heliyon*, **1**(4), e00050.
48. Mittal, V., and Krauss, L. (2014) Compatibilized polyethylene-thermally reduced graphene nanocomposites: Interfacial interactions and hyperspectral mapping for component distribution. *Colloid and Polymer Science*, **292**(10), 2509-2518
49. Castelain, M., Martinez, G., Marco, C., Ellis, G., and Salavagione, H. J. (2013) Effect of click chemistry approaches for graphene modifica-

tion on the electrical, thermal, and mechanical properties of polyethylene/graphene nanocomposites. *Macromolecules*, **46**, 8980-8987.

50. Vasileiou, A. A., Kontopoulou, M., and Docoslis, A. (2014) A noncovalent compatibilization approach to improve the filler dispersion and properties of polyethylene/graphene composites. *ACS Applied Materials & Interfaces*, **6**, 1916-1925.
51. Tarani, E., Wurm, A., Schick, C., Bikiaris, D. N., Chrissafis, K., and Vourlias, G. (2016) Effect of graphene nanoplatelets diameter on non-isothermal crystallization kinetics and melting behavior of high density polyethylene nanocomposites. *Thermochimica Acta*, **643**, 94-103.
52. Cheng, S., Chen, X., Hsuan, G., and Li, C. Y. (2012) Reduced graphene oxide-induced polyethylene crystallization in solution and nanocomposites. *Macromolecules*, **45**, 993-1000.
53. Avrami, M. (1939) Kinetics of phase change. I General theory. *The Journal of Chemical Physics*, **7**, 1103-1112.
54. Hu, W., Zhan, J., Hong, N., Wang, B., Song, L., Stec, A. A., Hull, T. R., Wang, J., and Hu, Y. (2014) Effect of functionalized graphene oxide with hyper-branched flame retardant on flammability and thermal stability of cross-linked polyethylene. *Industrial & Engineering Chemistry Research*, **53**, 3073-3083.
55. Han, Z., Wang, Y., Dong, W., and Wang, P. (2014) Enhanced fire retardancy of polyethylene/alumina trihydrate composites by graphene nanoplatelets. *Materials Letters*, **128**, 275-278.
56. Ran, S., Guo, Z., Han, L., and Fang, Z. (2014) Effect of Friedel-Crafts reaction on the thermal stability and flammability of high-density polyethylene/brominated polystyrene/graphene nanoplatelet composites. *Polymer International*, **63**, 1835-1841.
57. Carotenuto, C., Nicola, S. D., Ausanio, G., Massarotti, D., Nicolais, L., and Pepe, G. P. (2014) Synthesis and characterization of electrically conductive polyethylene-supported graphene films. *Nanoscale Research Letters*, **9**, 475.
58. Wang, B., Li, H., Li, L., Chen, P., Wang, Z., and Gu, Q. (2013) Electrostatic adsorption method for preparing electrically conducting ultrahigh molecular weight polyethylene/graphene nanosheets composites with a segregated network. *Composites Science and Technology*, **89**, 180-185.
59. Du, J., Zhao, L., Zeng, Y., Zhang, L., Li, F., Liu, P., and Liu, C. (2011) Comparison of electrical properties between multi-walled carbon nanotube and graphene nanosheet/high density polyethylene composites with a segregated network structure. *Carbon*, **49**, 1094-1100.

9

Polyimide Graphene Nanocomposites

9.1 Introduction

A large variety of polymers is generally employed for different purposes and their applications as engineering plastics is especially interesting. Such engineering polymers are used to replace metals in many appliances due to their high strength, stability, insulating properties, etc. Over the recent years, even more interesting property improvements have been achieved due to the addition of fillers to the polymers [1,2]. One of the reinforcing filler of significant research interest is graphene, which has superior electrical, mechanical and thermal properties [3]. Different forms of graphene such as graphene oxide, reduced graphene oxide, functionalized graphene sheets and fluorinated graphene sheets have been employed to incorporate into the polymer matrices. It has been generally reported that the incorporation of graphene-based nano-fillers into the polymer matrices provided significant improvements in thermal stability, mechanical properties, gas barrier, electrical properties and resistant to thermal expansion (lower coefficient of thermal expansion (CTE)), thereby expanding their domain for various applications.

An important category among the engineering polymers is polyimides (PI) which have a useful combination of many structural properties which lead to their use in many applications demanding high performance [4]. Polyimides are generally produced in a 2-step process, firstly reacting a di-anhydride and a di-amine at ambient conditions in a dipolar aprotic solvent (N,N-dimethylacetamide or N-methylpyrrolidinone) to generate corresponding acid, which is polyamic acid. Subsequent step is the cyclization of the acid into a polyimide. Majority of the polyimides are infusible due to their planar-aromatic and hetero-aromatic structure and, therefore, need to be processed through solvent route. The polyimide resins having high aromaticity possess high thermal stability (>300 °C), low creep, excellent

Omar B. Wani, Mujeeb O. Kareem and Vikas Mittal, The Petroleum Institute (part of Khalifa University of Science and Technology), Abu Dhabi, UAE*
**Current address: Bletchington, Wellington County, Australia*

radiation shielding capability, high glass transition temperatures of higher than 200 °C, good flexibility and low color. Thus, aromatic polyamides are among the finest performing materials in the fields of microelectronics and aircraft industries, due to their high mechanical properties, thermal stabilities and glass transition temperatures as well as resistance to solvents.

Graphene is a thin layer of pure carbon in the form of 2D platelets, and has extraordinary properties. For instance, it is about 200 times stronger that steel, with very high efficiency of heat and electricity (6000 S/cm) conduction. A defect free graphene layer is reported to have an elastic modulus of about 1 TPa and ultimate strength of over 100 GPa, thus, confirming it to be one of the strongest known materials [5,6]. These properties of graphene, added to its extremely high surface area and gas impermeability, are therefore used to improve mechanical, electrical, thermal and permeation characteristics of polymers [7,8]. However, the economics of generating it in large scale as well as reproducibility remain the biggest challenges. In addition, graphene in its original form is incompatible with most polymers, thus, the platelets of pure graphene usually form stacks and agglomerates within the polymer material matrix. It indicates that the optimal surface modification of graphene is required to achieve its nanoscale dispersion in the polymer matrices. The most emerging method to manufacture graphene on a large scale is based on exfoliating and reducing graphite oxide. Graphite oxide can be chemically or thermally reduced depending upon the conditions required for nanocomposite manufacture [9,10].

As mentioned earlier, the modification of the graphene surface (using both physical and chemical methods) is needed to enhance its dispersion in the polymer matrices, as the degree of dispersion of a filler strongly affects the properties of the resulting composite. In the case of carbon nanotubes (CNTs), functionalizing the surface using fluorine or acids as well as by generation of radicals are known to exhibit improved dispersion. In the case of graphene synthesis from graphite oxide, some epoxides and hydroxyl groups remain on the surface, which greatly facilitate functionalization [11]. Entanglement is not a problem in graphene due to flat platy morphology, however, restacking of the modified sheets can take place, thereby reducing the effectiveness. Use of surfactants and/or blending with polymer before chemical reduction are also useful methods to achieve good degree of dispersion in the polymer matrix. As graphite oxide exfoliates in water and other protic solvents using hydrogen bonding, polymer composites can also be generated using *in-situ* polymerization methodology [12].

This study reviews the recent reports on the synthesis and characterization of polyimide/graphene nanocomposites. Mainly, the mechanical and thermal properties of these composites have been studied and correlated with the filler content as well as synthesis strategy. In addition, from the application perspective, the studies reporting the use of PI/graphene nanocomposites for the development of fuel cell membranes, foams, anti-corrosion coatings, etc., have also been reviewed.

9.2 Nanocomposite Properties

In a study reporting the use of functionalization of filler surface, Dong *et al.* [13] modified the graphene oxide (GO) surface with 4,4'-diaminodipheny ether (ODA) to obtain GO-ODA filler. The crystal interlayer spacing for GO-ODA was observed to be 1.3 nm ($2\theta = 6.3°$), which was larger than GO. This was due to the intercalation of ODA between GO layers, which due to longer molecular chains in comparison to oxygenated functional groups pushed the GO platelets apart. The polyimide/graphene nanocomposite fibers were generated using *in-situ* polymerization. The mechanical properties like tensile strength and modulus were observed to be higher than pure polymers in the composites with GO, however, these were lower than the corresponding values in the GO-ODA containing composites. This was due to the reason that unmodified GO did not disperse uniformly in the polymer matrix and there was a possibility that a fraction of GO sheets might get pulled out, therefore, generating cracks within the polymer matrix. Also, the presence of functional groups on graphene oxide-ODA improved its compatibility with the polymer matrix, as a result of which better interfacial load transfer took place through bonding and entanglement. Specifically, the nanocomposite with 0.8 wt% of GO-ODA was observed to exhibit a tensile strength of 2.5 GPa, which was 1.6 times higher than the corresponding fiber of pure polymer. Similarly, the tensile modulus of the composite was 126 GPa, which was 223% higher as compared with the pure PI. The storage modulus of the composites also increased with increase in GO-ODA content below the glass transition temperature.

Ramakrishnan *et al.* [12] reported graphene oxide based polyimide (GO-PI) nanofiber composites by electrospinning poly(amic acid) (PAA) containing graphene oxide (GO), which was subsequently followed by thermal imidization of PAA. The electrospun PAA exhibited –OH (alcohol) stretching (at 3500 cm^{-1}), –C=O (ketone) stretching (at 1719 cm^{-1}) and vibrations of amide group (at 1660 and 1544 cm^{-1}), which were consistent with the presence of polyacrylic acid. Heat treatment led to amidation of PAA at temperatures between 50 °C and 125 °C, and further led to imidization at temperature of about 350 °C, as the stretching band for carboxylic –OH was not present in the material, thus, indicating that PAA was converted into PI. The formation of polyimide was also observed through Raman spectroscopy. A Raman shift was observed at around 1390 cm^{-1}, which confirmed the C-N stretching of the imide ring of the polymer. The authors also observed a shrinkage in the average fiber diameter after thermal imidization (polymerization). The observed effect was attributed to the shrinkage due to the elimination of water molecules during imidization. The authors reported that the incorporation of 2 wt% GO resulted in an enhancement of storage modulus from 1.4×10^8 to 3.8×10^8 Pa, as well as glass transition temperature from 317 to 323 °C as compared to pure polymer. In addition, the thermal stability of the composites was also observed to increase with the addition of filler.

Lim *et al.* [14] used aminophenyl functionalized reduced graphene oxide AP-rGO to generate polyimide nanocomposites using *in-situ* polymerization. The Raman spectra exhibited a higher D-band intensity for the AP-rGO than for GO as well as higher D- to G-band ratio, thus, suggesting that AP-rGO had a more disordered structure and defects than GO. The authors observed that due to the strong interfacial interaction between the polymer matrix and AP-rGO, the nanocomposite had significant reinforcement in the mechanical and oxygen barrier properties. For instance, the composite films with 3 wt% of AP-rGO exhibited increment in the tensile strength and modulus of 106% (262 MPa) and 52% (9.4 GPa), as compared to pure polymer. Figure 9.1 also shows the TGA thermograms of the composites with different filler fractions. The pure polymer and composites did not have variation in the thermal degradation behavior, however, the weight loss decreased in the composites owing to the presence of filler. In addition, the authors observed that the decrease in weight loss was higher than the corresponding increase in the filler content due to the thermal stability reinforcement on the incorporation of AP-rGO. The authors observed that the nanocomposites with 5 wt% of AP-rGO exhibited oxygen diffusivity and solubility values which were 99.6%, 86.6% and 96.2% lesser than that of pure PI. The authors also

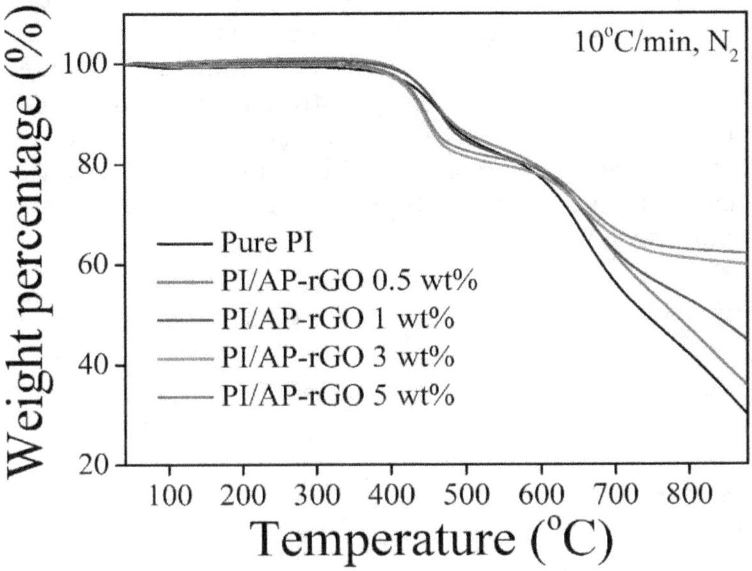

Figure 9.1 TGA thermograms of PI/AP-rGO composites. Reproduced from Reference 14 with permission from American Chemical Society.

generated phenyl functionalized graphene nano-sheets (PrGO)-PI nanocomposites and observed an improvement in the mechanical and oxygen barrier properties, however, the observed improvements were much smaller as compared to AP-rGO/PI nanocomposites. The morphological characterization of the (PrGO)-PI nanocomposites also revealed non-uniform filler dispersion, filler aggregation as well as phase separation, which probably resulted in the lower degree of property enhancement compared to AP-rGO/PI nanocomposites.

Yoonessi *et al.* [15] reported flexible polyimide graphene nanocomposites, generated by solution blending. The authors generated imide moieties on the amine-functionalized graphene for the surface functionalization of filler. As observed in Figure 9.2, the storage moduli of the composites with imidized graphene were 25-30% higher than the corresponding composite with unmodified graphene. The authors attributed the enhanced distribution of graphene in polyimide resin as well as existence of surface modifiers containing stiff benzene rings to the observed higher values of the storage modulus for the composite with modified graphene. The modulus was also observed to increase linearly with enhancing the content of both unmodified and imidized graphene in the polymer matrix. The thermal stability of the composites was also observed to be higher due to the addition of the filler, though no difference was observed in the thermal stability of the composites with unmodified and modified graphene. A significant enhancement in strain recovery rate and strain fixity rate was observed on adding graphene nanoparticles to PI as compared to neat polyimide resin. The improvement was observed to be large initially (for 0.5 wt% graphene), subsequently lower degree of improvement.

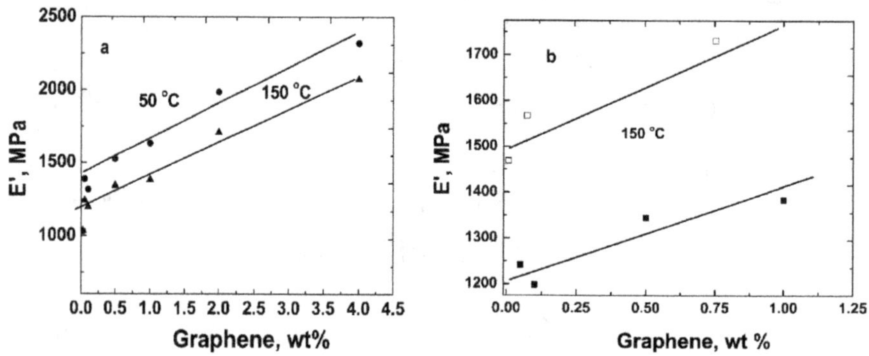

Figure 9.2 (a) Storage moduli of the polyimide graphene nanocomposites at 50 (●) and 150 °C (▲) as a function of filler content; (b) storage moduli of the nanocomposites containing graphene (■) and rigid imidized graphene (□) at 150 °C. Reproduced from Reference 15 with permission from American Chemical Society.

Ha *et al.* [16] reported the dispersion of chemaicllay modified graphene into PI by using polyamic acid (PAA)/graphene nanocomposite as a precursor. Subsequent Thermal imidization of the PAA resulted in the formation of PI-graphene nanocomposites. The authors mentioned that the degradation of the polymer by heat initiates due to the chain cleavage and formation of radicals, and the addition of graphene delays the thermal degradation as the surface of graphene sheets acts as radical scavenger. Thus, if the polymer/nano-sheet interfacial interaction can be tuned by the means of covalent bonding, the thermal conductivity of graphene can be imparted to the nanocomposite, therefore, increasing the overall thermal stability. The authors also reported that rGO containing composites had higher thermal stability as compared to GO based nanocomposites. Among the three rGOs namely DMF-reduced GO, hydrazine-reduced GO and ascorbic acid-reduced GO, the last one was the most thermally stable.

The reason for higher thermal stability of rGOs was due to the removal of the thermally liable oxygen groups during chemical reduction, thus, resulting in the decrease in defect density and corresponding increase in thermal stability. Addition of 5% GO was observed to enhance the tensile strength and tensile modulus of PI by 12 and 18 times respectively, which was attributed to good degree of filler dispersion and stress transfer from the polymer chains to the filler platelets. On the other hand, the composites with rGO exhibited lower extent of enhancement in the mechanical performance as compared to the GO based composites, as seen in Figure 9.3. The authors observed that the rGO (reduced by LAA) increased the electrical conductivity

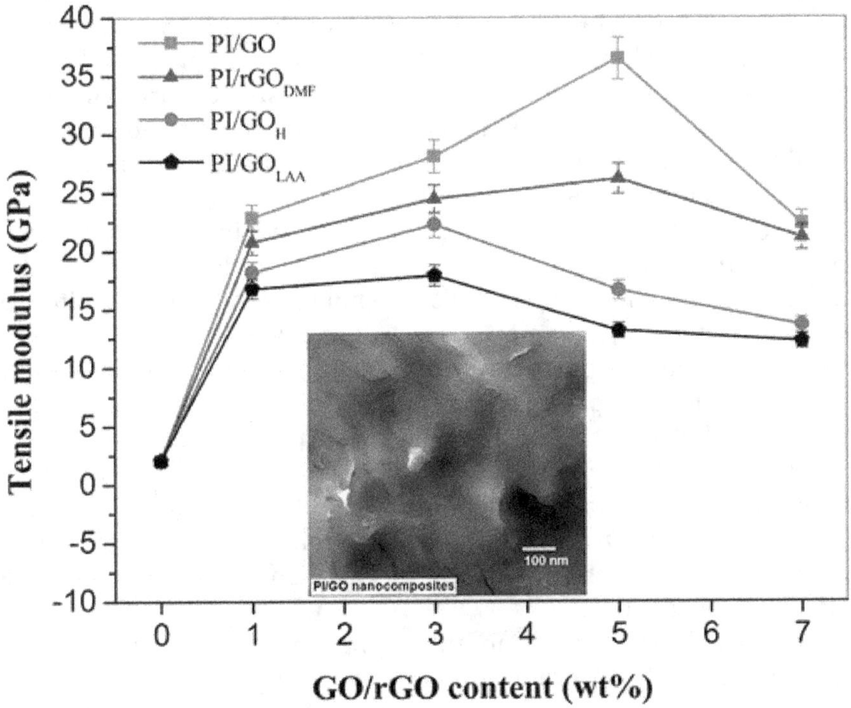

Figure 9.3 Tensile modulus of the PI-graphene nanocomposites with varying amounts of GO and rGO. Reproduced from Reference 16 with permission from American Chemical Society.

of the nanocomposite by 14 times, in comparison to pure PI. On the other hand, the composite with GO as filler exhibited an increased in the electrical conductivity by 10 times. The effectiveness in increase of electrical conductivity by rGOs was due to recovery of sp2 linkage of carbon by reducing with chemicals as well as uniform particle to particle link in the nanocomposite. The high electrical conductivity by LAA reduced GO was due to effective deoxygenation of GO during reduction.

In another study, Xu *et al.* [17] reported close-celled foams of aromatic polyimide (PI)/graphene composites, which had low density and improved flexibility. These materials were generated by thermal foaming of poly(amic ester)/graphene oxide (PAE/GO) precursor powders. The authors observed that the foaming capability of PAE was enhanced on the addition of GO, which resulted in larger cell size and decreased foam density. As observed in Figure 9.4, the cell size was observed to increase and become uniform as the fraction of GO was increased. In addition, the quantity of the unfoamed particles was also observed to decrease with increasing GO content. The surface of the cells also became rough with increasing GO fraction in the composites. The composite foams were also observed to possess high flexibility and flexural strength. For instance, an increase of 22.5% in flexural strength was obtained in the composite with the addition of 0.5 wt % GO in the precursor. The composites exhibited high gas barrier resistance due to the presence of GO. The authors observed the glass transition temperature (T_g) of the polymer in the composites to gradually reduce as the amount of graphene was enhanced. The observed effect was in contrast to the occasionally observed increment in the transition temperature due to the hindrance to the segmental mobility in the chains in the presence of graphene platelets. The authors suggested the reduction in the molecular weight of the polymer in the presence of graphene platelets to result in such degrease in the transition temperature. In addition, the weight loss in the composites in the temperature range of 50-170 °C was also observed to increase with graphene content.

Zheng *et al.* [18] used a modified hybrid of graphene and CNT as filler into PI, and the thermal stability was observed to decrease. This was deduced from the observation from TGA analysis, where the initiation of degradation in the composites occurred at a lower temperature as compared to pure PI. The decrease was suggested to be caused by the loss of carboxyl functional groups from graphene surface due to the reaction involved in the formation of the GS-CNT hybrid. Interestingly, the reduction in thermal properties was observed only at lower concentrations of the filler in the PI. On increasing the filler content in the polymer, an increase in thermal stability was observed. This was suggested to occur due to the high thermal stability

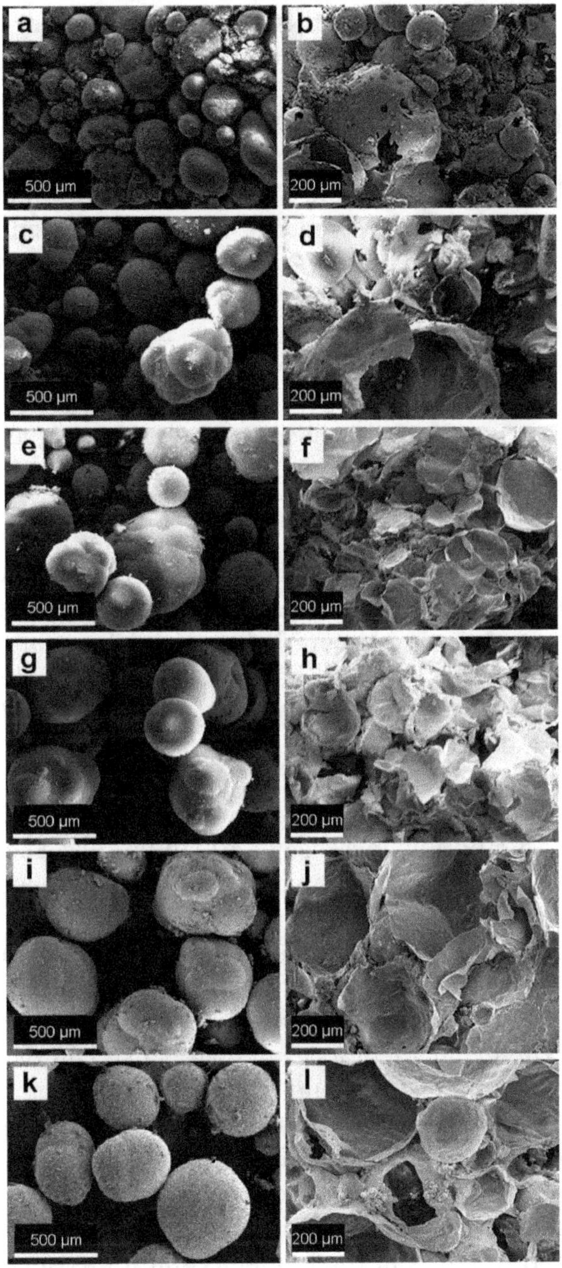

Figure 9.4 SEM images of individual cells (left column) and fractured surfaces (right column) of PI and the PI/graphene composite foams: (a, b) F0, (c, d) F0.1, (e, f) F0.2, (g, h) F0.5, (i, j) F1, and (k, l) F2. Reproduced from Reference 17 with permission from American Chemical Society.

of pure graphene which was in this case present in very substantial amount relative to the polymer [18].

In another study, Hu *et al.* [19] aimed to achieve effective dispersion of graphene oxide sheets in PI by *in-situ* polymerization. TGA analysis of the composites with different amounts of filler exhibited similar trends. Thus, the possible effect of increasing amounts of filler on the PI could not be observed and it was concluded that the GO sheets did not disturb the high thermal stability of the polymer. It was observed that the tensile strength of the nanocomposites decreased with increase in amount of filler added in small quantity (less than 1%). However, the authors observed that the addition of 10% filler produced a composite with better mechanical properties compared to that with 1%. Based on this observation, it was suggested that GO enhanced the properties of the polymer matrix only when added in high concentrations. In the study conducted by Longun and Iroh [20], the improvements in properties of PI due to the addition of nano-graphene sheets (NGS) were studied. It was observed from dynamic mechanical analysis that the glass transition temperature of the polymer in the composite improved remarkably with increasing amounts of the filler (greater than 1 wt%). There was, however, a decrease in T_g in the composites with very low amounts of filler (0.1-0.5 wt%). It was observed that addition of NGS to PI enhanced the storage modulus, especially in the rubbery phase. For instance, an increase of 500% were observed in the glassy phase for a 28.08% loading of NGS as filler, whereas a modulus increase of 400,000% was reported in the rubbery phase for the same composition of nanocomposite. The authors attributed the immense change in the modulus in the rubbery region to the high rigidity of the graphene sheets which expectedly caused the restrictions to the polymer chain movements. Thus, the presence of NGS in the rubbery polymer enhanced its stiffness to a high degree. The experimental results were also compared with the results from two models: the rule of mixtures (RM) model equation as well as the Halpin-Tsai (HT) micro-mechanical model. The models also exhibited the same trend of increase in the modulus of the nanocomposite with increase in filler composition although there were significant variations with the experiment results. These deviations were attributed to faults in the model assumptions such as uniform dispersion of fillers in the HT model and uniform alignment of fillers in the RM equation. Both of these assumptions (and a few others) are not necessarily achieved perfectly in real nanocomposites. In a similar work, Longun *et al.* [21] added clay to the PI-NGS matrix to study the effect of different amounts of clay and NGS on the thermal properties of the nanocomposite. It was observed that the T_g of the composite was higher than that of the pure polymer, similar to the finding observed in the previous study [20]. Also, the T_g remained fairly constant with varying amounts of clay, suggesting that the effect of NGS was more significant than that of clay in improving T_g. Also, the TGA revealed that the peak degradation temperature (T_d) of the composite improved significantly in the presence of NGS and clay and increased with increasing amounts of clay. Thus, it could be deduced that the incorporation of clay in the PI-NGS composite was effective in improving T_d but not significant in improving T_g [21]. As this report studied the effects of varying amounts of clay with a fixed amount of NGS on possible enhancements of PI-NGS system, it was observed that the increase in modulus with enhancing clay content was about 300% higher in the glassy phase and about 200,000% higher in the rubbery phase. It can, thus, be deduced by comparing the modulus enhancement in this work [21] and the previous one [20] that the property enhancement due to the addition of clay is significant, but not as high in magnitude as the enhancement resulting from the addition of NGS to PI. In addition, the authors observed an increase in storage modulus of the composite compared to pure polymer with increasing amount of clay (where NGS amount was fixed). This increase in modulus was, however, only up to a composition of about 4%, as addition of clay beyond this amount led to a decrease in storage modulus of the composite. In a recent work, Xu *et al.* [22] also exfoliated graphene oxide and reduced graphene oxide in PI, and observed improvement in the T_g of the polymer in the composite.

Kong *et al.* [23] added graphene to the PI matrix with Mg ions used as cross-linkers, which resulted in the improvement in the exfoliation of the filler. It was observed that the T_g of the polymer in the nanocomposites with different filler amounts did not vary significantly as compared to pure PI. Thus, the filler was observed to have no negative impact on the T_g of the polymer. The authors reported improvements in optical clarity, while the thermal stability of the polymer was also maintained in the nanocomposites. The authors also compared the improvements in tensile properties of Mg-ions cross-linked nanocomposite (called XPIG) with the one without Mg ions (called PIG). XPIG exhibited higher improvements in tensile properties as compared to both pure PI and PIG. The observed improvement was attributed to the better filler exfoliation achieved in the XPIG due to the effect of Mg ions. Furthermore, the authors observed that for both

XPIG and PIG, the highest property achievements occurred when (about) 1 wt% of filler was used. For higher weight fractions, the tensile properties reduced, which was attributed to agglomeration and non-uniform distribution of filler in the polymer matrix. The HT model was also used to predict the tensile properties of the polymer at different filler concentrations with the assumption (among others) that the graphene sheets were dispersed randomly inside the polymer. Comparison of the model results with the experimental data revealed a good fit. It was also observed that the elongation at break decreased with increasing amounts of GO filler in both XPIG and PIG suggesting increased rigidity of the composite. The observed results were also similar to the findings by Tsai *et al.* [24], where GO and reduced graphene (RG) were separately dispersed in varying amounts in PI to observe possible improvements in permeability and other properties. In the DMA analysis, the T_g of the composites did not change significantly. The authors observed that there was better dispersion of RG than GO in the polymer matrix. It was noted that the storage modulus of the PI-RG nanocomposite increased as the amounts of filler was enhanced. However, the trend shown in tensile strength increase for both nanocomposite systems was similar. In another study by Heo and Chang [25], functionalized graphene sheets were incorporated into the cross-linked PI matrix, and the thermal properties of the composites were studied using DSC and TGA for the determination of T_g and T_d, respectively. The authors observed that the T_g and T_d of the composites decreased with increasing the amounts of graphene from 1 to 10 wt%. There was, however, an increase in the T_g and T_d in the composite with 1 wt% filler as compared to pure PI. The authors suggested the existence of a threshold amount of filler loading beyond which stacking and agglomeration of the fillers in crosslinked polymer network possibly caused the decrease in T_g and T_d values.

Chen *et al.* [26] exfoliated graphene oxides in N,N-dimethylformamide by sonication and mixed the poly(amic acid) solution with graphene oxide suspension. Subsequently, polyimide nanocomposite films with different loadings of functionalized graphene sheets (FGS) were generated by reduction and imidization of the GO/PAA composites. Figure 9.5 depicts the SEM micrographs of the fracture surfaces of the sample after tensile test. The fractured surface of pure polymer was observed to be smooth, which indicated that the matrix was stretched due to the deformation. On the other hand, in the composite with 0.5 wt% FGS, the rough surface was attributed to the interfacial

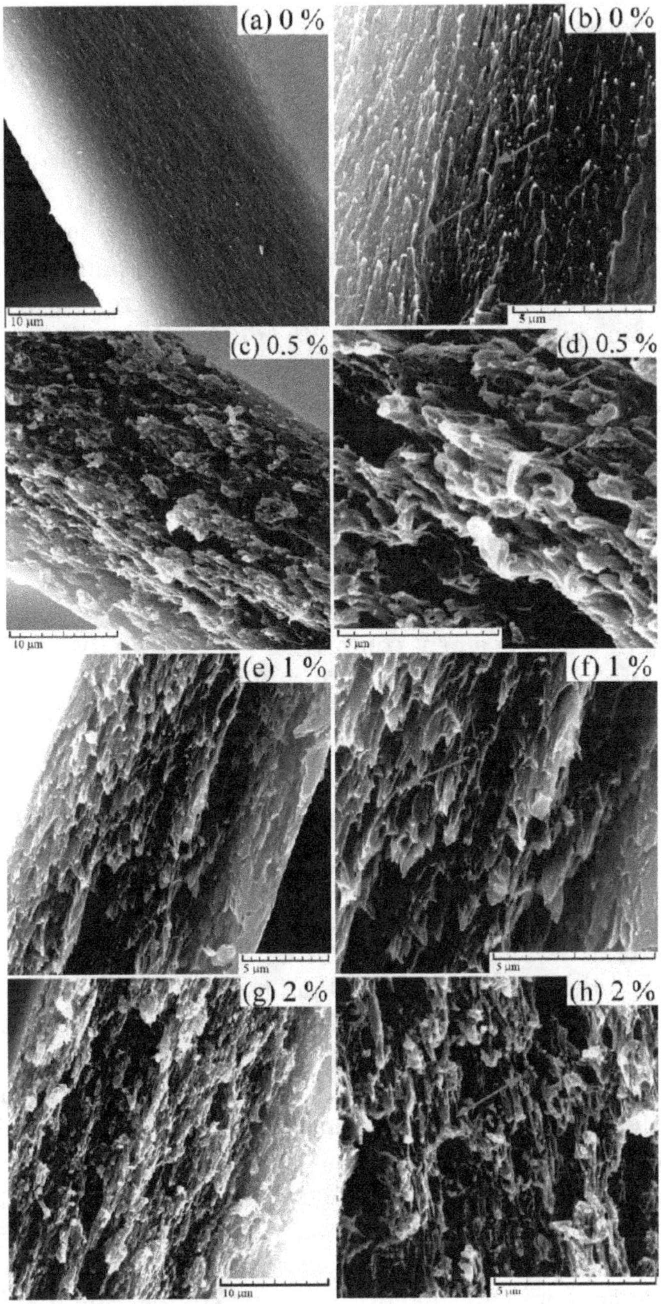

Figure 9.5 SEM micrographs of the fractured surfaces of neat PI and its nanocomposites containing different FGS loadings. Reproduced from Reference 26 with permission from American Chemical Society.

adhesion between the polymer and filler phases. The surfaces were also observed to become brittle with increasing filler content. The authors attributed the interfacial adhesion and the uniform filler dispersion to the significant enhancement in the tensile strength and modulus of the polymer. The authors also deduced that the method of synthesis allowed good dispersion of GO and alignment of the sheets along the surface direction of the polymer. The observed alignment (in addition to good interfacial bonding) thus also led to a significant enhancement of the modulus and tensile strength of the composites, which were recorded up to 2% addition of graphene. This improvement was followed with reduction in strain before rupture indicating an increased tendency towards brittleness, as also observed in morphology analysis in Figure 9.5.

Dai *et al.* [27] studied the thermal properties of the PI composites generated with increasing amounts of filler (graphene sheets modified with silicon carbide nanowires, SiC-GS). The authors reported the development in the thermal conduction with added amount of filler. The observed property improvement was suggested to result due to SiC-GS forming a heat conduction pathway in the cross-linked structure of PI [27]. Furthermore, the authors deduced that the thermal conductivity of the nanocomposites was stable within the temperature range of 25-175 °C. It was confirmed by conducting three heating-cooling cyclic experiments of thermal conductivity measurements within this temperature range. In addition, the TGA studies also revealed that the nanocomposites had a higher thermal stability than pure PI and was attributed to the anti-ablative nature of SiC (due to its low tendency to degrade in harsh external conditions). Interestingly, the final weight of materials remaining after decomposition was lower in the nanocomposites than pure PI. There was generally a reduction in tensile strength and storage modulus as temperature increased, however the mechanical properties did not show a steady pattern for temperature regions below the T_g of the polymer. In another study, Gong *et al.* [28] incorporated into the PI matrix about 12 wt% of 10-layer laminates of graphene woven fabrics (GWFs). The authors observed over 1400% increase in thermal conductivity of the pure PI, along with significant increase in the thermal diffusivity. It was also observed that GWFs/PI composites showed a higher increase in anisotropic thermal properties in comparison to other polymer-graphene composites with nano-fillers in the form of ribbons and platelets. The observed improvements were attributed to the peculiar arrangement of the GWFs layers within the PI matrix, which provided effective conduction paths due to their 3D structure, as well as the superior quality of the graphene used by virtue of its preparation by chemical vapor deposition (CVD).

As mentioned earlier, Ramakrishnan *et al.* [12] reported the increase in the T_g and storage modulus (E') of the nanocomposites with the increase in GO content, which indicated good degree of interaction between the polymer and filler phases. A similar observation was also made by Shukla *et al.* [29], where phosphorylated GO's dispersion in sulphonated PI led to improvements in the storage modulus. A reason for the observed good interaction could have been due to the fact that polymerization was carried out in the presence of graphene oxide which may have led to potential hydrogen bonding with the polyimide matrix. Tseng *et al.* [30] also reported that by adding 0.001 wt% of GO to PI, the storage modulus increased from 1487 for pure polymer to 2474 MPa at 60 °C. In addition, the composite had much lower water-vapor transmission rate of 30 g.mil.m^{-2}.day^{-1} as compared to 181 g.mi.m^{-2}.day^{-1} for pure PI. The composite had also lower coefficient of thermal expansion (CTE) resulting from the homogeneous distribution of GO nanosheets in the polymer and enhanced chain orientation of the polymer chains. The authors also reported that 2 μm thick films of polyimide/graphene oxide nanocomposite had marginal reduction in transparency on increasing the GO content.

Huang *et al.* [31] generated PI nanocomposites with modified graphene (MG/PI) using *in-situ* polymerization. The authors observed that the composite with 2 wt% MG had 20-fold increase in the wear resistance and 12% reduction in the friction coefficient. Incorporation of modified graphene was also observed to enhance thermal stability, electrical conductivity, and mechanical properties. As shown in Figure 9.6, the authors observed that the modified graphene existed on the worn surface of the nanocomposite, which resulted in the protective effect against the wearing force.

In another study, Kwon and Chang [32] produced PI nanocomposites using hexadecylamine-graphene sheets (HDA-GS) as filler. The authors observed a decrease in the T_g from 259 °C for the neat PI to 219 °C for the composite. On incorporating organically modified hectorite clay (STN)/HDA as filler, the T_g of the polymer in the composite decreased to 237 °C. In both cases, the filler content was increased from 0 to 20 wt%. It was explained that decrease in T_g was due to low heat resistance of the long alkyl chains that were present on the surface of the chemically modified graphene sheets [33]. In addition, the

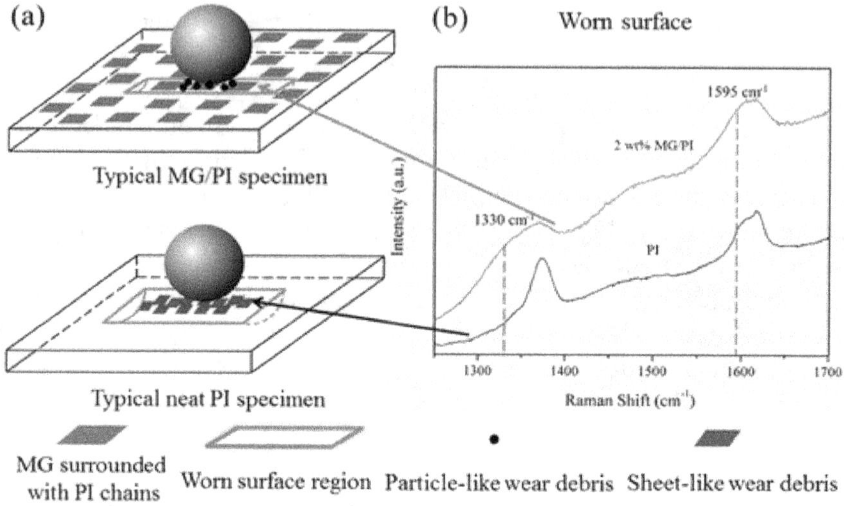

Figure 9.6 (a) Typical proposed model for the friction and wear mechanism of neat PI and MG/PI specimens. (b) Raman spectra of worn surface of neat PI and 2 wt% MG/PI specimens. Reproduced from Reference 31 with permission from American Chemical Society.

authors also reported that the thermal stabilities of nanocomposites having HDA-GS and STN/HDA-GS complex as fillers showed decreasing trend on increasing the filler content. The decrease in degradation temperature was attributed to unstable organic modifications. It was observed that by using HDA-GS as filler, the oxygen permeability decreased linearly. The authors also reported that nanocomposites with higher quantities of HDA-GS were stiffer which as a result added to lower gas penetrability. It was also observed that by adding 20 wt% of STN/HDA-GS as filler, the gas permeability decreased by an 80% in comparison to the pure PI film. Overall, the permeability was, thus, observed to decrease with increase in the length of diffusion path that the gas molecules have to travel in the filler. Thus, sheet like filler structures are beneficial to decrease gas permeability as these maximize the path distance due to high aspect ratio as compared to fillers with spherical or cubical shapes.

Ye *et al.* [34] employed fluorinated graphene (FG) as a filler to generate the nanocomposites with polyimide. The storage modulus of the PI/FG nanocomposites increased only after a critical filler content of 0.5 wt%. There was no improvement in T_g with FG as filler in the nanocomposites. For instance, in comparison to pure PI film, the composites exhibited only a slight increase of 1-3° in T_g. The authors also confirmed that FG can be used like any other graphene derivative to increase the thermal stability of the PI nanocomposites. The increase in thermal stability was due to the fact that fluorine did not remain on the filler surface, when FG was treated at high temperatures. Consequently, at high temperatures, FG in the composites was expected to behave similarly as graphene and facilitate heat dissipation in the nanocomposite, thereby, increasing overall thermal stability. The authors observed that the addition of 0.5 wt% FG in the polymer resulted in an increment of 30% in tensile stress, along with enhancement of 115% in elongation at break. The observed improvement in the mechanical performance was attributed to the good degree of filler dispersion in the polymer matrix, which resulted in effective stress transfer.

In a recent study, Ling *et al.* [35] developed microcellular polyetherimide (PEI)/graphene nanocomposite foams by employing a phase separation process. The authors observed the electrical conductivity percolation to decrease from 0.21 vol % for PEI/graphene nanocomposite to 0.18 vol % for PEI/graphene foam. The nanocomposite foams had enhanced electromagnetic interference (EMI) shielding effectiveness as well as low thermal conductivity of 0.065-0.037 W/m·K even at higher temperatures. Figure 9.7 also shows the

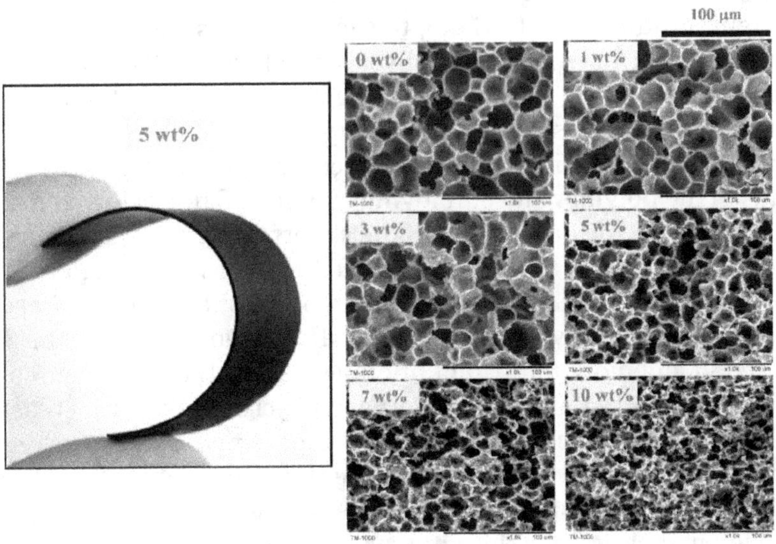

Figure 9.7 Optical and SEM micrographs of PEI/graphene nanocomposite foams as the function of graphene content. Reproduced from Reference 35 with permission from American Chemical Society.

morphological characterization of PEI/graphene nanocomposite foams as the function of graphene content. The pure PEI foam exhibited microcellular cell structure with call size of 15.3 μm, which was increased to 16.3 μm in the composite with 0.5 wt % filler loading. The size was increased further with the filler content of 1 wt%, however, the cell size of nanocomposite foams were observed to decrease with filler loading of 5 wt% onwards due to the higher viscosity of dispersion. In addition, the nanocomposite foams also exhibited uniform cell size distributions.

Park *et al.* [36] reported mechanically strong polyimide nanocomposites, generated using *in-situ* polymerization, with aminophenyl functionalized graphene nano-sheets (APGNS) as filler. Significant enhancement in the composites properties were reported on the incorporation of APGNS. For instance, the electrical conductivity of the composite with 3 wt% APGNS was 10^{11} times than that of pure PI. Similarly, the modulus of the 3 wt% composite was 610% higher than pure polymer. The transmission rate of water vapor was also observed to decrease by 74% in the composite. As also observed in Figure 9.8, the imidization temperature was observed to strongly affect the mechanical properties of the composites. The authors observed the composite generated at 250 °C to exhibit the highest tensile strength. The properties were observed to deteriorate at the treatment temperature of 300 °C due to the decomposition of functional groups on the graphene surface observed in Figure 9.8 c and d.

Yoonessi *et al.* [37] observed the effect of arrangement or orientation of the graphene sheets in the composite on its properties. This was achieved by connecting the graphene particles with nickel (Ni) strands in the polymer and employing magnetic fields of low and high intensity to vary the orientation of the sheets. Firstly, it was observed that the presence of Ni further increased the tensile modulus of the nanocomposite (in addition to the effect of graphene). The observed improvement was, however, limited to a Ni composition less than 3%. The second and more important finding was that the tensile modulus of the Ni connected graphene composite increased with increasing alignment of the graphene sheets in the polymer matrix. However, the modulus was not affected by increasing concentrations of graphene, when the graphene phase was aligned in the direction of the employed magnetic field. Notwithstanding, the tensile modulus of the composite exhibited a larger value compared to pure PI.

Wang *et al.* [38] observed that on addition of GO to polyimide, the tensile stress of the polymer increased up to 844 MPa. In addition, the

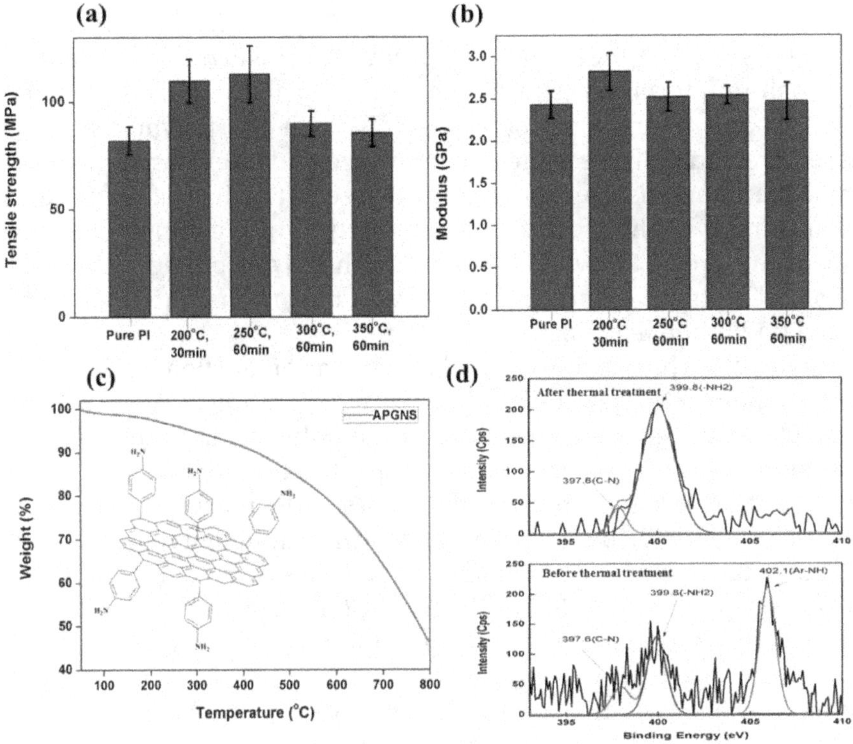

Figure 9.8 (a) Tensile strength and (b) modulus the PI composites with 1 wt% APGNS at various imidization temperatures, (c) TGA thermogram of APGNS, and (d) XPS N 1s spectra before and after thermal treatment (300 °C, 1 h, air conditioning) of APGNS. Reproduced from Reference 36 with permission from American Chemical Society.

Young's modulus was observed to increase to 20.5 GPa. The authors also reported that the use of NH_2-functionalized GO (ODA-GO) promoted dispersion of GO with the polymer matrix significantly. As a result, the nanocomposite with 3 wt% ODA-GO was observed to have Young's modulus and tensile strength to be 15 and 9 times greater than the corresponding pure PI film respectively. The authors also reported that the dielectric constant of the polymer in the composites decreased with increasing GO content. In another study, Luong *et al.* [39] reported the chemical modification of graphene oxide with ethyl isocyanate to generate ethyl isocyanate-treated graphene oxide (iGO). It was observed that the incorporation of 0.38 wt% of ethyl isocyanate treated graphene in polyimide resulted in the tensile strength to increase from122 MPa to 131 MPa. In addition, the

Young's modulus also increased from 1.8 GPa to 2.3 GPa. The increase in the tensile strength as well as modulus for the PI/graphene nanocomposite materials was attributed to uniform dispersion of filler platelets and their anisotropic positioning in the polymer matrix, thereby resulting in the achievement of a good degree of stress transfer. The authors also reported more than eight orders of magnitude enhancement in the electrical conductivity of the composite at the filler content of 0.38 wt%. In addition, the PI/graphene composite films had higher conductivities than that of carbon nanotubes reinforced polyimide nanocomposites.

Huang *et al.* [40] achieved planar orientation of chemically modified graphene (CMG) in the PI, along with generating strong covalent interactions at the interface between the polymer and filler phases, as shown in Figure 9.9. The authors reported that by adding a small amount of CMG, significant improvements in the mechanical properties, as well as thermal stability, electrical conductivity, and hydrophobic behavior were achieved. For instance, by the addition of 0.2 wt% of graphene to polyimide, the tensile strength increased by 30%, whereas the increment in the Young's modulus was 46%. In addition, the authors also achieved hydrophilic-to-hydrophobic transition and the electrical percolation at only 0.2 wt% filler content.

Kim *et al.* [41] studied the effect of graphene transferred on polyimide substrate which was pretreated using inductively coupled O_2, Ar, CH_4 and CHF_3 plasma. The authors measured the surface roughness as well as surface energies in each of the four cases and observed that for the improvement of electrical properties in graphene transferred to PI, substrate morphology and hydrophobicity were important parameters. The PI samples which were pretreated with CH_4 and CHF_3 displayed enhanced electrical conductivity of graphene device in comparison to the untreated PI substrate. The CH_4 and CHF_3 treated films also showed an increase in contact angles which suggested an increase in surface energy. The results of the PI samples treated with Ar and O_2 were not favorable as these exhibited degraded electrical conductivities, as well as a decrease in contact angle which implied an increase in surface energy. The authors concluded that pretreatment of PI films with CHF_3 and CH_4 leads to the removal of C-O and O-H chemical species by hydrophobization, which results in the improved electrical properties of graphene on these substrates. In another study, Liu *et al.* [42] investigated the method for the development of flexible conductive fibers by making use of interfacial modifications. The commercially obtained PI yarns were washed with DI

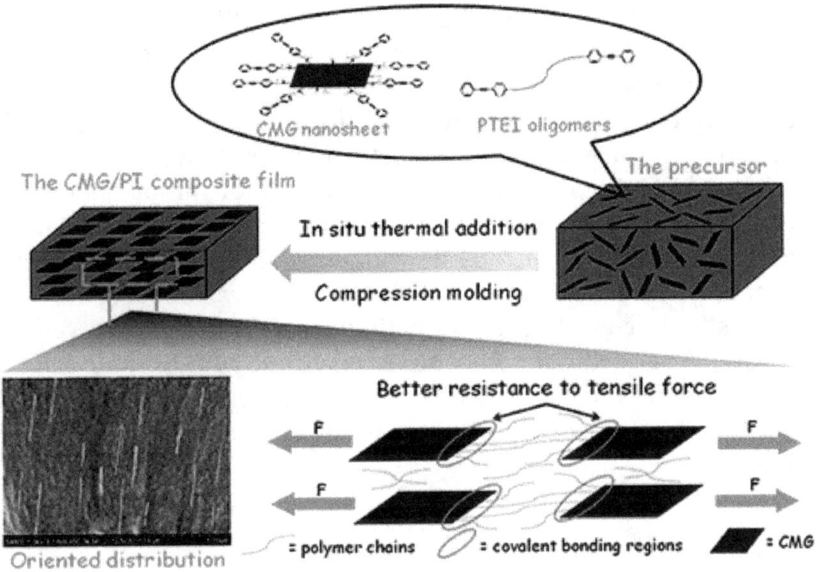

Figure 9.9 Schematic of the achievement of enhanced performance in the nanocomposites. Reproduced from Reference 40 with permission from American Chemical Society.

water and acetone in order to remove impurities. Subsequently, to modify the surface functional groups as well as the roughness, the yarns were immersed in an aqueous solution of sodium hydroxide and finally washed with DI water till the pH became neutral. These were then dipped into aqueous GO suspension. The amount of GO loading was controlled by the dipping timing. The PI/graphene yarns were finally dried and immersed into aqueous HCl in order to reduce the graphene oxide. The resulting yarns were washed with DI water and analyzed for different properties. It was observed that the conductivity reached a peak value of 1.02×10^3 $S.m^{-1}$ which was high for polymer fibers comprising conductive coatings. The authors also reported a current density of 90 $mA.cm^{-3}$ for fiber shaped supercapacitors in which the earlier mentioned PI-GO alkali treated yarns were used. The yarns also achieved a high electrochemical performance with a capacitance of 22.89 $F.cm^{-3}$.

Pandey *et al.* [43] reported the composites of sulfonated polyimide (SPI) with sulfonated propylsilane graphene oxide (SPSGO) (Figure 9.10) for use as polymer electrolyte membranes (PEMs). The incorporation of functionalized SPSGO (-SO_3H and -COOH) resulted in enhanced proton conductivity and thermal, mechanical, and chemical stabilities. For instance, the SPSGO containing PEM with 8 wt% content exhibited 10-fold enhanced bound water content, proton conductivity enhancement from 6.0 to 9.62 S·cm^{-1}, increment in the electron conductivity by 3 times and reduction in methanol permeability, as compared to pure SPI membrane. In another study, Park *et al.* [44] also reported highly conductive polyimide nanocomposites with reduced graphene oxide, generated using *in-situ* thermochemical synthesis. The thermo-chemically developed reduced GO from the phenyl functionalized GO (I-Ph-GO) (Figure 9.11) was observed to have five times higher electrical conductivity than conventional reduced GO (R-GO). As a result, the PI composite with R-I-Ph-GO had 4 orders magnitude higher conductivity than the corresponding composite with R-GO, thus, indicating the impact of filler surface modification strategy.

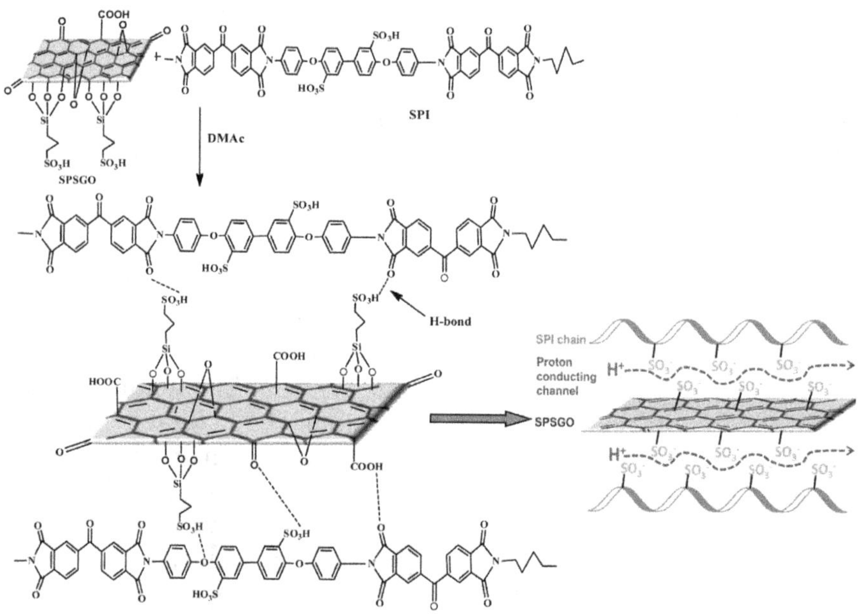

Figure 9.10 Schematic route for the preparation of SPI/SPSGO composite membranes. Reproduced from Reference 43 with permission from American Chemical Society.

Similar to the earlier study, Park *et al.* [45] further investigated the performance of iodo-functionalized graphene oxide (I-Ph-GO)/PI nanocomposites and reduced iodo-functionalized graphene oxide (R-I-Ph-GO)/PI nanocomposites. It was observed that the water vapor transmission rate for both the polymer nanocomposites was reduced by about 67% in comparison to pure PI, whereas for R-GO/PI nanocomposites, it was less than 20% of PI.

Dai *et al.* [46] studied the performance of different polyimide membranes containing different Ag-GO which were prepared by impregnation reduction. It was observed that small silver nanoparticles which were generated by the impregnation reduction using $Ag(NH_3)^{2+}$ and poly(ethylene glycol) were evenly distributed on the graphene oxide as well as in the hybrid membranes. The hybrid membrane containing Ag-GO-C (generated using the method mentioned above) reinforcement displayed the highest separation capability for benzene among different membranes. The authors attributed the enhanced performance to the cooperative effect of small silver particles and graphene oxide on the benzene molecules. It was also observed that by increasing the mass percent of silver beyond 15% in Ag-GO samples, the separation factor decreased because large silver particles were formed, which aggregated during polymerization, thereby reducing the separation performance of the membrane.

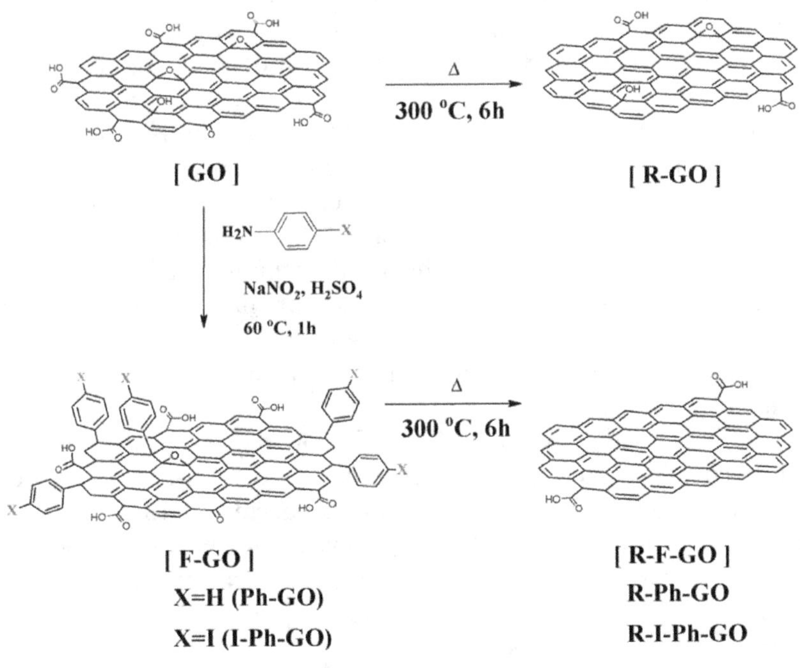

Figure 9.11 Schematic of the generation of functionalized graphene oxide (F-GO) and reduced graphene oxide (R-F-GO). Reproduced from Reference 44 with permission from American Chemical Society.

Hocker *et al.* [47] compared the water vapor permeability of unloaded polyimide, GO-PI nanocomposite and GO functionalized with 4-4' oxydianiline (ODAGO)-PI nanocomposite. The authors observed that the composite with 0.01 wt% GO exhibited five times decrease in water vapor transmission as compared to pure PI. The reduction was 10 times when 0.1 wt% GO in PI was used. Also, the nanocomposite with 0.1 wt% GO had a greater permeability than the nanocomposite with 0.01 wt% ODAGO. In other words, the ODAGO nanocomposite achieved the same extent of permeability reduction at much lower loadings than the GO nanocomposite.

In another study reporting the use of polyimide-graphene composite as proton exchange membranes, He *et al.* [48] studied the effect of different sizes of GO on the performance of fuel cells, by incorporating into sulfonated polyimide (SPI). The composite membrane containing smallest sized GO exhibited a highly uniform structure in comparison to the membrane having GO with larger sizes. Apart from improving the mechanical properties of the membrane, the SPI-GO membranes exhibited higher proton conductivity along with lower methanol permeability as compared to pure SPI film. The proton conductivity was observed to increase due to the formation of hydrogen bonds between GO and SPI, as the protons move through the hydrogen bond network. However, increase in the loading of even small sized GO particle after a certain limit decreased the proton conductivity due to blocking effect. The strong bonding between SPI and GO reduced membrane swelling and also increased the tortuosity as well as barrier effect for methanol crossing. The membranes with small particle size of GO were also observed to maintain higher ion exchange and water uptake. Thus, it can be concluded from these findings that it is important to choose the proper size of GO for polymer incorporation, as the filler size significantly affects the performance of nanocomposite materials.

In another study, Akhtar *et al.* [49] reported the development of polyetherimide-graphene nanocomposites as anti-corrosion coatings. Graphene in the polymer acted as a barrier for the diffusion of oxygen, water and chloride ions by increasing the tortuosity of the diffusion pathway when added in large concentrations of 5 wt%. The coated metal surfaces were also analyzed by scribing the 5 wt% graphene coating after 10 weeks of exposure to 3.5 wt% NaCl, as shown in Figure 9.12. The optical and SEM images of the metal surface at the scribed area didn't show corrosion products (Figure 9.12 a & b). It was confirmed further by the EDX analysis that the observed signals

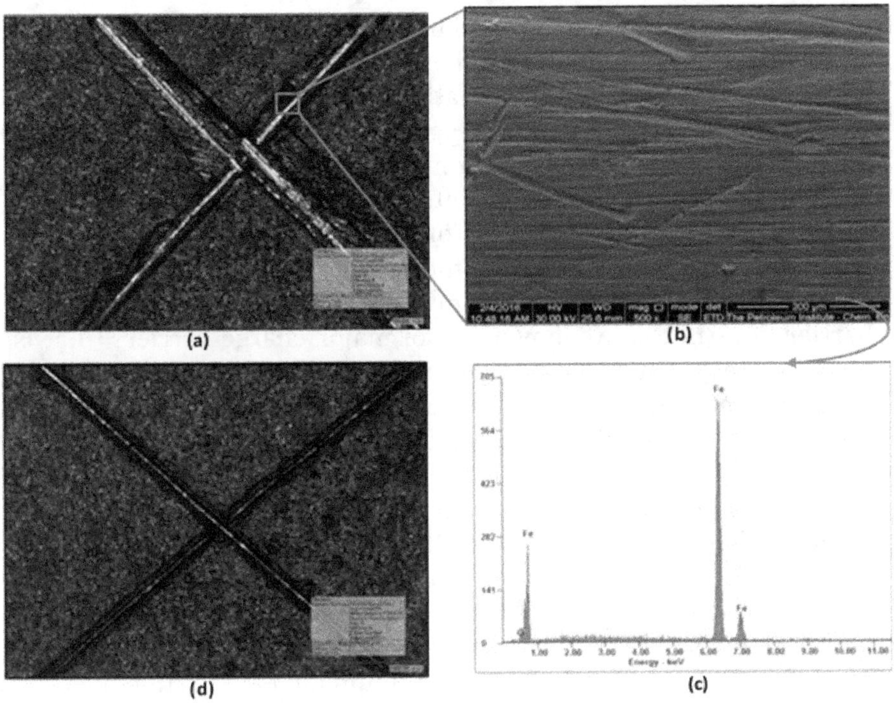

Figure 9.12 Optical image of scribed area on 5 wt% G-PEI coating (a), SEM image (b) and EDX (c) of the metal surface on the scribed area marked with red square. Optical image of 5 wt% G-PEI coating scribed and immersed for 6 weeks in 3.5 wt% sodium chloride solution.

corresponded to only metal ions (Figure 9.12 c). In order to analyze the stability of adhesion, a scribed 5 wt% G-PEI coating was immersed in 3.5 wt% NaCl over 6 weeks. No coating disbonding was seen even after immersion of scribed coatings showing that it maintained an excellent adhesion (Figure 9.12 d). The coatings did not allow the entry of water from the scribed region and did not allow lateral diffusion indicating the good interaction of graphene with PEI matrix and metal surface. The authors also presented mechanism of the protection provided by the graphene in PEI matrix, as shown in Figure 9.13. During the initial hours of immersion in NaCl solution, the 5 wt% G-PEI coatings prevented water and oxygen permeation through the coating due its hydrophobic nature and platelet structure with high aspect ratio which increased the diffusion pathway of the corrosive species and inhibited the corrosion reaction by barrier effect. However, at prolonged immersion, the protection was obtained by a different mechanism of steel passivation at the interface of steel and coating promoted by graphene. This noble behavior of steel might have occurred by the formation of Schottky barrier at the steel graphene junction. The insert in Figure 9.13 shows the protection obtained by the band gap formation at the metal-graphene interface. The graphene is n-doped at the interface when contact with metal is established. Thus, the interface of the steel-graphene forms a p-n junction with metal surface as the p-type semiconductor and graphene as the n-type semiconductor. The depletion region or potential barrier does not allow movement of major charge carriers, that is, electrons (e) from the graphene to the metal surface and holes (h) from the metal surface to graphene. Electron depletion occurs at the metal substrate (anode) and no electrons are available for transfer to the cathode. This reduces the tendency of metal to oxidize. The metal acts more nobly due to passivation by graphene contact.

In another study, Ahmad *et al.* [50] synthesized a nanocomposite electrode for lithium ion batteries using few layers of exfoliated graphene and polyimide (PI-FLEG). Due to the addition of exfoliated graphene, the electrical properties of PI-FLEG were enhanced. On employing PI-FLEG as a cathode in the lithium ion battery, it delivered a discharge capacity of 177 mA.h.g^{-1} at 0.1 °C and retained 80% of its initial discharge capacity after testing for 200 cycles. The synergistic effects of vertically grown polyimide sheets were attributed to the enhanced performance of the cathode in the lithium ion battery.

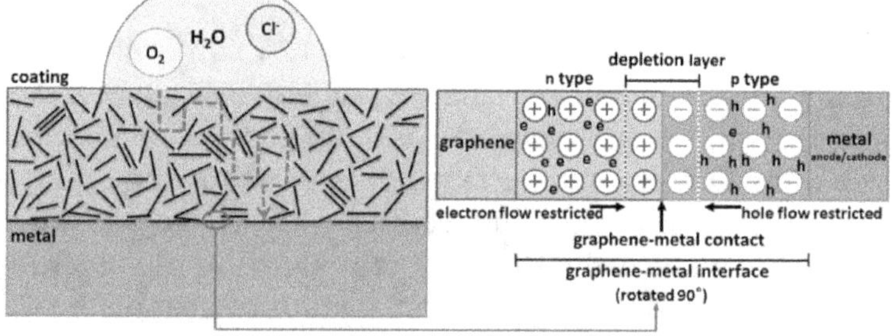

Figure 9.13 Pictorial representation of protection provided by graphene in 5 wt% G-PEI composite coating.

9.3 Conclusions

The polyimide-graphene nanocomposites are a novel and promising class of nano-materials with large number of excellent properties. Specifically, graphene has been reported to enhance the mechanical, thermal, viscoelastic, gas barrier, thermal and electrical conduction performance of the polyimide nanocomposites at low filler loadings. Owing to these enhancements in the properties, these polyimide nanocomposite materials have strong potential of use in many applications such as anti-corrosion coatings, fuel cell membranes, foams, conducting films, etc.

References

1. Mittal, V. (2009) Polymer layered silicate nanocomposites: A review. *Materials*, **2**, 992-1057.
2. Mittal, V. (2010) Polymer nanocomposites: Synthesis, microstructure, and properties. In: *Optimization of Polymer Nanocomposite Properties*, Mittal, V. (ed.), Wiley VCH, Germany, pp. 1-20.
3. Mittal, V. (2014) Functional polymer nanocomposites with graphene: A review. *Macromolecular Materials and Engineering*, **299**, 906-931.
4. Kirby, A. J. (1992) *Polyimides: Materials, Processing and Applications*, iSmithers Rapra Publishing, USA.
5. Lee, C., Wei, X., Kysar, J. W., and Hone, J. (2008) Measurement of the elastic properties and intrinsic strength of monolayer graphene. *Science*, **321**(5887), 385-388.
6. Tombros, N., Veligura, A., Junesch, J., Guimaraes, M. H. D., Vera-Marun, I. J., Jonkman, H. T., and van Wees, B. J. (2011) Quantized conductance of a suspended graphene nanoconstriction. *Nature Physics*, **7**(9), 697-700.
7. Bunch, J. S., Verbridge, S. S., Alden, J. S., van der Zande, A. M., Parpia, J. M., Craighead, H. G., and McEuen, P. L. (2008) Impermeable atomic membranes from graphene sheets. *Nano Letters*, **8**(8), 2458-2462.
8. Lee, Y., Bae, S., Jang, H., Jang, S., Zhu, S. E., Sim, S. H., Song, Y. I., Hong, B. H., and Ahn, J. H. (2010) Wafer-scale synthesis and transfer of graphene films. *Nano Letters*, **10**(2), 490-493.
9. Kim, H., Abdala, A. A., and Macosko, C. W. (2010) Graphene/polymer nanocomposites. *Macromolecules*, **43**(16), 6515-6530.
10. Bonaccorso, F., Lombardo, A., Hasan, T., Sun, Z., Colombo, L., and Ferrari, A. C. (2012) Production and processing of graphene and 2d crystals. *Materials Today*, **15**(12), 564-589.
11. Herrera-Alonso, M., Abdala, A. A., McAllister, M. J., Aksay, I. A., and Prud'homme, R. K. (2007) Intercalation and stitching of graphite oxide with diaminoalkanes. *Langmuir*, **23**(21), 10644-10649.
12. Ramakrishnan, S., Dhakshnamoorthy, M., Jelmy, E. J., Vasanthakum-

ari, R., and Kothurkar, N. K. (2014) Synthesis and characterization of graphene oxide-polyimide nanofiber composites. *RSC Advances*, **4**(19), 9743-9749.

13. Dong, J., Yin, C., Zhao, X., Li, Y., and Zhang, Q. (2013) High strength polyimide fibers with functionalized graphene. *Polymer*, **54**(23), 6415-6424.
14. Lim, J., Yeo, H., Goh, M., Ku, B.-C., Kim, S. G., Lee, H. S., Park, B., and You, N.-H. (2015) Grafting of polyimide onto chemically-functionalized graphene nanosheets for mechanically-strong barrier membranes. *Chemistry of Materials*, **27**(6), 2040-2047.
15. Yoonessi, M., Shi, Y., Scheiman, D. A., Lebron-Colon, M., Tigelaar, D. M., Weiss, R. A., and Meador, M. A. (2012) Graphene polyimide nanocomposites; Thermal, mechanical, and high-temperature shape memory effects. *ACS Nano*, **6**(9), 7644-7655.
16. Ha, H. W., Choudhury, A., Kamal, T., Kim, D.-H., and Park, S.-Y. (2012) Effect of chemical modification of graphene on mechanical, electrical, and thermal properties of polyimide/graphene nanocomposites. *ACS Applied Materials & Interfaces*, **4**(9), 4623-4630.
17. Xu, L., Jiang, S., Li, B., Hou, W., Li, G., Memon, M. A., Huang, Y., and Geng, J. (2015) Graphene oxide: A versatile agent for polyimide foams with improved foaming capability and enhanced flexibility. *Chemistry of Materials*, **27**(12), 4358-4367.
18. Zheng, Z., Wang, Z., Feng, Q., Du, Y., and Wang, C. (2013) Preparation of surface-silvered graphene-CNTs/polyimide hybrid films: Processing, morphology and properties. *Materials Chemistry and Physics*, **138**(1), 350-357.
19. Hu, N., Wei, L., Wang, Y., Gao, R., Chai, J., Yang, Z., Kong E. S.-W., and Zhang, Y. (2012) Graphene oxide reinforced polyimide nanocomposites via in situ polymerization. *Journal of Nanoscience and Nanotechnology*, **12**(1), 173-178.
20. Longun, J., and Iroh, J. (2012) Nano-graphene/polyimide composites with extremely high rubbery plateau modulus. *Carbon*, **50**(5), 1823-1832.
21. Longun, J., Walker, G., and Iroh, J. (2013) Surface and mechanical properties of graphene-clay/polyimide composites and thin films. *Carbon*, **63**, 9-22.
22. Xu, L., Chen, G., Wang, W., Li, L., and Fang, X. (2016) A facile assembly of polyimide/graphene core–shell structured nanocomposites with both high electrical and thermal conductivities. *Composites Part A: Applied Science and Manufacturing*, **84**, 472-481.
23. Kong, J.-Y., Choi, M.-C., Kim, G. Y., Park, J. J., Selvaraj, M., Han, M., and Ha, C.-S. (2012) Preparation and properties of polyimide/graphene oxide nanocomposite films with Mg ion crosslinker. *European Polymer Journal*, **48**(8), 1394-1405.
24. Tsai, M.-H., Chang, C.-J., Lu, H.-H., Liao, Y.-F., and Tseng, I.-H. (2013)

Properties of magnetron-sputtered moisture barrier layer on transparent polyimide/graphene nanocomposite film. *Thin Solid Films*, **544**, 324-330.

25. Heo, C., and Chang J.-H. (2013) Polyimide nanocomposites based on functionalized graphene sheets: Morphologies, thermal properties, and electrical and thermal conductivities. *Solid State Sciences*, **24**, 6-14.
26. Chen, D., Zhu, H., and Liu, T. (2010) In situ thermal preparation of polyimide nanocomposite films containing functionalized graphene sheets. *ACS Applied Materials & Interfaces*, **2**(12), 3702-3708.
27. Dai, W., Yu, J., Liu, Z., Wang, Y., Song, Y., Lyu, J., Bai, H., Nishimura, K., and Jiang, N. (2015) Enhanced thermal conductivity and retained electrical insulation for polyimide composites with SiC nanowires grown on graphene hybrid fillers. *Composites, Part A: Applied Science and Manufacturing*, **76**, 73-81.
28. Gong, J., Liu, Z., Yu, J., Dai, D., Dai, W., Du, S., Li, C., Jiang, N., Zhan, Z., and Lin, C.-T. (2016) Graphene woven fabric-reinforced polyimide films with enhanced and anisotropic thermal conductivity. *Composites, Part A: Applied Science and Manufacturing*, **87**, 290-296.
29. Shukla, G., Pandey, R. P., and Shahi, V. K. (2016) Temperature resistant phosphorylated graphene oxide-sulphonated polyimide composite cation exchange membrane for water desalination with improved performance. *Journal of Membrane Science*, **520**, 972-982.
30. Tseng, I. H., Liao, Y.-F., Chiang, J.-C., and Tsai, M.-H. (2012) Transparent polyimide/graphene oxide nanocomposite with improved moisture barrier property. *Materials Chemistry and Physics*, **136**(1), 247-253.
31. Huang, T., Xin, Y., Li, T., Nutt, S., Su, C., Chen, H., Liu, P., and Lai, Z. (2013) Modified graphene/polyimide nanocomposites: reinforcing and tribological effects. *ACS Applied Materials & Interfaces*, **5**(11), 4878-4891.
32. Kwon, K., and Chang, J.-H. (2014) Comparison of the properties of polyimide nanocomposites containing three different nanofillers: organoclay, functionalized graphene, and organoclay/functionalized graphene complex. *Journal of Composite Materials*, **49**(24), 3031-3044.
33. Xu, H., Kuo, S.-W., Lee, J.-S., and Chang, F.-C. (2002) Preparations, thermal properties, and Tg increase mechanism of inorganic/organic hybrid polymers based on polyhedral oligomeric silsesquioxanes. *Macromolecules*, **35**(23), 8788-8793.
34. Ye, X., Gong, P., Wang, J., Wang, H., Ren, S., and Yang, S. (2015) Fluorinated graphene reinforced polyimide films with the improved thermal and mechanical properties. *Composites, Part A: Applied Science and Manufacturing*, **75**, 96-103.
35. Ling, J., Zhai, W., Feng, W., Shen, B., Zhang, J., and Zheng, W. G. (2013)

Facile preparation of lightweight microcellular polyetherimide/graphene composite foams for electromagnetic interference shielding. *ACS Applied Materials & Interfaces*, **5**(7), 2677-2684.
36. Park, O.-K., Hwang, J.-Y., Goh, M., Lee, J. H., Ku, B.-C., and You, N.-H. (2013) Mechanically strong and multifunctional polyimide nanocomposites using amimophenyl functionalized graphene nanosheets. *Macromolecules*, **46**(9), 3505-3511.
37. Yoonessi, M., Gaier, J. R., Peck, J. A., and Meador, M. A. (2015) Controlled direction of electrical and mechanical properties in nickel tethered graphene polyimide nanocomposites using magnetic field. *Carbon*, **84**, 375-382.
38. Wang, J.-Y., Yang, S.-Y., Huang, Y.-L., Tien, H.-W., Chin, W.-K., and Ma, C.-C.M. (2011) Preparation and properties of graphene oxide/polyimide composite films with low dielectric constant and ultrahigh strength via in situpolymerization. *Journal of Materials Chemistry*, **21**(35), 13569-13575.
39. Luong, N. D., Hippi, U., Korhonen, J. T., Soininen, A. J., Ruokolainen, J., Johansson, L.-S., Nam, J.-D., Sinh, L. H., and Seppala, J. (2011) Enhanced mechanical and electrical properties of polyimide film by graphene sheets via in situ polymerization. *Polymer*, **52**(23), 5237-5242.
40. Huang, T., Lu, R., Su, C., Wang, H., Guo, Z., Liu, P., Huang, Z., Chen, H., and Li, T. (2012) Chemically modified graphene/polyimide composite films based on utilization of covalent bonding and oriented distribution. *ACS Applied Materials & Interfaces*, **4**(5), 2699-2708.
41. Kim, J.-K., Kang, K. H., and Lee, J.-M. (2016) Electrical properties of transferred graphene films on pre-treated polyimide substrate by inductively coupled plasma. *Carbon*, **109**, 282-289.
42. Liu, L., Weng, W., Dai, X., Liu, N., Yang, J., Liang, Y., and Ding, X. (2016) Highly conductive graphene-bonded polyimide yarns for flexible electronics. *RSC Advances*, **6**(110), 108362-108368.
43. Pandey, R. P., Thakur, A. K., and Shahi V. K. (2014) Sulphonated polyimide/acid functionalized graphene oxide composite polymer electrolyte membranes with improved proton conductivity and water retention properties. *ACS Applied Materials & Interfaces*, **6**, 16993-17002.
44. Park, O.-K., Hahm, M. G., Lee, S., Joh, H.-I., Na, S.-I., Vajtai, R., Lee, J. H., Ku, B.-C., and Ajayan, P. M. (2012) In situ synthesis of thermochemically reduced graphene oxide conducting nanocomposites. *Nano Letters*, **12**(4), 1789-1793.
45. Park, O.-K., Kim, S.-G., You, N.-H., Ku, B.-C., Hui, D., and Lee, J. H. (2014) Synthesis and properties of iodo functionalized graphene oxide/polyimide nanocomposites. *Composites, Part B: Engineering*, **56**, 365-371.
46. Dai, S.-q., Jiang, Y.-y., Wang, T., Wu, L.-g., Yu, X.-y., and Lin, J.-z. (2016)

Enhanced performance of polyimide hybrid membranes for benzene separation by incorporating three-dimensional Ag-graphene oxide. *Journal of Colloid and Interface Science*, **478**, 145-54.

47. Hocker, S., Hudson-Smith, N., Schniepp, H. C., and Kranbuehl, D. E. (2016) Enhancing polyimide's water barrier properties through addition of functionalized graphene oxide. *Polymer*, **93**, 23-29.
48. He, Y., Tong, C., Geng, L., Liu, L., and Lu, C. (2014) Enhanced performance of the sulfonated polyimide proton exchange membranes by graphene oxide: Size effect of graphene oxide. *Journal of Membrane Science*, **458**, 36-46.
49. Akhtar, T., Luckachan, G. E., and Mittal, V. Inhibition of corrosion by graphene in polyetherimide composite coating for long term protection of carbon steel, unpublished.
50. Ahmad, A., Wu, H., Guo, Y., Meng, Q., Meng, Y., Lu, K., Liu, L., and Wei, Z. (2016) A graphene supported polyimide nanocomposite as a high performance organic cathode material for lithium ion batteries. *RSC Advances*, **6**(40), 33287-33294.

10

Poly(phenylene sulfide) Nanocomposites with Expanded Graphite/Graphene

10.1 Introduction

Engineering polymers are materials with superior structure-property correlations. These polymers have exceptional mechanical properties such as strength, stiffness, creep and dimensional stability, which are further complemented by superior thermal stability, flame retardancy, electrical conduction, etc. These properties enable the use of engineering polymers in advanced, high-end applications in structural, automotive and aerospace industries. For example, aromatic polyesters and polybenzamide have decomposition temperatures of around 480-500 °C, whereas polybenzimidazole, polypyrrole, and poly(*p*-phenylene) decompose around 650 °C. Various engineering polymers include polysulfones, polyimides, poly(phenylene oxide), aramids, polysulfides, polyketones, liquid crystalline polymers, poly(phenylene ether), poly(phenylene sulfide), etc.

Poly(phenylene sulfide) (PPS) is one of the most useful engineering polymers due to its extensive set of superior properties [1-9]. It was first discovered by Charles Friedel and James Mason Crafts in 1888. This form of regular PPS was found to have a linear high-performance thermoplastic structure with semi-crystalline nature [1-5]. Its structure mainly comprises of a benzene ring and sulfur. It has a glass transition temperature (T_g) of 85 °C (80-90 °C) and a peak melt temperature (T_m) of 285 °C (280-290 °C), which in turn allow PPS's heat deflection/service temperature at around 230 °C. It has a high density of 1350 kg.m^{-3} and a dielectric strength in the range of 22-28 kV.mm^{-1}. Its coefficient of thermal expansion is around 51.6 ($<T_g$) and 97.9 ($>T_g$). Its thermal and electrical conductivity values are approx. 0.22 W.(m.K)$^{-1}$ and 1.5x10^{-16} S.cm^{-1}. PPS has

Hersi M. Abdulkadir, Georges H. Faraj and Vikas Mittal, The Petroleum Institute (part of Khalifa University of Science and Technology), Abu Dhabi, UAE*
**Current address: Bletchinton, Wellington County, Australia*

a Vickers microhardness around 24 kg.mm^{-2} [1]. In addition, PPS has good mechanical properties (strong, rigid, tough, light weight), high dimensional stability and is resistant to high temperature, flames, long term creep, chemicals and solvents (insoluble in many organic solvents below 200 °C). Moreover, it is flexible to inject mold, blow mold and machine as well as retains its attributes at temperatures even above 200 °C. One of PPS's most interesting features is its ability to change structure and, in turn, properties using heat and oxygen (curing, crosslinking, chain extension). However, PPS has some limitations as it has a low elongation at break, is relatively brittle and has a high cost of production [5,6,9]. Different types of PPS have been developed, such as cured, linear and branched, which are briefly described below:

- ***Cured***: generated by heating the regular PPS polymer in the presence of oxygen. This leads to an increase in chain length and branching, thus, causing it to have a thermoset-like behavior.
- ***Linear***: has double the molecular weight of regular PPS. This is achieved by higher degree of polymerization. The increase in molecular weight increases tenacity (ability to grip), elongation and impact strength.
- ***Branched***: has a higher molecular weight than the regular PPS polymer with higher extent of short and long chain branching which leads to improvements in tenacity and ductility. Such materials have reduced extent of crystallinity due to the presence of branches which do not allow the polymer chains to crystallize easily.

The synthesis or preparation of PPS is usually achieved using different routes, some of which are mentioned below [10-13]:

- Polycondensation of 1,4-dichlorobenzene and sodium sulfide in a polar solvent (N-methylpyrrolidone) at a high temperature and pressure. However, as the temperature gets higher, the yield decreases. The yield is also affected by the ratio of acid-to-monomer.
- Polycondensation of metal *p*-halobenzenethiolate with Lewis acid at 350 °C. Here the ratio of acid to oxidant affects structure, morphology and transport properties. The polymerization proceeds in a stepwise fashion and consists

of an electrophilic reaction of cations with the diphenyl sulfide to form a trimer.

- Electrolysis using 2 platinum strips acting as the working and auxiliary electrodes in addition to an Ag/AgCl reference electrode. The setup is placed in a nitromethane solution of thiophenol contained in a Lewis acid at 20 °C temperature in an inert N_2 atmosphere.

Many recent developments in the design of molecular architecture of the PPS chains have been achieved. For instance, Li *et al.* [14] reported the synthesis and characterization of novel PPS containing a chromophore in the main chain. To achieve this, the reaction of dihalogenated monomer and sodium sulfide was performed via nucleophilic substitution polymerization under high pressure. The authors reported that the mechanical performance of modified PPS containing chromophore groups was higher than the pure PPS. It was, thus, concluded that the chromophore groups in the PPS main chain hindered their movement. Specifically, due to the strong polarity of the chromophore groups in the main chain, the tensile and compression strength were increased.

Like any other polymer, PPS is susceptible to degradation. Oxygen and nitrogen monoxide are mainly responsible for its degradation [11-13]. Although, oxygen may be used to tailor the properties of PPS, however, oxygen may also be detrimental to the polymer properties as well. A major impact of oxygen is the change in the crystallinity of PPS. In general, the crystallinity of PPS is affected by temperature, heating time and oxygen presence. This is because crystallinity is controlled by annealing and curing parameters which are affected by the stated parameters. During curing, several structural changes occur that include chain extension, oxygen linkage branching, formation of crosslinks as well as sulfur dioxide generation due to oxygen uptake. All these variations during the curing process affect mechanical properties of PPS. To summarize, the strength of PPS depends on the degree of crystallinity as well as any flaws/imperfections in its amorphous phase. Nitrogen monoxide (NO) increases amorphous and crystalline region deterioration, but does not affect the crystallization process.

PPS has a broad range of applications and uses, such as nonwoven bag filters, light reflectors, flow measuring units, valves, nozzles, fuel manifolds, photovoltaic cells, gas sensors, and ultra-capacitors. Moreover, high strength machine parts employed under very harsh environmental conditions, such as engine components, are generally made from PPS [1-8]. The uses are also classified based on the polymer type as follows:

- Regular PPS is typically used as coatings, which have high thermal and chemical stability. Cured PPS is also used as coatings and injection molding compounds, among other applications.
- Linear PPS is used to make fibers and tougher injection molding compounds since it has better tenacity, elongation and impact strength.
- Branched PPS is used to make films and several injection molding compounds.

The development of polymer nanocomposites, where at least one of the dimension of the filler phase is less than 100 nm, have revolutionized the composites research due to the achievement of large polymer-filler interfacial contacts which result in significant enhancement in the polymer properties at much lower filler volume fractions. Fillers such as layered silicates, graphene or carbon nanotubes (CNTs) are commonly used to enhance the properties of polymers [8,12,13]. For any noticeable improvements in the polymer properties to occur, the filler and the matrix phase must be compatible and have good interfacial attraction, else large amount of filler is needed to achieve required improvement in the properties, which make the composite bulky and results in loss of transparency. A large variety of nanocomposites with different fillers have also been developed using engineering polymers. For instance, Figure 10.1 demonstrates the various examples of the development of nanocomposites of polypyrrole, polyetherimides, polyaniline, polyamides, etc. [15-18]. The incorporation of uniformly distributed filler phase was observed to enhance the performance of various engineering polymers significantly. In other studies, Sangermano *et al.* [19] reported the synthesis of diacrylate polysulfone oligomer which was subsequently used as the acrylic oligomer for the *in-situ* synthesis of noble metal/PSU nanocomposites through UV-induced simultaneous radical polymerization of acrylic functionalities and NP formation by reduction of their precursors. The authors observed a slight decrease in the glass transition temperature of the cured films prepared in the presence of either silver or gold precursor. The observed behavior also agreed with the FTIR analysis of the

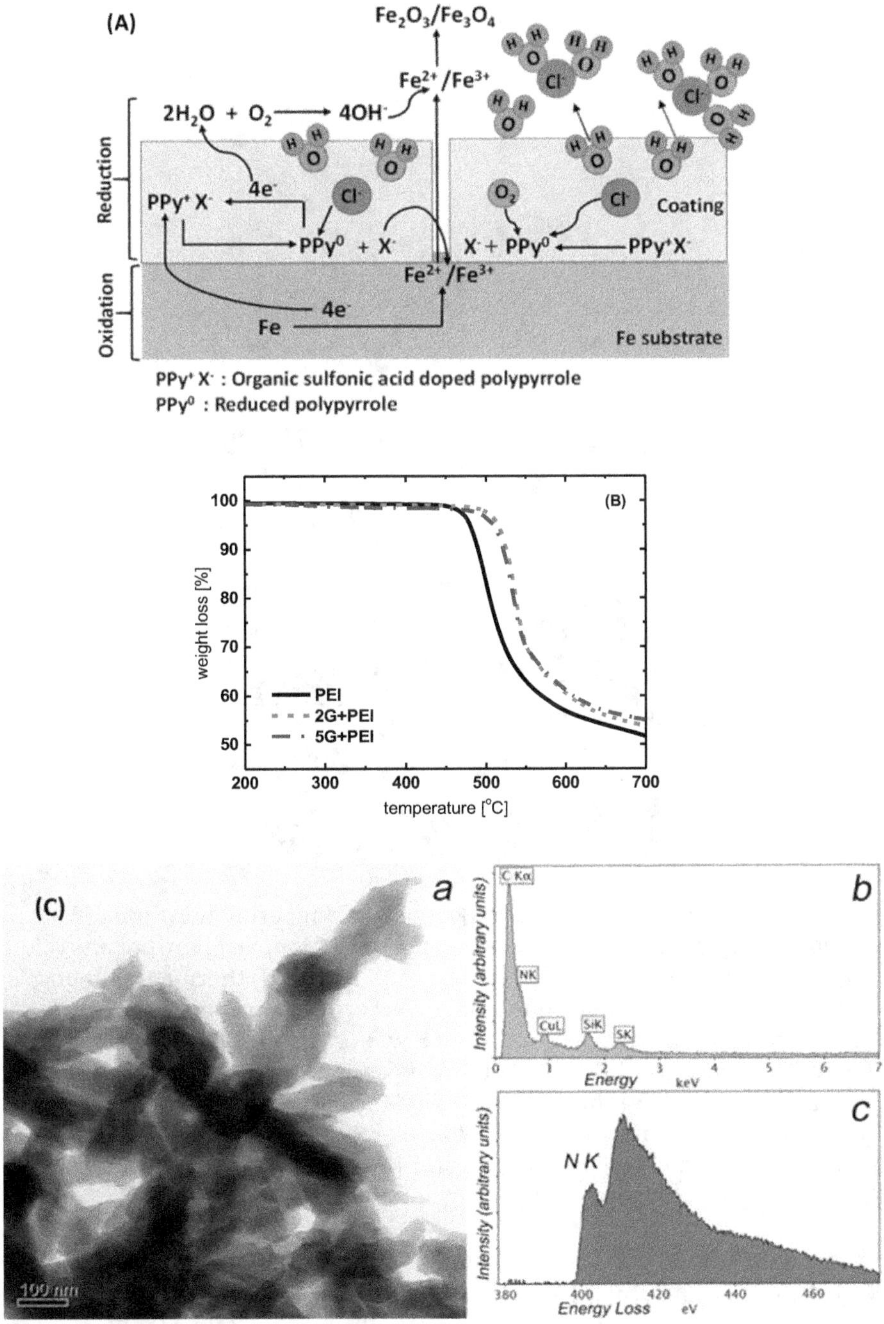

Figure 10.1 (continued on next page)

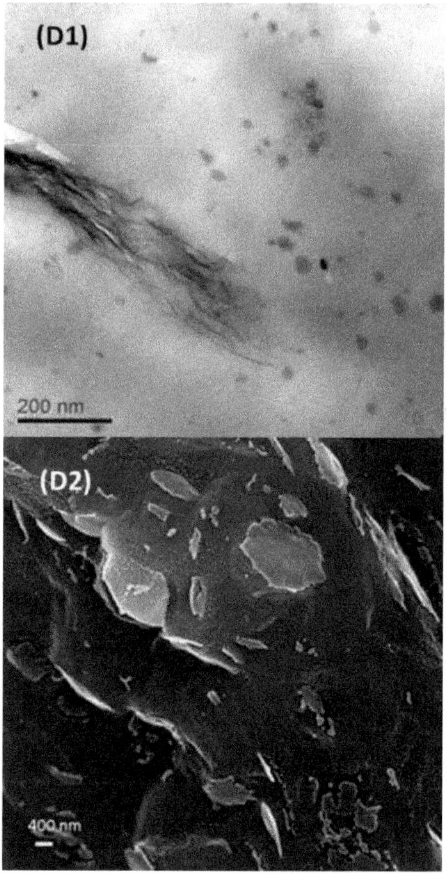

Figure 10.1 Examples of nanocomposites of engineering polymers: (A) schematic representation of corrosion protection imparted by polypyrrole pigments in composite coatings with carbon black, (B) thermogravimetric analysis of pure polyetherimide and polyetherimide-graphene nanocomposites films with 2 wt% and 5 wt% graphene, (C) (a) TEM micrograph and (b,c) EDX and EELS analysis of polyaniline modified thermally reduced graphene and (D) TEM (1) and SEM (2) images of polyamide nanocomposite with 5 wt% filler. Figures A, C and D are reproduced from References 15, 17, and 18 with permission from Royal Society of Chemistry, Taylor and Francis and Elsevier respectively.

decrease of double bond conversion and crosslink density. Gokceoren and Erbil [20] reported a series of poly(N-vinyl carbazole) (PNVCz)/sodium montmorillonite (Na-MMT) and PNVCz/organo-MMT composites prepared by *in-situ* free radical polymerization. The authors reported that the type of intercalation agent and mixing method had an impact on the structure and properties of obtained composite materials. It was observed that the interlayer spacing of clays and glass transition temperature of the polymer in the composites increased from 1.18 nm to 3.53 nm and from 140 °C to 185 °C respectively.

The nanocomposites of PPS with various fillers have also been similarly developed [21,22]. Figure 10.2 demonstrates an example of PPS nanocomposites with zinc oxide (ZnO), generated for use in medical applications [21]. For this purpose, aminated poly(phenylene sulfide) derivative (PPS-NH_2) was melt-mixed with different ZnO concentrations. The authors observed a gradual increment in the crystallization temperature and the degree of crystallinity with ZnO content due to the nucleation effect of the particles. The authors observed that the properties like stiffness, strength, toughness, glass transition, and heat distortion temperature were

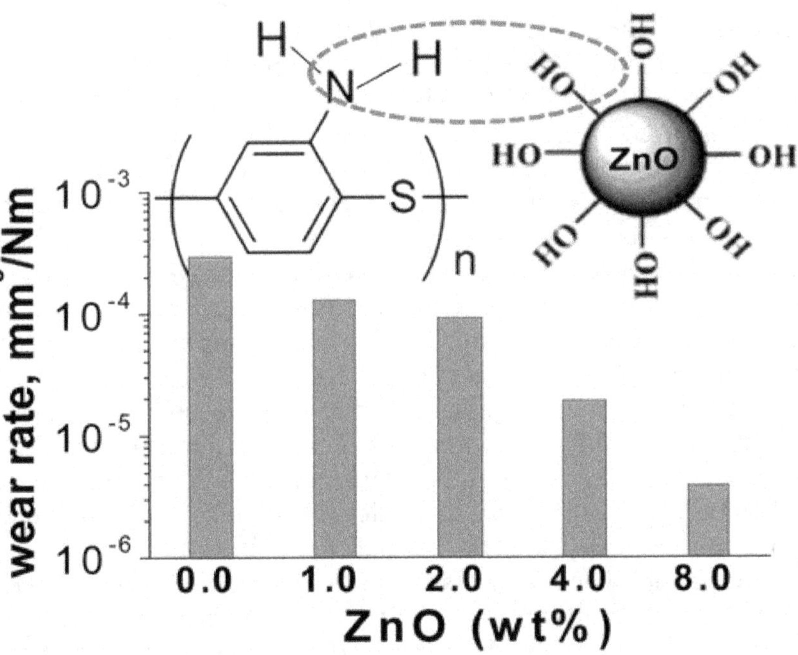

Figure 10.2 Wear rate of the PPS nanocomposites as a function of ZnO content. Reproduced from reference 21 with permission from American Chemical Society.

significantly improved in the PPS nanocomposites. A significant drop in wear rate of nearly 100-fold was observed in the nanocomposite with highest loading of ZnO (Figure 10.2). The nanocomposites revealed active inhibition against both Gram-positive and Gram-negative bacteria. Flores *et al.* [22] also evaluated the reinforcement of PPS with inorganic fullerene-like tungsten disulfide nanoparticles. The authors observed an increase of 60% in hardness in the composites with 1% filler content. The nature of the polymer amorphous phase, whether in the glassy or rubbery state, was observed to play an important role in defining the nanocomposite performance. It should also be noted that during injection molding process to generate the PPS nanocomposites, complex interaction between solid filler/fiber and flow field might lead to issues such as product anisotropy, weld line formation as well as poor surface appearance. These issues are generally resolved using *in-situ* composite manufacturing.

Carbon based fillers (CBFs), superficially graphite and graphene, are multifunctional fillers with superior electrical and thermal conductivity, high aspect ratio and low density. Enhanced thermal, electrical and mechanical properties of the composites are, thus, achievable if CBFs are employed as fillers. Graphite's structure consists of a hexagon of covalently bonded carbon atoms forming a several crystal structured layers. Moreover, the layers are bonded together via van der Waal's forces. As these forces are weak, therefore, it is possible to intercalate atoms/molecules/ions in the interlayers which result in graphite intercalation compounds (GICs). If the level of intercalation is sufficient, exfoliated graphite or graphene can be attained using this technique. Furthermore, graphene, which consists of one atomic thick sheets of covalently sp^2-bonded carbon atoms in a hexagonal arrangement, has already received attention of researchers for the generation of polymer nanocomposites to achieve a large increase in polymer properties at low weight or volume fractions [23]. Its choice as a filler is due to its excellent electrical and mechanical properties, which are significantly better than other inorganic filler materials. A single defect-free graphene layer has Young's modulus of ~1.0 TPa, intrinsic strength ~42 N/m, thermal conductivity ~4840-5300 W/(m.K), electron mobility exceeding 25,000 $cm^2/V.s$, excellent gas impermeability and specific surface area of ~2630 m^2/g [23]. All these properties make this material even superior to carbon nanotubes for use in polymer nanocomposites. Graphene and its derivatives, when added to a polymer, do not significantly affect the density ifn added in low concentrations. Graphene has a low bulk density and is difficult to feed to a compounder. However, dry blending before melt compounding solves this issue. Due to these properties, the development of the graphite and graphene based nanocomposites with PPS are also expected to enhance its spectrum of properties as well as applications further. Also, CNTs have issues when it comes to dispersion, alignment and compatibility with polymers. Furthermore, breakage of the CNTs during processing (high shear stress) or acid treatment decreases its aspect ratio. Thus, expanded graphite and graphene platelets represent superior alternatives to enhance the polymer properties.

In the current review, the mechanical, thermal and electrical properties of PPS nanocomposites reinforced with graphite, graphene and other derivatives have been reviewed. The properties have also been correlated with the composite generation methodologies as well as filler content in the nanocomposites. The mechanical properties included Young's modulus, tensile strength, elongation at break, flexural strength, etc. On the other hand, the various thermal properties included glass transition temperature, onset and peak melting temperatures as well as degradation temperatures. Some of the main electrical properties included electrical conductivities, percolation thresholds, dielectric constants and electrical resistances. Percolation threshold is an important property since it is the point at which a material's properties are greatly enhanced. Regarding dielectric constant, it is the point at which the polymer transits from an insulator to a semi-conductor.

10.2 Review of the PPS Nanocomposite Properties

Chang *et al.* [24] studied the effect of defective graphene (DG) and graphene oxide (GO) fillers in PPS. Computer simulations with a NEMD (non-equilibrium molecular dynamics) model were used to determine Young's modulus and thermal conductivity of the polymer nanocomposites. A significant change in properties was predicted as the weight fraction reached above 10% due to filler particles forming aggregates by clumping together. The authors also mentioned that the efficient dispersion of the filler in the polymer was achieved only when the filler weight fraction was strictly less than 10%. The results indicated that GO was uniformly dispersed in the polymer matrix during the simulations, as the parameter of solubility of GO is very close to that of PPS, making filler dispersion in the matrix easier. It was observed from the results that increasing the filler weight fraction led to the enhancements in the thermal conductivity of the polymer. Furthermore, the thermal conductivity of PPS/GO nanocomposite was suggested to be greater than the corresponding PPS/DG nanocomposites due to the stronger interactions between the PPS and graphene oxide phases. In addition, the defect graphene particles were observed to start aggregating, once a weight fraction larger than 20% was achieved. The authors observed that the Young's modulus increased constantly with increasing the filler content. This behavior was observed for both DG and GO containing composites, however, the PPS/GO composite was stiffer and had better interactions compared to the PPS/DG system. Also, as mentioned earlier, the graphene oxide (GO) filler was more uniformly dispersed in the matrix phase, due to which the PPS/GO system had noticeably higher modulus than that of the PPS/DG system. Thus, the DG and GO enhanced the overall mechanical and thermal properties of the polymer, though their effects on the composite performance were different owing to their molecular nature and corresponding interactions with the polymer.

Gu *et al.* [25] also generated the PPS nanocomposites with graphene nano-platelets (GNP) and concluded that the addition of graphene increased the thermal conductivity of the composites. Both pristine and functionalized GNPs were incorporated into PPS. The functionalization of the pristine/neat GNPs was expected to result in stronger interfacial bonds and better adhesion of the filler with the matrix phase. The PPS-GNP nanocomposites were prepared by utilizing a ball mill machine followed by compression molding at moderate conditions (295 °C/10 MPa). As expected, functionalized graphene platelets enhanced the composites properties more significantly. The nanocomposite was observed to achieve percolation at low concentrations of filler particles. A noticeable increase in the thermal conductivity was observed with increasing filler volume fraction. At roughly 29 vol% mass fraction of functionalized GNPs, a thermal conductivity of 4.414 W/mK was achieved, which was 19 times higher than that of pure PPS matrix. This was attributed to the higher extent of thermal transition properties through the GNP network in the PPS polymer following percolation behavior. The addition of functionalized GNPs was also observed to improve the degradation properties of the nanocomposites with increasing the filler fraction. These findings also confirmed the role of filler surface modification to achieve optimal enhancement in the nanocomposite performance.

Li *et al.* [26] reported the process to reduce weakly exfoliated graphite oxide (WEGO) using Grignard reagent phenylmagnesium bromide (PhMgBr), which acted both as a reducing agent as well as a chemical functionalization agent, as shown in Figure 10.3. The PPS nanocomposites with 2.5 wt% of the modified fillers exhibited higher electrical conductivity of 10.5 S/m. the tensile strength of the composites was 88.9 MPa, whereas a value of 740 MPa was measured for the Young's modulus. The authors attributed the observed enhancements to the directional homogeneous dispersion of GS in the polymer as well as the π-π conjugated interaction between the polymer and filler phases.

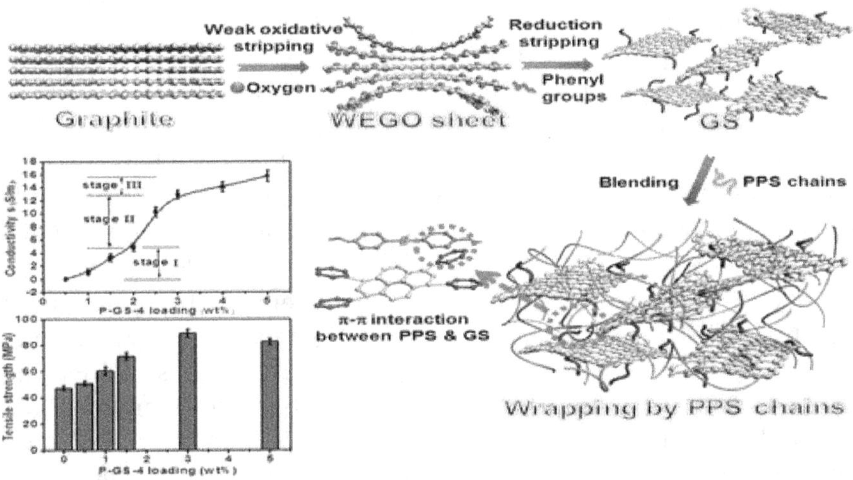

Figure 10.3 Representation of the nanocomposite synthesis as well as conductivity as well as mechanical properties as a function of filler content. Reproduced from Reference 26 with permission from American Chemical Society.

Khan *et al.* [27] compared the property enhancement achieved by different CBFs when added to PPS. The CBFs included carbon fibers (CF), multi-wall nanotubes (MWNT) and graphite nanoplatelets (GNP). The PPS composites were prepared using the melt mixing technique. Specifically, using a twin screw microcompounder, the initially blended powdered PPS and fillers were melt compounded. The effective thermal conductivity (K_{eff}) increased with increasing fractions of all 3 CBFs. The greatest improvements were observed using GNP followed by MWNT and CF. The reason for the observed behavior was attributed to the difference in the aspect ratios. Higher aspect ratio platelets helped to achieve better interconnectivity of particles as well as a large surface area which improved polymer-filler interactions. The coefficient of thermal expansion (CTE) was observed to decrease with increasing filler content irrespective of the filler type. This was attributed to the enhanced restrictions to the polymer chain mobility as the filler content increased. Finally, the glass transition temperature (T_g) of the composites exhibited no significant effect due to the incorporation of different fillers, irrespective of the filler contents. The authors suggested this observation to be unexpected as an increase in T_g is usually expected on the addition of the filler to the polymer. It was opined that weak interfacial bonding between the polymer and filler may have resulted in the observed finding. The authors also studied the effect of the CBFs on the compressive elastic modulus (E_{comp}). It was observed that the E_{comp} increased with increasing content for any of the fillers. However, MWNT and CF provided a greater enhancement that GNP. This could be explained due to the difference in the structure of these fillers, MWNT and CF were 1-D nano-fillers and offered higher rigidity than the 2-D GNP. It was reported that all fillers enhanced the electrical properties, however, GNP exhibited the greatest improvement. The percolation threshold of GNP was approximately 5 wt%, however, it was around 10-20 wt% for CF and MWNT. This could be because GNPs had higher aspect ratios compared to the other CBFs, which enabled these platelets to form a better network of conduction in the polymer. Several factors were suggested to affect this property, including aspect ratio, dispersion of filler, interfacial interaction of the polymer and matrix, etc. Moreover, polymers with high tendencies to crystallize have the possibility to expel the dispersed phase in the molten polymer to the amorphous phase, which leads to agglomeration.

In another study, Zhao *et al.* [28] used two forms of expanded graphite, prepared using different preparation methods, as fillers for PPS. One of the filler was prepared using rapid thermal treatment of expandable graphite with sulfuric acid, thereby, producing EG. On the other hand, the other filler was generated by immersing EG in an alcohol solution and ultrasonication to form ultrasonciated EG (S-EG). The forms of EG were observed to have differences in their aspect ratio. The composites were generated using melt blending method. The authors observed that an increase in the content of the fillers led to corresponding increase in the onset of crystallization temperature (T_{co}) as well as temperature of peak crystallization

(T_c) along with narrowing of the peak. Furthermore, T_{mo} (onset of melting), T_m (peak melting) and T_{max} (maximum degradation rate) temperatures were also observed to increase, which indicated an enhancement in the polymer's thermal stability. In addition, the melt enthalpy of the polymer in the composites increased, which indicated higher degree of crystallinity in the composites. The authors also observed the percolation threshold to be 1 wt% and 2 wt% for PPS/S-EG and PPS/EG nanocomposites, respectively, thus, indicating the better filler dispersion and network formation in the case of S-EG containing composites. The flexural strength of the PPS composites with EG and S-EG was observed to decrease with increasing the filler content. This was probably due to the poor filler interactions with the matrix and stress concentrations which resulted from filler aggregation. There was also a critical point where the flexural strength and flexural modulus started to increase with weight fraction of filler in both fillers, however, the flexural strength of both PPS/EG and PPS/S-EG nanocomposites was still lower than that of pure PPS. Use of both forms of filler was observed to improve the electrical conductivity of PPS, however, S-EG incorporation achieved a lower percolation threshold and higher conductivity. This could be justified by the fact that EG had un-exfoliated sheets which led to micro-pores in the structure of the surface, which hampered the conductive network. On the other hand, the S-EG polymer composite resulted in a better dispersed system with enhanced bridging/network structure.

Liu *et al.* [29] generated PPS nanocomposites with cetyltrimethylammonium bromide modified graphene oxide (CGO). It was observed that the modification of GO with cetyltrimethylammonium bromide resulted in the impairment of the ordered structure of GO as the interlayer distance was observed to increase. The authors observed that the addition of the surface modified graphene oxide filler improved the thermal properties of the composites. For instance, the glass transition temperature was increased by roughly 10 °C by the addition of 0.5 wt% CGO. This was attributed to the combined effect of the chemical bonding between CGO filler and polymer chains as well as the hindered segmental motion caused by the structure of the surface modified filler particles. Moreover, CGO also significantly improved the thermal stability of the polymer. In the composite with 1 wt% CGO, the temperature at which 30% weight loss occurred was increased by 10 °C and the heat resistance index was offset by 2 °C. Moreover, the presence of evenly dispersed CGO fillers in the PPS matrix phase enhanced char formation which was favorable in improving heat stability. The addition of cetyl-trimethylammonium bromide modified graphene oxide to PPS also resulted in significant improvements in Young's modulus and tensile strength, as a function of filler content. The composite with 0.1 wt% CGO exhibited an increase of 45.4% and 171.8% respectively in the tensile strength and Young's modulus as compared to pure PPS. This was attributed to the entanglement of the PPS chains with cetyl-trimethylammonium bromide. The tensile strength was observed to decrease rapidly as the CGO content was enhanced in the nanocomposites. At lower filler concentrations, a good degree of filler dispersion existed in the matrix phase. Furthermore, the presence of the surface modification increased the surface contact area and thereby greatly enhanced the aspect ratio. However, at higher filler concentrations, agglomerates were formed which induced additional stresses, thereby resulting in a lowered tensile strength.

Pan *et al.* [30] studied the methods to improve wear performance of the polymer coating generated using a blend of PPS and polytetrafluroethylene wax (PFW) reinforced with graphene. Composites with varying mass ratios of the three components were generated, and the wear life of the blend/composite was tested. The authors observed that the composite with mass ratio 70/30/0.8 (PPS/PFW/graphene) resulted in the maximum increase in the coating wear life. Adding higher extent graphene reinforcement was observed to hamper the coating wear life, as the test conducted on the nanocomposite sample with 1.5 graphene mass ratio resulted in a reduced wear life. The observed decline in the wear life with increasing graphene content was attributed to the aggregation of graphene phase, thereby destroying the uniform structure of the composite coating. Furthermore, adhesive wear was found to be the major form of damage in pure PPS coating, however, abrasive wear became the major form of damage in the composite coatings. Overall, the coating with 0.8 mass ratio of graphene resulted in a 750% increase in wear life compared to that of neat PPS coating. As high strength contributed by graphene resulted in the improvement in the load carrying capacity and wear life of coatings, thus, it indicated the potential use of graphene for such applications.

Goyal *et al.* [31] studied the electrical properties of PPS nanocomposites with expanded graphite, prepared using ultrasonication. The electrical conductivity percolation threshold of 1 wt% was observed in the PPS/expanded graphite nanocomposites, which was much lower than the 5 wt% measured for the PPS composite generated with micrometer-sized graphite flakes. An increase of 14 orders of magnitude was observed in the electrical conductivity of the composite with 2 wt% filler content. Similarly, the dielectric constant (DC) increased with increasing EG content at any frequency, with a sharp increase at 1 wt% filler content. As the frequency increased, the DC values decreased at a fixed EG content. However, at low EG content, the DC was independent of the frequency. This was due to the Maxwell-Wagner polarization effect. Another property that was studied was the dissipation factor (DF), which is the ratio of the lost electrical energy over the stored electrical energy. As the EG content increased, DF increased at any constant frequency. Also, at low EG content, DF increased with increasing frequencies. However, above 1 wt% EG content, DF was observed to decrease with increasing frequencies.

Ramanujam and Radhakrishnan [32] studied the property enhancement of PPS using a hybrid composite system of graphite and EG. EG was prepared using acid intercalation at high temperature followed by ultrasonication. Electrical properties improved noticeably upon the addition of EG to the PPS/graphite (7 wt%) composite. The authors observed the percolation threshold at a content of 0.25 wt% EG in the PPS/graphite (7 wt%) composite. In addition, incorporation of 1 wt% EG in the PPS/graphite composite was observed to increase the conductivity by six orders of magnitude. Moreover, the DC increased as the fraction of EG was enhanced in the composite. This was possibly due to the EG entering the inter-graphite layers and, hence, reducing the barrier between adjacent graphite sheets. The DC was observed to decrease with increasing frequencies. In addition, bulk and contact resistances decreased with increasing the EG loading. A possible explanation for the observed behavior was that EG distributed among the PPS matrix and graphite and acted as a charge carrier between both phases.

Ramanujam *et al.* [33] also investigated the effect of polymerization route on the electrical properties of PPS/EG nanocomposites. PPS/EG nanocomposites were prepared by two different polymerization routes, viz. powder mixing and *in-situ* polymerization. It was observed that the electrical conductivity was improved by the addition of expanded graphite as filler to PPS. Furthermore, the *in-situ* polymerization route resulted in higher electrical conductivity compared to powder mixing route. Moreover, the percolation threshold by the powder mixing route was found to be higher than that of the *in-situ* polymerization route. Therefore, it could be concluded that the *in-situ* polymerization route resulted in a better dispersion of expanded graphite filler in the polymer phase, which resulted in higher electrical conductivity at the same EG filler concentration.

10.3 Conclusions

Poly(phenylene sulfide) is a versatile polymer and the use of fillers such as graphene and expanded graphite can be very useful in significantly improving a broad range of its properties. In terms of thermal properties, the addition of graphene was observed to result in an overall improvement in the thermal conductivity, thermal stability and degradation temperatures of the PPS composite, as a function of filler content. Functionalization of the graphene and expanded graphite based fillers also improved these properties further as compared to the composites generated with neat or pristine filler. In addition, the improvement in heat stability and heat resistance of the polymer were observed at much lower graphene/expanded graphite concentrations as compared to other conventional fillers. The mechanical properties of PPS composites e.g. Young's modulus and tensile strength were also observed to be the affected by the amount of filler. This was attributed to the rigid structure of the filler particles and their ability to absorb stress. However, the impact and flexural strength of PPS/graphene composites were generally observed to decrease with the addition of filler. This was due to the filler aggregates, especially at higher filler concentrations, which made the polymer more brittle. The wear life of the PPS based composites was also reported to increase significantly with the addition of graphene filler. Similar to other properties, the optimum filler quantity led to the maximum increment in performance, as the additional filler content subsequently reduced the wear life due to agglomeration. Electrical conductivity, dielectric constant and dissipation factor were also observed to increase with the addition of graphene and expanded graphite. Furthermore, bulk & contact resistances along with percolation threshold decreased with increasing the filler content. Overall, PPS plays an important role in many applications and is a vital engineering polymer. This review has reviewed many instances where the addition of graphene and expanded graphite has yielded significant improvements in the thermal, mechanical and electrical properties of PPS. Due to their superior properties, PPS nanocomposites with graphene and graphite thus have the usefulness in a large variety of applications, where these materials can successfully replace metallic materials. Table 10.1 also presents the summary of the thermal, mechanical and electrical properties of the PPS nanocomposites with increasing filler content.

Table 10.1 Summary of the thermal, mechanical and electrical properties of the PPS/filler nanocomposites

Thermal properties	Mechanical properties	Electrical properties
Increasing filler content increases thermal conductivity. However, aggregates can occur at higher concentrations.	Increasing filler concentration increases Young's modulus, whereas tensile strength is observed to increase only at low filler concentrations.	Electrical conductivity, dielectric constant and dissipation factor increase by adding small amounts of graphene/expanded graphite.
Surface modifications of filler platelets yields enhanced properties.	Flexural and impact strength exhibit maximum at very low filler fraction. Increasing the filler content further reduces these properties because of strain hardening and aggregate formation, which make the system more brittle.	Bulk & contact resistances and percolation threshold decrease with increasing graphene content.
Degradation and thermal stability are improved as a function of filler fraction.	Wear life of coatings is enhanced significantly by incorporating only a small fraction of filler.	Increasing the frequency at fixed filler content decreases the dielectric constant.

References

1. Rahate, A. S., Nemade, K. R., and Waghuley, S. A. (2013) Polyphenylene sulfide (PPS): state of the art and applications. *Reviews in Chemical Engineering,* **29**, 471-489.
2. Caglar, B., Fischer, P., Kauranen, P., Karttunen, M., and Elsner, P. (2014) Development of carbon nanotube and graphite filled polyphenylene sulfide based bipolar plates for all-vanadium redox flow

batteries. *Journal of Power Sources,* **256**, 88-95.

3. Jang, H., Hong, T., Yoo, J., Lee, S., Pyo, J., Sutradhar, S. C., Ju, H., and Kim, W. (2015) Preparation and characterization of sulfonated poly(phenylene)s membranes containing conjugated moiety via nickel catalyzed carbonecarbon coupling polymerization. *International Journal of Hydrogen Energy,* **40**(41), 14364-14370.
4. Young, R. T., and Baird, D. G. (2000) The influence of processing variables on injection molded in situ composites based on polyphenylene sulfide and a melt processable glass. *Composites: Part B,* **31**, 209-221.
5. Zhang, R.-C., Lu, A., Xu, Y., Min, M., Xia, J.-Q., Zhou, J.-H.,Huang, Y.-G., and Li, Z.-M. (2009) Equilibrium melting temperature and spherulitic growth rate of poly(phenylene sulfide). *European Polymer Journal,* **45**, 2867-2872.
6. Ju, S.-P., Haung, T.-J., Liao, C.-H., and Chang, J.-W. (2013) Investigation of thermal conductivity of graphite flake/poly (p-phenylene sulfide) composite by experimental measurement and non-equilibrium molecular dynamics simulation. *Polymer,* **54**, 4702-4709.
7. Lee, T. H., Boey, F. Y. C., and Khor, K. A. (1995) On the determination of polymer crystallinity for a thermoplastic PPS composite by thermal analysis. *Composites Science and Technology,* **53**, 259-274.
8. Yylmaz, T., and Synmazcelik, T. (2007) Investigation of load bearing performances of pin connected carbon / polyphenylene sulphide composites under static loading conditions. *Materials and Design,* **28**, 520-527.
9. Sergeyev, V. A., Shitikov. V. K., Nedel'kin, V. I., Askadskii, A. A., Bychki, Slonimskii, G. L., and Korshak, V. V. (1977) The effect of heat on some of the physico-mechanical properties of polyphenylene sulphide. *Polymer Science U. S. S. R.,* **19**(6), 1494-1498.
10. Sergeyev, V. A., Shitikov, V. K., Nedel'kin, V. I., and Korshak, V. V. (1975) Synthesis and properties of soluble polyphenylene sulphides. *Polymer Science U. S. S. R.,* **17**(11), 2783-2789.
11. Tanthapanichakoon, W., Furuuchi, M., Nitta, K.-h., Hata, M., Endoh, S., and Otani, Y. (2006) Degradation of semi-crystalline PPS bag-filter materials by NO and O_2 at high temperature. *Polymer Degradation and Stability,* **91**, 1637-1644.
12. Tanthapanichakoon, W., Mitsuhiko Hata, K.-h. N., Furuuchi, M., and Otani, Y. (2006) Mechanical degradation of filter polymer materials: Polyphenylene sulfide. *Polymer Degradation and Stability,* **91**, 2614-2621.
13. Huang, X., Qi, X., Boey, F., and Zhang, H. (2012) Graphene-based composites. *Chemical Society Reviews,* **41**, 666-686.
14. Li, H., Lv, G.-y., Zhang, G., Ren, H.-h., Fan, X.-x., and Yan, Y.-g. (2014) Synthesis and characterization of novel poly(phenylene sulfide)

containing a chromophore in the main chain. *Polymer International*, **63**, 1707-1714.

15. Niratiwongkorn, T., Luckachan, G. E., and Mittal, V. (2016) Self-healing protective coatings of polyvinyl butyral/polypyrrole-carbon black composite on carbon steel. *RSC Advances*, **6**, 43237-43249.
16. Akhtar, T., Luckachan, G. E., and Mittal, V. Inhibition of corrosion by graphene in polyetherimide composite coating for long term protection of carbon steel, unpublished.
17. Mittal, V., Chaudhry, A. U., and Matsko, N. B. (2016) Organic functionalization of thermally reduced graphene oxide nanoplatelets by adsorption: structural and morphological characterization. *Philosophical Magazine*, **96**(20), 2143-2160.
18. Mittal, V., Chaudhry, A. U., and Luckachan, G. E. (2014) Biopolymer - thermally reduced graphene nanocomposites: Structural characterization and properties. *Materials Chemistry and Physics*, **147**(1-2), 319-332.
19. Sangermano, M., Roppolo, I., Alves Camara, V. H., Dizman, C., Ates, S., Torun, L., and Yagci, Y. (2011) Polysulfone/metal nanocomposites by simultaneous photoinduced crosslinking and redox reaction. *Macromolecular Materials and Engineering*, **296**, 820-825.
20. Gokceoren, A. T., and Erbil, C. (2011) Morphological characterization and electrochemical/electrical properties of poly(N-vinyl carbazole)/montmorillonite composites. *Polymer Composites*, **32**(9), doi:10.1002/pc.21153.
21. Diez-Pascual, A. M., and Diez-Vicente, A. L. (2014) High-performance aminated poly(phenylene sulfide)/ZnO nanocomposites for medical applications. *ACS Applied Materials & Interfaces*, **6**(13), 10132-10145.
22. Flores, A., Naffakh, M., Diez-Pascual, A. M., Ania, F., and Gomez-Fatou, M. A. (2013) Evaluating the reinforcement of inorganic fullerene-like nano-particles in thermoplastic matrices by depth-sensing indentation. *Journal of Physical Chemistry C*, **117**(40), 20936-20943.
23. Mukhopadhyay, P., and Gupta, R. K. (2011) Trends and frontiers in graphene-based polymer nanocomposites. *Plastics Engineering*, **67**, 32-42.
24. Chang, C.-Y., Ju, S.-P., Chang, J.-W., Huang, S.-C., and Yang, H.-W. (2014) The thermal conductivity and mechanical properties of poly(p-phenylene sulfide)/oxidedgraphene and poly(p-phenylene sulfide)/defectgraphene nano-composites by molecular dynamics simulation. *RSC Advances*, **4**, 26074-26080.
25. Gu, J., Xie, C., Li, H., Dang, J., Geng, W., and Zhang, Q. (2014) Thermal percolation behavior of graphene nanoplatelets/polyphenylene sulfide thermal conductivity composites. *Polymer Composites*, **35**,

1087-1092.

26. Li, C., Li, Z., Cao, L., and Cheng, B. (2016) Graphene preparation by phenylmagnesium bromide and its excellent electrical conductivity performance in graphene/poly(p-phenylene sulfide) composites. *Industrial & Engineering Chemistry Research*, **55**(41), 10860-10867.
27. Khan, M. O., Leung, S. N., Chan, E., Naguib, H. E., Dawson, F., and Adinkrah, V. (2013) Effects of microsized and nanosized carbon fillers on the thermal and electrical properties of polyphenylene sulfide based composites. *Polymer Engineering and Science*, **53**(11), 2398-2407.
28. Zhao, Y. F., Xiao, M., Wang, S. J., Ge, X. C., and Meng, Y. Z. (2007) Preparation and properties of electrically conductive PPS/expanded graphite nanocomposites. *Composites Science and Technology*, **67**, 2528-2534.
29. Liu, Q., Luo, W., Chen, Y., Zou, H., and Liang, M. (2016) Enhanced mechanical and thermal properties of CTAB-functionalized graphene oxide-polyphenylene sulfide composites. *High Performance Polymers*, doi: DOI: 10.1177/0954008316663810.
30. Pan, B., Zhao, J., Zhang, Y., and Zhang, Y. (2012) Wear performance and mechanisms of polyphenylene sulfide/polytetrafluoroethylene wax composite coatings reinforced by graphene. *Journal of Macromolecular Science, Part B*, **51**(6), 1218-1227.
31. Goyal, R. K., Samant, S. D., Thakar, A. K., and Kadam, A. (2010) Electrical properties of poly-mer/expanded graphite nanocomposites with low percolation. *Journal of Physics D: Applied Physics*, **43**, 365-404.
32. Ramanujam, B. T. S., and Radhakrishnan, S. (2010) Polyphenylene sulfide-graphite hybrid composites: Charge transport and impedance characteristics. *International Journal of Plastics Technology*, **14**, 37-44.
33. Ramanujam, B., Radhakrishnan, S., and Deshpande, S. (2016) Polyphenylene sulfide-expanded graphite nanocomposites. *Journal of Thermoplastic Composite Materials*, doi: 10.1177/0892705716646420.

Index

A

B

C

D

E

F

G

H

I

L

M

O

P

R

S

T

U

V

W

X

Y

Z

www.ingramcontent.com/pod-product-compliance
Lightning Source LLC
LaVergne TN
LVHW011527200926
848484LV00001B/177